Established and Emerging Practices for Soil and Crop Productivity

The book explains the various existing, emerging and environmentally viable technologies for the sustainable and profitable crop productivity from the soil to the undergraduate and graduate students, young scientists, extension workers and farmers. In this book discussed as one of the important objective of green technology, waste should be reduced, which a crucial component of sustainable, which will help to promote healthy environment and good habits for the coming generations. The waste production adds to emissions of carbon dioxide, methane, leaking dangerous materials into the soil and waterways. Compost, FYM, cover crops, green and brown manuring are the most important supplement depleted soil and add nutrient-rich humus, which fuels plant growth and restores vitality to depleted soil. There is also focused on in this book climate change, hurricanes and tropical storms, natural resources management, crop diversification, crop resource management, cropping systems, farming system, management of land use resources, conservation agriculture, crop residue management, renewable energy, precision agriculture, integrated nutrient management, integrated pest management.

Dr. Avtar Singh Bimbraw joined service in April, 1986 as Assistant Professor of Agronomy and retired as Senior Agronomist/Professor of Agronomy in June, 2014 after 28 years experience in teaching, research and extension work in agronomy from Punjab Agricultural University, Ludhiana. He secured B.Sc. Agri. (Hons.) and Ph.D Agronomy degrees from Punjab Agricultural University, Ludhiana and M.Sc. Forage Production from National Dairy Research Institute, Karnal, Haryana. He developed the agronomic technologies and evaluated the genotypes of Medicinal and Aromatic Plants, forage and fodder crops, sugarcane, pulses, cotton rice-wheat cropping and also worked on conservation agriculture.

Established and Emerging Practices for Soil and Crop Productivity

Avtar Singh Bimbraw

Senior Agronomist (Retd)
Punjab Agricultural University
Department of Agronomy
Ludhiana - 141004

CRC Press is an imprint of the
Taylor & Francis Group, an **informa** business

NARENDRA PUBLISHING HOUSE

First published 2021
by CRC Press
2 Park Square, Milton Park, Abingdon, Oxon, OX14 4RN

and by CRC Press
6000 Broken Sound Parkway NW, Suite 300, Boca Raton, FL 33487-2742

© 2021 Narendra Publishing House

CRC Press is an imprint of Informa UK Limited

The right of Avtar Singh Bimbraw to be identified as author of this work has been asserted by him in accordance with sections 77 and 78 of the Copyright, Designs and Patents Act 1988.

Reasonable efforts have been made to publish reliable data and information, but the author and publisher cannot assume responsibility for the validity of all materials or the consequences of their use. The authors and publishers have attempted to trace the copyright holders of all material reproduced in this publication and apologize to copyright holders if permission to publish in this form has not been obtained. If any copyright material has not been acknowledged please write and let us know so we may rectify in any future reprint.

All rights reserved. No part of this book may be reprinted or reproduced or utilised in any form or by any electronic, mechanical, or other means, now known or hereafter invented, including photocopying and recording, or in any information storage or retrieval system, without permission in writing from the publishers.

For permission to photocopy or use material electronically from this work, access www.copyright.com or contact the Copyright Clearance Center, Inc. (CCC), 222 Rosewood Drive, Danvers, MA 01923, 978-750-8400. For works that are not available on CCC please contact mpkbookspermissions@tandf.co.uk

Trademark notice: Product or corporate names may be trademarks or registered trademarks, and are used only for identification and explanation without intent to infringe.

Print edition not for sale in South Asia (India, Sri Lanka, Nepal, Bangladesh, Pakistan or Bhutan).

British Library Cataloguing-in-Publication Data
A catalogue record for this book is available from the British Library

Library of Congress Cataloging-in-Publication Data
A catalog record has been requested

ISBN: 978-1-032-06020-0 (hbk)
ISBN: 978-1-003-20034-5 (ebk)

Contents

	Preface	vii
1	Importance, impact and objectives of green technology	1
2	History, importance, preparation and organisms in composting	8
3	Preparation, requirement and maintenance of vermi-compost and farm yard manure	42
4	Reuse, type, treatment of grey water	46
5	Types, significance, choice of cover crops	73
6	Carbon farming: green and brown manuring	84
7	Climate change and global warming impact on agriculture	122
8	Causes and impact of hurricanes and tropical storms	168
9	Classification, depletion, management of natural resources	184
10	Crop diversification for maximizing productivity	231
11	Concept, significance and sustainability of cropping system	247
12	Approach, objectives and components of farming system	290
13	Degradation and management of land use resource	301
14	Principles, significance and future of conservation agriculture	348
15	Policy challenges and potential solutions of residue burning in India	361

16	Type, importance, history, emerging technology of renewable energy	390
17	Elements, technologies, approaches of precision farming	407
18	Challenges, concept, approaches, management of site-specific nutrient	429
19	Concept, advantages, components of integrated nutrient management	441
20	Concept and function of integrated pest management	460
	Challenges of established and emerging technologies	*471*
	References	*473*

Preface

For agriculture to continue along a sustainable path of economic development further production increases must be generated by technologies that are both profitable and more environmentally benign. In this context, we assess the role of these green or sustainable technologies in steering agriculture along a more sustainable path. However, the lack of markets for the environmental attributes associated with green technologies can limit their development. In addition, simply making a technology available does not mean it will be adopted. Experience with various green technologies demonstrates that even when technologies are profitable, barriers to adopting new practices can limit their effectiveness. In this book discussed as one of the important objective of green technology, waste should be reduced, which a crucial component of sustainable is living, which will help to promote healthy environment and good habits for the coming generations. The waste production adds to emissions of carbon dioxide, methane, leaking dangerous materials into the soil and waterways. Compost, FYM, cover crops, green and brown manuring are the most important supplement depleted soil and add nutrient-rich humus, which fuels plant growth and restores vitality to depleted soil. They are also free, easy to make and good for the environment. Grey water is called as any domestic wastewater produced excluding sewage. Water from dishwashers and kitchen sinks is often referred to as dark grey water and water from toilets is called black water. People are now waking up to the benefits of grey water re-use. The impact of climate change on agriculture could result in problems with food security and may threaten the livelihood activities upon, which much of the population depends. Climate change can affect crop yields (both positively and negatively), as well as the types of crops. Both the population and the natural and built environment are also vulnerable to the impacts of hurricanes and Tropical Storms. The conservation of natural resources is the fundamental problem. Unless it solves that problem, it will avail little to solve all others. Depletion of natural resources is associated with social inequity. Considering most biodiversity are located in developing countries, depletion of this resource could result in losses of ecosystem services for these countries. Diversification is the process to take advantage of emerging opportunities created by technology, new markets, changes in policy etc. to meet certain goals, challenges and threats and to reduce risk. Cropping system refers to the crops, crop sequences and management techniques used on a particular agricultural field over a period of years. Historically, cropping systems have been designed to maximise yield, but modern agriculture is increasingly concerned with promoting environmental sustainability in cropping systems. Orientation from cropping system research to farming system research focusing on multiple uses of inputs and recycling for reduction in cost and resource conservation. Integrated farming systems are of sequential production of crops, trees, livestock and fisheries after their proper assessment and evaluation that is expected to expand livelihood basket. Land-use Planning is a systematic and iterative procedure carried out in order to create an enabling environment for sustainable development of land resources which meets people's needs and demands. Conservation agriculture is a concept for resource-saving agricultural crop production that strives to achieve acceptable profits together with high and sustained production levels while concurrently conserving the environment. The crop residues, specifically as a field residue is a natural resource that traditionally contributed to the soil stability and fertility through ploughing directly into the soil, or by composting etc. Renewable and non-renewable energy: Energy exists freely in nature. Some of them exist infinitely (never run out, called renewable, the rest have finite amounts (they took millions of years to form and will run out one day called non-renewable is generally defined as energy that comes from resources such as sunlight, wind, rain, tides, waves and geothermal heat. Renewable energy replaces conventional fuels in four distinct areas: electricity generation, hot water/space heating, motor fuels, and rural (off-grid) energy services. Precision farming is a current buzz word among agricultural circles. The term precision farming means carefully tailoring soil and crop management to fit the different conditions found in each field. In lay man language it can be defined as a need based or variable rate application system. Site-specific nutrient management as developed in Asian rice producing countries provides an approach for 'feeding' rice with nutrients as and when needed. Integrated Nutrient Management refers to the maintenance of soil fertility and of plant nutrient supply at an optimum level for sustaining the desired productivity through optimization of the benefits from all possible sources of organic, inorganic and biological components in an integrated manner. Integrated pest management (IPM) is a decision support system for the selection and use of pest control tactics, singly or harmoniously coordinated into a management strategy, based on cost/benefit analyses that take into account the interests of and impacts on producers, society and the environment.

Avtar Singh Bimbraw

Importance, Impact And Objectives Of Green Technology

As the name implies green technology is one that has a "green" purpose. By green do not mean the colour, however, Mother Nature is quite green and the long and short term impact an invention has on the environment. Green inventions are eco- friendly inventions that often involve: energy efficiency, recycling, safety and health concerns, renewable resources and more. One of the best known examples of green technology may be the solar cell. A solar cell directly converts the solar energy into electrical energy through the process of photovoltaic. Generating electricity from solar energy means less consumption of fossil fuels, reducing pollution and greenhouse gas emissions. Another simple invention that can be considered green is the re-usable water bottle. Drinking lots of water is healthy. Reducing plastic waste is great for the environment. Hence, trendy reusable water bottles that you can refill yourself are health-promoting, eco-friendly, and green. The word «Green Technology» is relatively new. Having been adopted just over the last couple of decades, Green is the way to go today. In order to understand the term green technology, let us first know the definition of green technology.

DEFINITION OF GREEN TECHNOLOGY

In simple words, it means the technology, which is eco-friendly, developed and used in such a way so that it doesn't disturb our environment and conserves natural resources. The green technology is being referred to as environmental technology and clean technology. In the nascent stages of its development, the future only promises to bring bigger and better things for this field. It will in fact be a necessity of the future.

The term 'green technology' is now becoming quite popular. It appeared just over the last two decades. Green technology means this type of technology, which is environmentally friendly. Such technology was developed and is being used in a way, which does not disturb the environment and does not destroy natural resources. Green technology is also known as 'environmental technology' or 'clean technology'.

Unlike the technological waves in recent decades, Green Technology is almost entirely materials science based. Relying on the availability of alternative sources of energy, the purpose of this technology is to reduce global warming as well as the green house effect.

Its main objective is to find ways to create new technologies in such a way that they do not damage or deplete the planets natural resources. It also expresses less harm to human, animal, and plant health, as well as damage to the world, in general. Going on with the green technology definition, our environment needs immediate recoup from pollution. With the help of green technology, one can reduce pollution and improve the cleanliness as well. Today developed as well as developing countries are turning to green technology to secure the environment from negative impacts. The green technology definition explained here basically gives you an idea about the messing up of the environment due to human intrusion and the important need to slow down and adopting healthier ways towards life. By adopting green technology wisely, the earth can be protected against environmental pollution.

The main purpose of green technology is to slow down global warming and reduce the green house effect. The main idea is the creation of new technologies which do not damage the natural resources. This should result into less harm to people, species and the general health of our planet.

Importance of Green Technology

Using green technology has become imperative today. It is important to stay current with what's happening around. Today's technology and conveniences are more focused on making more luxuries available to more people. The ugly side of this is the harm these products or new technologies are doing to our environment. This raises the importance of using green technology all the more.

> » As about using green technology, industries and regulatory bodies are already taking steps in the right direction. But one need not wait for others to come up with all the answers. By using green technology one can make their effort to help the environmental heal. One need not be a scientist to understand all about using green

technology. Some easy environmentally friendly solutions are there for everybody to follow. Developing a planned community, making less packaging for frozen foods, using Environmental friendly machines are just a few examples. One can go out and seek products that support green technology.

- » By using green technology, we will march a step ahead in saving the environment and make the earth free from any danger. Destroying environment any further can lead to situations which can be irreversible. As green technology will help to reduce the use of fossil fuel, it is expected that energy production from green technology will be higher than fossil fuel sources of energy like oil, and gas in the future.

- » Using green technology will become not only important but mandatory too in the coming years. With the earths' energy resources depleting fast, we have to rely on alternative sources of energy. Green technology encourages the concept of cleanliness, freshness as well as promotes new dimensions. The sooner we realize the importance of using green technology, the better it will be for our planet and its environment.

- » It now becomes obvious that our planet starts to suffocate from all the pollution we create. But if there is a will, there is a way to make this problem much smaller. The active use of green technology can help significantly reduce pollution. That is why the developed and some of the developing countries are now turning to this type of technology to help them protect the environment from aggressive impacts. Although the problems with pollution are old enough, green technology is a relatively new idea. It is now becoming more popular because people started to realize that we are really destroying our planet.

Effect of Green Technologies

Green technology, an environmentally friendly technology is developed and used in a way that protects the environment and conserves natural resources. A part of the renewable energy branch of the environmental technology movement, the green technology importance cannot be ignored. Therefore, there is need to pause and reflect on the growing green technology importance and why it is going to be important for humanity.

There are several reasons in respect of green technology importance, perhaps volumes may be written and spoken on the subject. Whether it is the growing importance of green technology in the industry or at homes, it is obvious that things need to be done fast to clean environment and save energy resources. Going green can only help us come out of the present grim situation. Before things turn for the worst, we should realize the green technology importance to solve this problem.

Important Objectives of Green Technology

The objectives of green technology are many and some are given below:

- » The main objective of green technology is to meet the needs of society in any way without damaging or depleting natural resources on earth. The idea is to meet present needs without making any compromises.
- » Focusing on making products that can be fully reclaimed or re-used.
- » Changing patterns of production and consumption as steps to reduce waste and pollution.
- » It is essential to develop alternative technologies to prevent any further damage to health and the environment. Speeding their implementation can benefit our environment and truly protect the planet.

Explore the objectives of green technology by introducing sustainable living, developing renewable energy and reducing waste.

Ways to Reduce Waste and Promote Green Technology

With the ever rising populations and the subsequent climb in resource demands, the waste production has also increased. So, there is need to reduce waste. As one of the important objective of green technology, waste should be reduced, which is a crucial component of sustainable living. In this way, it can help to promote healthy environment and good habits for the coming generations. One of the basic ways to reduce waste is by reusing commodities and recycling. The waste production adds to emissions of carbon dioxide, methane, leaking dangerous materials into the soil and waterways. In the coming years, trillion pounds of natural resources are feared to get transformed into non productive wastes and gases. So, it has become very essential to take some steps immediately.

Reduce paper waste: Paper wastage can be reduced with the move of online document. Another means to reduce waste is to avoid buy in bulk, which lessens the use of packaging materials.

Food wastage: Organic waste keeps piling up, leading to costly methane emissions. Food waste can be re-introduced into the environment through composting, which can easily be carried out at home or with community composting.

Avoid buying materials in excess means more than requirement. It should be based on non-hazardous or less hazardous items. In general, addition of waste to landfills is practised. In order to curb this, the waste materials should be reused. Reuse not only saves natural resources, but also lessens our dependency on them. Recycling is another process to reduce waste. It primarily involves three primary processes; collection and processing, manufacturing. Purchase recycled products and promotes the goals of green technology

Green Sustainable Living

In simple words, sustainable living communities mean living a lifestyle that uses to reduce the depletion of the natural resource. Followers of sustainable living make efforts to shrink their carbon footprint by making changes in their lifestyle, like the way they use the transport, how they eat and the energy use in their homes. There is no shortage of ideas and tips for sustainable living and make the environment happier. To live a simpler, more sustainable life, just follow the following simple ideas for green sustainable living:

Recycle

Perhaps you already know about this are already recycling. But look around you and you may find a lot more that can be recycled. Reuse and recycle regularly.

Use a reusable bag

If one is casual, it is amazing to see how on ends up with piles and piles of plastic bags around them. Plastic bags are difficult to recycle and extremely toxic to our environment. Get a tote bag whenever you go shopping.

Eat sustainable

Cook from scratch and avoid preserved or pre cooked food from the market. If you eat more vegetarian diet, you'll achieve the goals of green technology a lot faster by buying sustainable living products.

Drive less

Keeping in view environment, there are some trips, which are completely avoidable. Walk or use bicycle whenever possible and avoid becoming too reliant on the automobile. You will be doing yourself as well as the environment a lot good.

Make less waste

Avoid wastage wherever you can. Buy less, eat less, and spend less. The impulse decisions and impulsive purchases build only more wastage around us. In order to promote sustainable living communities around you, get out of this habit fast and buy only what you really need.

Feature of Sustainable Living Communities

Sustainable communities generally strive to minimize waste, reduce consumption and preserve open space. Ideally, they don't use resources faster than they can be replenished. Sustainable communities don't produce waste faster than it can be assimilated back into the environment. Sustainable communities are designing the neighbourhood to encourage walking or bicycling. Less deriving means less gas and reduced emissions. Many eco-villages also incorporate work space into homes or encourage telecommuting. They also might zone part of the development for commercial use, essentially making the community a self contained environment where residents don't even have to leave for shopping or entertainment. This design sometimes is called a live–work-play lifestyle.

Using green building techniques is another staple of sustainable communities. For ex. Architects design buildings to take advantages of the sun's lighting and heating capabilities. They install energy-efficient appliances. They try to use local sources of materials as much as possible to cut back on the environmental costs of transport. They build with durable, non-toxic materials that have either been recycled or sustainably harvested. Houses made of straw, which essentially use bales of straw as the structural building blocks; cob houses, which are a mix of straw, clay and sand or earth bag homes, which are exactly what they sound like, homes made out of bags of dirt.

Along with green building techniques, sustainable communities rely on green gardening methods. They landscape with native, drought-tolerant plants and raise them organically to reduce water and keep pesticides and herbicides out of the environment. Some settlements, like Serenbe, also maintain sizable organic vegetable gardens to provide a local food source. Serenbe is a progressive community connected to nature on the edge of Atlanta. A neighbourhood full of fresh food, fresh air and focused on wellbeing. This community is set among acres of preserved forests and meadows with miles of nature trails that connect homes and restaurants with arts and businesses. Serenbe's architectural planning sets a new standard for community living.

Another way sustainable communities reduce their ecological footprint is by capturing and recycling their wastes, often creating their own contained natural cycles. Instead of treating normally perceived waste products such as rainwater and sewage as pollution to be gotten rid of, residents turn them into resources. Sewage, for example, is turned into compost that fertilizes plants and increases soil productivity, while captured rainwater is cleansed through innovative filtering systems and reused for watering plants. Composting is also other sustaining technique to depleted soil and grey water, it can solve the global water crisis.

History, Importance, Preparation And Organisms In Composting

HISTORY OF COMPOSTING

Occasionally, curious individuals want to know the origins of composting. It is difficult to attribute the birth of composting to a specific individual or even one society. The ancient Acadian Empire in the Mesopotamian Valley referred to the use of manure in agriculture on clay tablets 1,000 years before Moses was born. There is evidence that Romans, Greeks and the Tribes of Israel knew about compost. The Bible and Talmud both contain numerous references to the use of rotted manure straw, and organic references to compost are contained in tenth and twelfth century Arab writings, in medieval Church texts, and in Renaissance literature. Notable writers such as William Shakespeare, Sir Francis Bacon and Sir Walter Raleigh all mentioned the use of compost.

On the North American continent, the benefits of compost were enjoyed by both Native Americans and early European settlers of America. Many New England farmers made compost as a recipe of 10 parts muck to 1 part fish, periodically turning their compost heaps until the fish disintegrated (except the bones). One Connecticut farm, Stephen Hoyt and Sons, used 220,000 fish in one season of compost production. Other famous individuals that produced and promoted the use of compost include George Washington, Thomas Jefferson, James Madison, and George Washington Carver.

The early 20th century saw the development of a new "scientific" method of farming. Work done in 1840 by a well-known German scientist, Justus von Liebig, proved that plants obtained nourishment from certain chemicals in solution. Liebig dismissed the significance of humus, because it was insoluble in water. After that discovery, agricultural practices became increasingly chemical in nature. Combinations of manure and dead fish did not look very effective beside a bag of fertilizer. For farmers in many areas of the world, chemical fertilizers replaced compost.

Sir Albert Howard, a British agronomist, went to India in 1905 and spent almost 30 years experimenting with organic gardening and farming. He found that the best compost consisted of three times as much plant matter as manure, with materials initially layered in sandwich fashion, and then turned during decomposition known as the Indore method. In 1943, Sir Howard published a book, An Agriculture Testament, based on his work. The book renewed interest in organic methods of agriculture and earned him recognition as the modern day father of organic farming and gardening.

J.I. Rodale carried Sir Howard's work further and introduced American gardeners to the value of composting for improving soil quality. He established a farming research centre in Pennsylvania and the monthly Organic Gardening magazine. Now, organic methods in gardening and farming are becoming increasingly popular. A growing number of farmers and gardeners who rely on chemical fertilizers are realizing the value of compost for plant growth and restoring depleted soil.

COMPOST

Compost is the single most important supplement depleted soil. Composting is a simple way to add nutrient-rich humus which fuels plant growth and restores vitality to depleted soil. It's also free, easy to make and good for the environment.

Composting for Kitchen Gardening

Composting is the biological decomposition of organic material into a humus-like substance called compost. The process occurs naturally but can be accelerated and improved by controlling environmental factors.

If raw wastes are put directly into the soil, the decomposition process will rob the soil of nitrogen, an important nutrient for plants. (Soil incorporation is one method of composting but requires leaving the area fallow.) Finished compost from a pile is typically a more uniform product with a better balance of nutrients. It can be used throughout the growing season in many different types of applications.

With a pile, composters have more control over adding and mixing the amount of carbon and nitrogen rich materials used to make the end product. In addition, a properly controlled composting environment can ensure production of high temperatures needed for killing weed seeds, diseased plant tissue, and pathogenic organisms.

Materials for Composting

The materials used in compost pile have a major impact on how well the composting process works and the quality of the final compost. The key to good composting is to have a variety of materials and a balanced carbon to nitrogen ratio. Variety increases the types of microorganisms at work in the pile and chances to obtain nutrient rich compost. Some folks (Compost preparing group) think they don't have enough organic material to build and maintain a compost pile. In addition to the leaves and grass clippings that we usually think of composting, there are numerous other suitable organic materials. Most of these materials are easy to find at home. Occasionally, it may be helpful to find free or cheap local sources of organics to add to a pile.

In contrast to those who worry about having enough materials, some folks want to put almost any type of organic material into their pile. While anything organic will eventually decompose, it may not belong in a backyard composting pile. It is important to be aware of these materials and the reasons they should be avoided. New and potential composters often have questions about what materials can be composted.

Commonly used compostable materials

As you are collecting materials around your yard and home, it may not be easy to determine if materials are higher in carbon or nitrogen. Carbon to nitrogen ratios for particular materials are helpful, but they usually only show a limited number of materials (Table 1). A simple method for differentiating between materials is to remember that fresh, juicy materials are usually higher in nitrogen. In addition, materials of animal origin (such as feathers, manure, and blood meal) are typically higher in nitrogen. Drier, older, or woody vegetable and plant tissues are usually higher in carbon. The presence of a carbon, nitrogen, or oxygen in the C/N column indicates whether a material's effect on compost would be carbonaceous (C), nitrogenous (N), or other (O). Materials designated as other (O) do not affect the C: N.

Table 1. List of compostable materials

S. No.	Material	Contain C or N
1.	Bedding, herbivorous	C & N
2.	Blood meal	N
3.	Bone meal	N
4.	Coffee grounds	N
5.	Crushed egg shells	O, alkalizer
6.	Feathers	N
7.	Fruit	N
8.	Fruit peels and rinds	N
9.	Garden debris, dried	C
10.	Garden debris, fresh	C & N
11.	Grass clippings, dried	C
12.	Grass clippings, fresh	N
13.	Hair	N
14.	Hay	C
15.	Lake weeds	N
16.	Leaves	C
17.	Lint	N
18.	Manure	N
19.	Paper(non-recyclable)	C
20.	Peanut shells	C
21.	Straw	C
22.	Pumpkins	N
23.	Vegetable scraps	N
24.	Tea grounds and leaves	N

There are a number of compostable materials that require special handling before they are put into a backyard pile. Some of the materials listed below may require extra preparation or they may need to be added in layers or small quantities. Other materials listed may cause difficulties with the composting process or negatively affect the final product. The comments are intended to help you decide whether to include these particular materials in your own pile.

Table 2. Materials used for making compost and require special handling

S. No.	Material	Contain C or N	Nature of material
1.	Cardboard (non-recyclable)	C	Slow to decompose. Shred into small pieces. If desired, put in water and add a drop of detergent to further speed decomposition.
2.	Corn cobs and stalks	C	Slow to decompose. Run through shredder or chop into very small pieces, mix with nitrogen rich material.
3.	Diseased plants	C	Diseases may be hard to eliminate. Sun-bake plants in plastic bag until thoroughly dried, or leave in hot pile at 55-60°C (131°-140°F) at least one week or burn and put ashes in pile, or omit from pile.
4.	Grass clippings with chemicals	C	Pesticides and herbicides are a concern, degradability ranges from one to twelve months. Leave grass clippings on the lawn (best) or add to pile if material composts for at least 12 months or wait 2-3 weeks before using clippings from lawn after chemicals applied. Do not use clippings as garden mulch for at least 2-3 weeks (or after 2 mowing) after chemical application.
5.	Hedge trimmings	C or N	Slow to decompose. Thin layers of hedge trimmings can be used occasionally for roughage; chop twigs and branches into small pieces.
6.	Lime	O, Alkalizer	Changes pile chemistry, causes nitrogen loss, and too much lime hurts bacteria and other microorganisms. Omit from pile or use very sparingly in thin layers if pile is going anaerobic (do not mix with manure).
7.	Nut shells -walnut, pecan	C	Slow to decompose. Pulverize with shredder.
8.	Peat moss	O, low in nutrients	Highly moisture absorbent, slow to decompose. Mix thoroughly with other materials, adds in small quantities. If possible, soak peat moss in warm water before adding to pile.
9.	Pine Cones	C	Slow to decompose. Shred or chop into very small pieces.

10.	Pine needles	C	Slow to decompose. Mix thoroughly with other materials, adds in small quantities.
11.	Rhubarb leaves	N	Contains oxalic acid which lowers pH and inhibits microbial activity. Add in very small quantities, mix thoroughly with other materials or omit from pile.
12.	Sawdust	C	Slow to decompose, can negatively affect aeration. Work into pile in thin sprinklings, mix with nitrogen rich material.
13.	Sod	N	Slow to decompose. Break into small clumps, mix thoroughly with other materials or cover top of the pile with roots up, grass down (better in fall), or compost separately with roots side up, water thoroughly, cover with a dark tarp.
14.	Soil	O, Activator source	Can make finished compost heavy. Add small quantities in thin layers as soil activator or omit from pile (finished compost produces the same results and typically weighs less).
15.	Walnut leaves	C	Contain jug lone which can be toxic to plants. Add in small quantities, mix thoroughly; toxins will biodegrade in 30 to 40 days.
16.	Weeds, pernicious	C	Rhizomatous root system hard to kill. Sun-bake in plastic bag until thoroughly dried or omit from pile.
17.	Weeds, other	N	Weed seeds hard to kill. Best to use when green and no seed heads present or leave in hot pile at 55-60°C (131-140°F) at least one week.
18.	Wood ashes	O, Alkalizer, potash	Changes pile chemistry, can cause nutrient imbalance. Use very sparingly in thin layers; do not use on top of pile or omit from pile.
19.	Wood chips	C	Slow to decompose. Shred or chop into very small pieces; mix with nitrogen rich material.

Unsuitable Organic Materials

Someday when your compost pile has shrunk and looks disappointedly small, you may scour your yard and home for organics to add to it. Some of those materials do not belong in your backyard compost pile (Table 3).

Table 3. Materials not used in a home compost pile

S. No.	Material	Nature of material
1.	Bones	Very slow to decompose; can attract pests.
2.	Cat litter	May contain pathogens harmful to humans; may also contain chemicals to perfume litter.
3.	Charcoal and briquettes	Contain sulfur oxides and other chemicals that are toxic to soil and plants.
4.	Cooked food waste	May contain fats which attract animals; slow to decompose.
5.	Dairy products	May smell, take a long time to decompose, and attract pests (butter, cheese, mayonnaise, salad dressing, milk, yogurt and sour cream).
6.	Dishwater	May contain grease, perfume, and sodium.
7.	Meat	Can attract pests; smells bad during decomposition.
8.	Paper, glossy colored	May contain inks that could contribute toxins to the pile.
7.	Peanut butter	Can attract pests; slow to decompose.
8.	Pet wastes, human excrement	May contain pathogenic bacteria, viruses, and parasites that require prolonged high temperatures to be destroyed
9.	Sludge (bio-solids)	Requires special handling and high temperatures to kill disease organisms and get rid of toxic metals; do not use unless product is sold in compliance with government regulations.

Additives for Successful Composting

There is a wide array of compost inoculants, starters, and activators sold in stores and mail order catalogs. Fortunately, compost additives are **not** required for successful composting. In some situations, certain additives can be helpful.

Inoculants contain special cultures of dormant bacteria and fungi. The theory behind using them is that they are supposed to introduce microorganisms, hasten the breakdown of materials in a compost pile and produce a better product. They are rarely needed because leaves, kitchen scraps, finished compost, and other organic materials already contain ample bacteria that work readily on their own.

Commercial "starters" or accelerators are supposed to help the decomposition process by adding nitrogen, enzymes, and bacteria to a pile. Some people feel better putting these products in their piles, but independent tests conducted to date have not shown significant benefits. Tests conducted at universities and private research stations showed that the best compost additives are finished compost or topsoil from your yard. (Store bought soil is sometimes sterilized so it does not always add microorganisms.)

Activators contain a nitrogen source. Activators include organic types (manure, blood meal, finished compost, soil) and artificial types (chemically synthesized compounds such as commercial nitrogen fertilizers). While activators are not necessary for successful composting, they can sometimes help if a pile is made from materials low in nitrogen. Nitrogen is usually the limiting nutrient in a pile that doesn't heat up or decay quickly enough. Some purists do not recommend using commercial nitrogen fertilizers as an activator, but if you have some readily available, it may be helpful. Avoid using ammonium sulfate as it may be toxic to earthworms. Keep in mind that chemical fertilizers are not as effective as organic sources because they contain no protein (which microorganisms use). Organic sources are better sources of nitrogen if you need to add an activator.

Table 4. Amounts of various nitrogen sources needed to apply 68 g (0.15 pounds or 2.4 oz) nitrogen

S. No.	Source (N)	N (%)	Application (g)
1.	Ammonium nitrate	33	198.5 (7.0)
2.	Calcium nitrate	15	453.6 (16.0)
3.	Urea	46	147.4 (5.2)
4.	Dried Blood	12	567.0 (20.0)
5.	Fish meal	10	680.4 (24.0)

Figures are in the parenthesis (Ounces)
Source: Dickson, *et al.* (1991)

If additional nitrogen is needed, apply approximately 68 g (0.15 pounds) actual nitrogen per 3 bushels (3 3/4 cubic feet) of carbon rich materials such as leaves. Authors of *The Rodale Book of Composting* recommend adding 907 to 1361 g (2 to 3 pounds) of organic nitrogen supplement (blood meal, manure, bone meal, alfalfa meal) per 45.4 kg (100 pounds) of low nitrogen materials (for example, straw or sawdust).

Composting Methods

The secret to successful composting is to select an approach and technique that suits needs and lifestyle. The choice will depend on a number of factors such as space and material available, plan to use the compost, time period for use and neat look of compost pile. For example, if you only need a little compost, want to expend minimal effort, and have a small area to do it in, and best choice might be a commercially available bin. If you have plenty of space and want large quantities of compost quickly, there is need to build a deluxe three bin unit. If required to compost vegetative food waste separately, it is easiest to directly incorporate them into the soil. Here will cover four methods of composting: holding units, turning units, heaps and sheet composting.

Holding units

Holding units are bins used to hold yard and kitchen materials until composting is complete. They need relatively little maintenance, and some models can be used by apartment dwellers for composting on balconies. Non-woody materials can be added to a holding unit as they are generated. (Many of the commercial one bin systems sold in stores and mail-order catalogs are holding units.) Using a holding unit is one of the easiest ways to compost but is generally slower. This type of enclosure makes it difficult to turn the heap as a way of increasing oxygen. No turning is required, but the lack of aeration causes the composting process to take from six months to two years.

The process can be hastened by using portable bins. Some lightweight units are designed to be taken apart and easily moved. These units can be removed from an existing heap and transferred to an adjacent location. The heap is then turned back over into the unit, mixing and aerating materials. Portable units can be purchased (usually plastic) or constructed from circles of wire fencing or hardware cloth, snow fencing, or wire framed in wood.

Other folks attempt to improve aeration in holding units by adding one or more ventilating stacks or by poking holes into the pile. Ventilating stacks need to be placed into the center of the bin prior to making a pile. Stacks can be made out of perforated pipe, a cylinder of wire mesh or even a bunch of twigs loosely tied together. PVC pipes should be at least one inch in diameter with holes drilled randomly along the length. They can be inserted vertically or horizontally. Another alternative to improve aeration is to place the holding unit on a wood pallet or plastic aeration mat (available from composting equipment dealers).

In holding units, stages of decomposition will vary from the top to the bottom of the heap since yard trimmings and other organics are added continuously. Typically, the more finished compost will be found near the bottom of a pile. Finished compost at the bottom can be removed and used. How easily one gets to the finished compost depends on the type of bin used. Some holding units are designed with a removable front or small doors at the bottom of the bin. With portable bins, finished and unfinished compost can be separated using a similar method to the one described previously. The portable bin should be removed and set nearby. Less decomposed materials from the top of the pile can be put into the empty unit until finished compost is uncovered. More effort is required for heavy or permanent holding units without removable doors. Unfinished compost must be removed and placed in an adjoining unit or temporary storage container. If you have room, it is helpful to have two or three stationary units. One bin can be used for fresh organics, another for maturing materials, and possibly, a third for finished compost.

In addition to the portable bins mentioned earlier, there are numerous other types of commercial and home-built units. Stores and mail order catalogs typically sell units made from plastic and occasionally wood. Home-built units can be constructed from pallets, lumber, hardware cloth, tires, and metal barrels, among other materials. Some people like the appearance of permanent structures which can be made from landscape timbers, concrete blocks, rocks, or bricks.

If plan to build a wood composting unit, avoid using this lumber treated with copper arsenate (CCA), creosote, and penta. (Avoid using the lumber around vegetable gardens.) Toxic compounds from the wood preservatives could leach into your compost. The compounds are harmful to humans and pets. They have been shown to cause cancer and skin and eye irritations. Use wood that is naturally resistant to decay such as cedar or untreated pine. Structures built from pine will probably have to be replaced within a few years. By then, you may be ready for a multiple bin unit or a new design.

Turning units

Turning units are systems designed to be turned or aerated. These units work faster than holding units, because aerobic bacteria are provided with the oxygen they need to break down materials. There are two general forms of turning units: either a series of bins, or a rotating barrel or rolling ball. When organic materials are turned and mixed on a regular basis (every five to ten days), compost can be made in two months or less (assuming a good carbon/nitrogen mix and proper moisture content). Frequent turning offers important advantages in addition to faster composting. Higher temperatures produced as a result of turning (32.2- 60°C or 90° - 140° F) will kill major disease organisms and fly larvae, help kill weed seeds, and provide a good environment for the most effective decomposer organisms.

Turning systems typically cost more than holding units and/or require greater effort to build. Turning composting materials in multiple bins and rolling balls may be difficult for people with back problems or limited physical strength. In contrast, some barrel units are designed for ease of turning and maintenance. These systems may actually be easier to use than holding units for older or physically challenged composters. Barrel units tend to have smaller capacities than most other bins, which make them better suited for people with small amounts of yard trimmings and food scraps.

Materials need to be carefully prepared and added to turning units in stockpiled batches. Materials should be saved until there is enough to fill one bin of a multiple unit, or to fill a barrel unit to the prescribed level. Food wastes can be accumulated in a pest-proof container such as a plastic, five gallon bucket. If necessary, sawdust can be added to the top of each day's scraps to reduce odor.

Heaps

Heap composting is similar to composting with holding and turning units except that it does not require a structure. Recommended dimensions for a heap are 150 cm (5 feet) wide by 90 cm (3 feet) high. Length can vary depending on the amount of materials used. Heaps take more space due to gravity. The wider width will help the pile retain heat better. Materials can be added as they are generated or they can be stored until enough are available to make a good sized heap. During fall months, making a good sized heap will help the composting process work longer into the winter season. Ideally, two heaps are better than one. When the first heap is large enough, it should be allowed to compost undisturbed. A second heap can be started with new materials.

Turning a heap is optional. The composting process will obviously take longer if the pile is not turned. Food scraps should not be thrown on an unturned pile because pests are likely to be attracted. Woody materials may also pose a problem. If woody materials are not cut up into small pieces, the pile may tend to become more of a brush pile than a composting pile. A woody pile decomposes extremely slowly, usually over a period of several years, and can become huge quickly.

Sheet composting

Sheet composting is a way to obtain the benefits of decayed organic material without building a composting pile. Sheet composting involves spreading a thin layer of organic materials, such as leaves, over a garden area. The materials are then tilled in with a hoe, spade, garden fork, or rotary tiller. Leaves, garden debris, weeds, grass clippings, and vegetative food scraps are examples of materials that can be easily tilled into the soil. To aid decomposition, materials should be shredded or chopped prior to layering.

The danger of sheet composting as a compost-making method is that carbon containing residues will call upon the nitrogen reserves of the soil for their decomposition. On the other hand, high-nitrogen materials may release their nitrogen too quickly in the wrong form. What may take a matter of weeks in a compost pile, given confined and thermophilic conditions, may take a full season in the soil.

To ensure adequate decomposition of organic materials before planting, it is best to do sheet composting in the fall. Spread a 5 – 10 cm (2 to 4 inch) layer of organic materials on the soil surface and till in. A rotary tiller will do the most thorough job of working materials into a vegetable garden. In a flower bed containing perennials and bulbs, it may be necessary to carefully work the organic material in with a garden fork or hoe.

Pit or trench composting

This is the simplest way for composting kitchen scraps. Dig a one-foot-deep hole. Chop and mix the food wastes into the soil then cover with at least 20 cm (8 inches) of additional soil. Depending on soil temperature, the supply of microorganisms in the soil and the content of the materials, decomposition will occur in one month to one year.

Food waste burial can be done randomly in unused areas of the garden or in an organized system. One system is to bury scraps in holes dug around the drip line of trees or shrubs. An English system, known as pit or trench composting, maintains three season rotation or soil incorporation and growing. Sometimes this is also called vertical composting. Divide garden space into 3' wide rows.

Firstt year – Dig a 30 cm (1 foot) wide trench on the left hand 1/3 of the 3' area (A). Add compostable materials in this trench and cover with soil when 1.25 cm (half an inch) full. Leave the center 1' section open for a path (B), and plant your crop in the remaining 1' strip along the right side (C).

Second year – Section A is a path for year 2 allowing time for the Materials to break down. Plant your crop in section B. Section C, where you planted last year, becomes the compost trench.

Third year – Section A is now ready for planting. Section B is your trench for composting. Section C is in the second year of composting is it will be the path.

Building of compost pile

Put a pile of leaves, a cardboard box and a watermelon in your back yard, exposed to the elements, and they will eventually decompose. Time taken to break down depends on a number of factors such as

- » Materials used in compost pile
- » Exposed surface area
- » The availability of moisture and air

Backyard composting is a process designed to speed up the breakdown or decomposing of organic materials. Insure the makeup of the material is a mixture that bacteria and other microorganisms can easily feed upon, breaking them down into compost. A proper C: N is the goal. Carbon in fallen leaves or woodier wastes serves as an energy source. Nitrogen in the greener materials provides microbes with the raw element of proteins to build their bodies.

The more surface area then the microorganisms has to work on, the faster the materials will decompose. It's like a block of ice in the sun: slow to melt when it is large, but melting very quickly when broken into smaller pieces. Chopping your garden wastes with a shovel or a machete, or running them through a shredding machine or lawnmower will increase their surface area, thus speeding up your composting.

All life on earth needs a certain amount of water and air to sustain itself. The microbes in the compost pile are no different. They function best when the compost materials are about as moist as a wrung out sponge and are provided with many air passages for aerobic breakdown. Adding water and turning the pile maintains efficient decomposition. Extremes of sun, wind, or rain can adversely affect this balance in your pile.

Understanding these key factors when composting allows for efficient, quick break down of kitchen and yard wastes, turning them into "Black Gold" ¾ compost.

Selection of Location, Time and Tools for Composting

Location

The right location is important for a successful compost pile. Choose a level area with good drainage. Standing water will slow down the pile. If possible avoid direct sunlight and areas exposed to strong winds, which can dry and cool the pile. A half day sun situation is ideal. Shaded areas are fine but pay attention to limited rainfall through a canopy of leaves, and slow drying out of a saturated pile. Some trees may send roots up into the pile in search of water and nutrients. When the pile is turned, these roots may be damaged. If your only location is near trees, you may want to consider setting a brick or stone foundation.

Select a convenient location. One that is easy to get to and not where you will have to trudge a long distance just to add your carrot peelings. Choose an area that does not interfere with family activities but is close to a water source and has enough space for temporary storage of excess organic wastes. Avoid placing your compost pile near dog areas or other animals. Animal urine and faeces may harbour unwanted pathogens.

Don't place your pile directly against wooden buildings, fences or trees, because wood in contact with compost will decay. Avoid placing under a wide overhang that would limit rainfall, or under a drippy eave or rain spout that would continually saturate your pile.

Size

The recommended size for a home compost pile is no smaller than 1.5 m x 1.5m x 1.5m, and no larger than 1.5 m x 1.5m x 1.5m. A smaller pile may not heat up high enough for efficient breakdown, or it may lose heat and quickly slow down the process. A larger pile may hold too much water not allowing air into the centre. This would create an anaerobic environment. Air naturally penetrates 45 to 60 cm into a pile from all directions. The biggest problem with a large pile is physically turning the pile. It can be too much for some people to manage.

Time of composting

A compost pile can start any time of the year but there are limitations during certain seasons. You can build your pile as materials become available. In the spring and early summer, high nitrogen materials are available, but very little carbon materials are available unless you stored leaves from the fall. In summer you start to have garden debris, but your mowing may be lessened due to high summer temperatures. Fall is the time of year when both nitrogen from cool season lawn mowing and carbon from fallen leaves are readily available.

Tools

Besides the type of bin use - if any at all - there are a few tools that will make your composting easier. A 4- or 5-tined pitch fork for turning pile is the main item. The compost turners do work especially for those who lack upper body strength or have back problems. A garden hose or watering can needs to be handy to water pile. Other items are also useful: Pruners, a machete or a shredder to cut up large pieces of organic waste to increase surface area.

A compost thermometer can monitor pile's temperature. A metal pole will also indicate heat. Insert the pole into the centre of the pile as a compost thermometer. Where the thermometer will give the exact temperature of the pile, a warm/hot to the touch metal pole will also indicate activity.

For kitchen scraps it is a good idea to keep a covered container in the kitchen to store scraps until you have time to take it to the pile. Often it is impractical to run out and dump every kitchen scrap as it is produced. Place the container under the sink or near the back door.

Compost Piling, Care and Problems

Layering

When starting a compost pile the recommended practice is to layer the materials thinly and uniformly, the same way lasagne (Italian food) is made with thin layers of pasta, cheese, and sauce. Never overdoing any one single ingredient and never skipping a layer in the construction process will prove successful! It will happen only when starting a new pile. Once the pile is active you add materials by either burying them in the centre or incorporating them when you turn your pile.

It is recommended to start your pile on bare ground. Don't place your pile on asphalt or concrete. This impedes aeration and inhibits microbial contact with the earth. If tree roots are a problem, a loosely laid brick foundation could be installed. Placing a pallet underneath the pile is a possibility if the area may be damp or holds water in the spring. This creates air channels from below. Starting with the bottom layer, continue to layer until you reach the top or (what happens most often) you run out of material. Firm and lightly water each layer as it is added but do not compact.

First layer - The organic materials layer can be vegetable wastes, sod, grass clippings, leaves, hay, straw, chopped corncobs, corn stalks, untreated sawdust, twigs less than 1.25 cm

(½ inch) in diameter, or garden debris. Remember the proper C: N and mix accordingly. The bulkier organic materials do best in the first ground level layer. As pile settles, these items tend to allow for more air spaces. Shred or chop up materials for greater surface area. The organic layers should be between 15- 20 cm (6-8 inches) thick. Materials that tend to mat such as grass clippings should be either mixed in or placed in 5 - 7.5 cm (2-3 inch) layers within this 15 - 20 cm (6-8 inch) layer.

Second layer - Animal manures, fertilizers or starters serve as activators that accelerate the ignition or initial heating of your pile. They all provide a nitrogen source for the microbial community. Some provide proteins and enzymes. If manure from a grain eating animal is available, add 2.5 – 5 cm (1-2 inch) layer. If this is not available, add one cup of 10-10-10 or 12-12-12 commercial fertilizer per 2.32 square meters (25 square feet). If using a commercial starter, follow label directions.

Third layer - Top soil or active compost introduce microorganisms. Plain garden soil is fine. Avoid soil that has been treated with insecticides recently and sterile potting soils, which lack these necessary microbes. A 2.5 to 5 cm (one to two inch) layer is enough.

Care

Temperature plays an important role in the composting process. Decomposition occurs most rapidly between 43.3 to 71.1 °C (110° to 160°F). Within two weeks, a properly made compost pile will reach these temperatures. At this time, pile is settling, which is a good sign of that the pile is working properly.

If there is needed to add material in pile, it can be added throughout the growing season and into the winter months. Add fresh material, there is need to turn and water to pile more often. Monitoring the temperature and turning whenever the piles temperature dips below 43.3 °C (110°F) keeps pile active at its highest level, and will have the fastest breakdown. This means turning the pile more often. This can be weekly and it is work. In reality, the average composter turns their pile once every 4 to 5 weeks. This mixes in the fresh material with the older, adds air to the pile and allows you to add water. With this method, a pile started in the fall, added to and turned the following summer will be ready in late fall of that year or the next spring.

If there is no need to add new material, turn and water the pile 35-42 days after initial heating. Make sure to turn the outside of the old pile into the centre of the new pile. The compost should be ready to use 90 to 120 days later.

The squeeze test is an easy way to gauge the moisture content of the pile. The organic material should feel damp to the touch, with just one or two drops of water expelled when squeezed tightly in the hand.

In the winter months when the temperatures lower, the pile cools down and eventually all activity ceases. Most people let the pile shut down and plan to reactivate it in the spring. During extremely cold weather, the task of getting to compost pile may be the most difficult. To keep the pile active in the winter, it will have to insulate. Covering the pile will help retain heat and prevent water build-up. During the winter, kitchen scraps are generally the only items added.

Problems and their Solution

» Not enough air due to overwatering and compaction. If odor of ammonia, too much nitrogen. If compost has unpleasant odor then add dry materials such as cornstalks, leave, or wood chips to soak up excess water. Turn the pile to aerate. Cover pile if rains continue. Turn the pile to aerate. Add carbon materials and turn the pile to aerate.

» Pile is too small if has insufficient moisture, not enough air and lack of nitrogen, compost may be finished but pile is not heating up. Its solution is make thepile bigger. Add water while turning and add water by sticking a garden hose into the centre in several locations. Poke holes into the pile and add water using a watering can. Turn the pile to aerate. Mix in nitrogen materials, add 10-10-10. Use it and start over.

» Pile is too small and compost is damp and warm only in the centre then adds more material.

» Not enough air, lack of carbon and pile temperature exceeds 71 ^{0}C (160°F). Then turn the pile to aerate. Mix in carbon materials.

» When low surface area then large, undecomposed items are still in the mix. Remove that items, chop or shred before adding.

» Presence of meat scraps. There will be rodents then only add items recommended for your pile and remove offensive material. Animal-proof bin.

» Good pile is composting correctly. Insects are a sign of a productive compost pile. The white cobweb material is actinomycete, part of the microbial community. Then compost pile has flies, earwigs, slugs and/or other insects. I find white material throughout my pile.If there is an abundance of flies, bury your food scraps as you turn the pile.

Benefits and Uses

Use of compost

Compost is ready to use when it is dark, brown, and crumbly with an earthy odour. It would not be mouldy and rotten. Crumbly compost will be sort of fluffy; it does not need to be decomposed to a point of being powdery. The original materials that went into the compost pile should no longer be recognizable in finished compost, except for some woody pieces. The temperature of the finished compost should be the same as the outside air temperature, and the material should not reheat. At this stage see earthworms and other insects now that the temperature is lower. If compost is still hot, smells like ammonia, or you can still recognize much of the original material, which went into the pile, then it is not ready to use yet. Once the compost appears finished, let it sit for at least 21 days to make sure the decomposition process has stabilized.

Compost may be tempted to use before it is ready. However, if incompletely decomposed material is added to the garden compost, bacteria may compete with plants for nitrogen in the soil. Plants will look stunted and yellow. Unfinished compost (Not fully prepared compost) has been found to also retard germination and growth of seedlings.

Benefits of compost

Whether a compost pile is quick and hot or slow and cool, when the decomposers have completed their work the contents of the pile have been transformed to an entirely new material. The volume of the finished compost has been reduced because of biochemical breakdown and water respiration to about 30 to 50 per cent into the pile. This finished product offers numerous benefits to soil, are given below:

» Compost will improve the quality of almost any soil, and for this reason it is most often considered a soil conditioner.

» Compost improves the structure and texture of the soil enabling it to better retain nutrients, moisture, and air for the betterment of plants.

Incorporating compost into soil dramatically improves soil structure. Soil structure refers to how inorganic particles (sand, silt, clay) combine with decayed organic particles (compost, humus). Soil with good structure has a crumbly texture, drains well, retains some moisture, and is easy to turn over. "Crumbly" means how the particles held together. A soil amended with compost shows that it is made up of many round, irregular aggregates. Aggregates are groups of particles loosely bound together by secretions of worms and compost bacteria giving it this crumbly appearance. If lightly crush one of these aggregates, it breaks down into smaller aggregates. Crumbly soil allows air to penetrate and holds moisture well but allows excess water to drain away. Tender young roots also have an easier time penetrating into the soil. A well-structured soil with lots of small aggregates stays loose and is easy to cultivate. Compost helps improve all soil types, especially sandy and heavy clay soils.

A garden with sandy soil has very little water and nutrient retention. Sandy soil feels loose and has coarse particles that won't hold their shape when squeezed in hand. Water and nutrients pass through quickly since there is nothing to hold them there. In loose, sandy soil compost helps to bind these particles together and increase the soil's ability to retain moisture and nutrients. In other words, there is now something to hold onto. Plant roots penetrate easily, finding moisture where there was none before.

Clay soils appear heavy and dense. The soil particles are small and tightly bound together. When wet, clay is sticky and easily holds together when squeezed in your hand. When compost is mixed with clay soils, it binds to the clay particles forming larger particles that now have larger air spaces between them. These spaces allow better surface water drainage and air penetration.

Compost also adds nutrients to your soil. Compost contains a variety of the basic nutrients that plants require for healthy growth. In addition to the main three; nitrogen, phosphorous, and potassium, of special importance are the micronutrients found in compost such as manganese, copper, iron, and zinc. Micronutrients are only needed in small doses, like vitamins in our diet, but they play an important role in the plant's ability to extract nutrients from other foods. In a commercial fertilizer, such as 10-10-10, micronutrients are often missing. Compost is basically a free nutrient boost for your plants.

Compost is made up of different ingredients, some of which rot more rapidly than others. As a result, nutrients are released over a long period of time. Call it a slow-release fertilizer. Actually, if everything decomposed at the same rate, compost would not be so valuable. The nutrient content of each batch of compost is impossible to predict because it depends on so many variables. The greater the variety of materials used in making compost, the greater the variety of nutrients in the finished product.

Adding compost to soil breaks down over time and provides nitrogen to garden soil and landscape plants. Sufficiently aged compost releases organic nitrogen after soils warm in the spring. It has been show that the breakdown of this organic material provides 25% of its nitrogen in the first year, 10% in the second and third year, and declines to 5% in the fourth and fifth year.

Compost attracts earthworms and provides them with a healthy diet. The presence of earthworms, red worms, centipedes, sow bugs, and other soil critters shows that compost is a healthy living material. The presence of these decomposers means that there is still some organic material being slowly broken down releasing nutrients as foods pass through their digestive tracts. They also represent balanced soil ecology.

Research showed that soil treated with compost tends to produce plants with fewer pest problems. Compost helps to control diseases and insects that might otherwise overrun a more sterile soil lacking natural checks against their spread. Leaf based compost is showing promise suppressing nematodes. Compost application to turf has suppressed many fungal diseases.

Soil pH also benefits with the addition of compost. This is a measure of soil acidity or alkalinity. Finished compost has a neutral pH. For the majority of ornamental plants, nutrients are available to them at a pH range of 5.5 to 7.5. When mixed into the soil, compost helps keep the pH at optimum levels for nutrient availability. Organic matter also has a high capacity to fix soil toxins. According to Dr. Selman Waksman in his book *Humus: Origin, Chemical Composition, and Importance in Nature,* high salt concentrations are less injurious, and high aluminium solubility and its specific injurious action are markedly decreased.

Other Uses of Compost

Mulch

In nature, plants and trees drop leaves that accumulate at their bases. Every year, a new layer is added while the old layers start to decompose. This is leaf mould and it is a form of compost. Nature is doing important role for providing a protective layer over the roots of plants. This layer of vegetative material protects the bare soil during the summer months by reducing soil temperature, suppressing weed growth and reducing soil moisture loss. The compost can do the same thing in the soils of gardens and landscapes.

To prepare any area for mulching, first clear away grass or weeds that might grow up through the mulch. Make sure to remove the roots of tough perennial weeds such as ground ivy. When using compost as mulch in flower beds, vegetable gardens, landscape beds, or lawns, screen the finished compost. A simple screen can be made using 1.25 cm (½-inch) mesh hardware cloth and attaching it to a wooden frame. Place the screen over a wheel barrow or other container and sift the compost into it. The large pieces left behind can go into next compost pile as an activator, introducing the necessary microorganisms. Cover the garden or bed area with screened compost to a depth of one to two inches. If apply compost on a lawn, be sure it is finely ground or sifted because have less of a chance of smothering the lawn. Use 0.6 cm (¼-inch) mesh hardware cloth. One way to incorporate the compost is to aerate the sod, then apply a 0.3 to 0.6 cm (1/8-inch to ¼-inch) covering of fine compost. Use a rake to distribute the compost into the crevices. When mulching around trees and shrubs, screening may not be necessary. This is really a matter of aesthetic desire.

Soil amendment

At the above already discussed about how compost helps soil, especially sandy and clay soils. When starting a new garden soil amending is recommended before you plant. It is so much easier to add compost now than it is after the garden is planted. Cover the garden area with 7.5 to 10.0 cm (3 to 4 inches) of compost and till it into the upper 15 cm (six inches) of the soil. If garden is already established and incorporation of compost in the deeper layers of soil, the options are limited.

With perennials, every time add a new plant to the garden or divide an existing one, add compost. In annuals can add compost every spring. Loosen up the entire area where annuals will be planted and work in compost. Around trees and shrubs add at planting time, mixing no more than 25 percent of soil volume. Some references say not to use any at all for fear that the roots will remain in the planting hole area and not grow out into the surrounding soil. Keeping the compost level at one-quarter of the total soil volume will not lead to this problem. If required, use the compost as mulch only.

Around existing trees it may be difficult to incorporate into the upper 15 cm (six inches) of the soil. Add compost by injecting nutrients the way professional arborists do. Drill 2.5 cm to 5.0 cm (1-to 2-inch) diameter holes 30 cm (12 inches) deep in the soil throughout the tree canopy and beyond at 45 cm (18 inch) spacing. Fill the bottom of each hole with recommended rates of dry fertilizer and then top off the holes with compost. For shrubs, the holes only need to be drilled 20 to 25 cm (8 to 10 inches) deep. This treatment should supply nutrients for two to three years.

Using compost in potting mixes

It can also blend fine-textured compost in potting mixtures. However, make sure the compost does not make up more than one quarter to one half of the potting mixture's volume. Plants growing in containers are entirely reliant on the water and nutrients provided in the potting mix. Compost is excellent for container growing mixes because it stores moisture effectively and provides a variety of nutrients not typically supplied in commercial fertilizers or soil-free potting mixes. There is still need to fertilize containers on a regular basis to provide the high volume of nutrients they need. Finely sifted compost can also be used in seed starting mixtures.

Compost tea

An old fashioned way of providing liquid fertilizer for plants is to brew compost tea. Similar to manure tea, compost tea gives your plants a good dose of nutrients. Compost tea works especially well for providing nutrients to new transplants and young seedlings. To make compost tea fill burlap sack or an old pillow case with finished compost and secure the open end. Place in a tub, barrel, or watering can filled with water. Agitate for a few minutes and then let it steep for a few days. Water will leach out nutrients from the compost and the mixture will take on the colour of tea. Spray or pour compost tea on and around plants. Use the bag of compost for several batches. Afterwards, simply empty the bag's contents onto the garden.

Reduce, reuse and recycle

Armed with a better understanding of the value of finished compost and the biological processes that transform excess organic material generated around our home into a usable product, it is easy to see the use of compost. Home composting not only provides a free valuable soil amendment but also eliminates dumping fees, landscape yard waste bags, and the work involved filling them and hauling them to the curb.

Many of the organic materials generated such as lawn clippings, landscape trimmings, kitchen scraps, leaves, and untreated cardboard can be managed through composting. They can be reused and recycled. But what about reducing the amounts produced in the first place. Proper landscape plant selection and recycling grass clippings are two ways can eliminate excess yard waste:

Landscape plant selection

The landscapes frame our homes. The intent of landscaping is to enhance these structures and their surrounding environment. Landscapes also can generate a lot of yard waste that has managed throughout the growing season. Reduce the excess of production of waste; consider the following points when selecting landscape material:

Maples are beautiful trees that with age generate large amounts of leaves in the fall. Rather than selecting an eventual 24 m (80-foot) giant, perhaps a smaller maple will do such as an Amur Maple or a Paper bark Maple, both reaching 6 to 7.5 m (20 to 25 feet). Your fall leaf pile will be much smaller.

Sycamore trees produce large dinner-plate size leaves. Honey locust trees produce small leaves. Remember, the greater the exposed leaf surface the faster the decomposition. Honey locust leaves break down and blow away in one season while sycamore leaves linger, smothering desirable plant material if not removed. Sycamore leaves will eventually breakdown.

Crabapple trees are beautiful in the spring but many of the older cultivars are prone to diseases and lose their leaves midsummer only to re-leaf and fall off again in autumn. It will generate the same amount of yard waste twice in the same season. Choose a disease resistant variety.

Ash trees can produce huge numbers of seeds every year. Select a seedless cultivar and eliminate this yard waste. Evergreen plant material doesn't annually shed all of its foliage.

When selecting a plant for your yard, find out the mature size. Take into consideration the eventual location of the plant. With careful consideration of mature size, yard trimmings can be reduced. If have a bush shaped like an elephant then have more yard waste.

Recycle grass clippings

The Illinois Department of Energy and Natural Resources estimates that lawn care for an average Midwestern residence generates some 340.19 kg (750 pounds) of grass clippings a year. Multiply that times the number of lawns of the community and how quickly the tons of just this one type of excess yard waste add up. Landfills no longer accept yard waste. If a regional composting facility is nearby, then bag the grass clippings and haul them there or community may have a yard waste pick-up day. What to do to reduce this large amount of material? An easy option is to leave the grass clippings on the lawn.

This recycling strategy has many different names as Don't Bag it, Just Say Mow and Grass cycling etc. The idea is the same: clippings left on the lawn will naturally breakdown. Contrary to what some folks believe, grass clippings will not damage lawns. In the 1960's, consumers were told that a bagger on their mower was a necessity to slow down thatch development. If not, a build-up of thatch would occur. Research has shown that leaving grass clippings on the lawn does not contribute to thatch.

Thatch: It is a tightly intertwined layer of dead and living grass stems and roots that can develop between the soil surface and green vegetation. This layer develops when dead organic matter accumulates faster than it decomposes. A thatch layer less than 1.25 cm (½ inch) is considered beneficial. It insulates roots, reduces soil compaction and serves as a mulch to prevent excessive water evaporation. A layer more than 1.25 cm (½ inch) increases the disease susceptibility of the turf and reduces tolerance to environmental extremes. An 11-year study at the USDA research station in Beltsville, Maryland, found that on an annual basis, leaving grass clippings on the lawn contributes only 0.08 cm (0 .03 inches) to the thatch layer. Grass recycling does not spread lawn disease either. Disease spores are present whether clippings are left or removed. Turf grass disease occurs when disease-causing spores contact susceptible grasses under certain environmental conditions.

Grass clippings are a valuable organic source of nutrients, especially nitrogen. As clippings decompose, these nutrients become available for use by the grass plant. Clippings also shade the soil surface and reduce moisture loss due to evaporation. In addition, leaving your clippings saves time, work, and money. A recent study was conducted in Fort Worth, Texas, with 147 homeowners who quit bagging their clippings. The homeowners mowed their lawns 5.4 times per month versus 4.1 times by homeowners who bagged their grass. However, the grass recyclers spent an average of seven hours less during the grass cutting season on yard work because they did not have to spend time bagging grass for disposal. Savings in money can be realized from reduced fertilizer applications. Leaving grass clippings can reduce yearly nitrogen applications by 25 percent. It also saves by not purchasing trash bags and with less wear and tear on mowers by not having a bag attachment full of heavy clippings.

Mulching mowers and mulching attachments for existing mowers are available but they do is reducing the grass clipping size thus increasing the rate at which clippings decompose. No need to purchase special equipment. A normal rotary mower is fine. Owners of old rear or side discharge mowers can leave the discharge opening closed, or they can remove the catching bag. Cover the discharge chute with a plate/cover that can be purchased at a local hardware store. Most new mowers are designed for improved grass recycling with decks that are shaped to cut clippings into smaller pieces so they will fall below the tops of the grass blades, helping to prevent clumping.

- » Keep your mower blade sharp. A dull mower blade tears grass which increases the chance of disease infestations.

- » Cut your lawn when the grass is dry. Wet grass is difficult to cut evenly, dulls blades, and tends to form clumps.

- » Clean the mower deck periodically. Wet clippings can become matted on the underside of the mower deck, resulting in clumping of clippings or mechanical failure. Proper mowing will insure quick breakdown of grass clippings and overall lawn quality.

- » For best results, do not cut more than one-third of the leaf surface at any one cutting (for example, if grass is three inches tall, only remove one inch). This practice allows you to leave clippings on the lawn without visible clumping and is healthier for the turf than cutting too short.

Grass recycling is not appropriate in every situation. Prolonged wet weather, mechanical breakdown of mowers, or infrequent mowing are situations where clippings should be bagged or collected.

Municipal Compost Facilities

Across Illinois, there is Environmental Protection Agency permitted compost facilities that will accept homeowner yard waste. Using the windrow system where a large machine mechanically turns the rows, compost is made in about 10 to 12 months. For a minimal fee, homeowners have this valuable soil amendment locally available. These sites are visited twice a year to insure that they are fulfilling regulated requirements.

Microorganisms in compost

Microorganisms such as bacteria, fungi, and actinomycetes account for most of the decomposition that takes place in a pile. They are considered chemical decomposers, because they change the chemistry of organic wastes. The larger decomposers, or macroorganisms, in a compost pile include mites, centipedes, sow bugs, snails, millipedes, springtails, spiders, slugs, beetles, ants, flies, nematodes, flatworms, rotifers, and earthworms. They are considered to be physical decomposers because they grind, bite, suck, tear, and chew materials into smaller pieces.

Of all these organisms, aerobic bacteria are the most important decomposers. They are very abundant; there may be millions in a gram of soil or decaying organic matter. You would need 25,000 of them laid end to end on a ruler to make an inch. They are the most nutritionally diverse of all organisms and can eat nearly anything. Bacteria utilize carbon as a source of energy (to keep on eating) and nitrogen to build protein in their bodies (so they can grow and reproduce). They obtain energy by oxidizing organic material, especially the carbon fraction. This oxidation process heats up the compost pile from ambient air temperature. If proper conditions are present, the pile will heat up fairly rapidly (within days) due to bacteria consuming readily decomposable materials.

While bacteria can eat a wide variety of organic compounds, they have difficulty escaping unfavorable environments due to their size and lack of complexity. Changes in oxygen, moisture, temperature, and acidity can make bacteria die or become inactive. Aerobic bacteria need oxygen levels greater than five percent. They are the preferred organisms, because they provide the most rapid and effective composting. They also excrete plant nutrients such as nitrogen, phosphorus, and magnesium. When oxygen levels fall below five percent, the aerobes die and decomposition slows by as much as 90 percent. Anaerobic microorganisms take over and, in the process, produce a lot of useless organic acids and amines (ammonia-like substances) which are smelly, contain unavailable nitrogen and, in some cases, are toxic to plants. In addition, anaerobes produce hydrogen sulfide (aroma-like rotten eggs), cadaverine, and putrescine (other sources of offensive odors).

There are different types of aerobic bacteria that work in composting piles. Their populations will vary according to the pile temperature. Psychrophilic bacteria work in the lowest temperature range. They are most active at 12.8°C (55° F) and will work in the pile if the initial pile temperature is less than 21.1°C (70° F). They give off a small amount of heat in comparison to other types of bacteria. The heat they produce is enough however, to help build the pile temperature to the point where another set of bacteria, mesophilic bacteria, start to take over.

Mesophilic bacteria rapidly decompose organic matter, producing acids, carbon dioxide and heat. Their working temperature range is generally between 21.1 to 37.8°C (70° to 100° F). When the pile temperature rises above 37.8°C (100° F), the mesophilic bacteria begin to die off or move to the outer part of the heap. They are replaced by heat-loving thermophilic bacteria.

Thermophilic bacteria thrive at temperatures ranging from 45 to 71.1°C (113° to 160° F). Thermophilic bacteria continue the decomposition process, raising the pile temperature 54.4 to 71.1°C (130° to 160° F), where it usually stabilizes. Unless a pile is constantly fed new materials and turned at strategic times, the high range temperatures typically last no more than three to five days. Thermophilic bacteria use up too much of the degradable materials to sustain their population for any length of time. As the thermophilic bacteria decline and the temperature of the pile gradually cools off, the mesophilic bacteria again become dominant. The mesophilic bacteria consume remaining organic material with the help of other organisms.

The drop in compost pile temperature is not a sign that composting is complete, but rather an indication that the compost pile is entering another phase of the composting process. While high temperatures (above 60°C or 140° F) have the advantage of killing pathogenic organisms and weed seeds, it is unnecessary to achieve those temperatures unless there is a specific concern about killing disease organisms and seeds. (You can greatly reduce the possibility of pathogens in a pile by excluding pet waste, diseased plants, and manure from diseased animals.) Many decomposers are killed or become inactive when pile temperatures rise above 60°C or 140° F. If the pile temperature exceeds 60°C or 160° F, you may want to take action and cool the pile by turning it. A number of research projects have shown that soil amended with compost can help fight fungal infestations. If the compost pile temperature goes above 60°C or 160° F, the composting material may become sterile and lose its disease fighting properties.

While the various types of bacteria are at work, other microorganisms are also contributing to the degradation process. Actinomycetes, higher-form bacteria similar to fungi and molds, are responsible for the pleasant earthy smell of compost. Greyish in appearance, actinomycetes work in the moderate heat zones of a compost pile. They decompose some of the more resistant materials in the pile such as lignin, cellulose, starches, and proteins. As they reduce materials, they liberate carbon, nitrogen, and ammonia, making nutrients available for higher plants. Actinomycetes occur in large clusters and become most evident during the later stages of decomposition.

Like bacteria and actinomycetes, fungi are also responsible for organic matter decay in a compost pile. Fungi are primitive plants that can be either single celled or many celled and filamentous. They lack a photosynthetic pigment. Their main contribution to a compost pile is to break down cellulose and lignin, after faster acting bacteria make inroads on them. They prefer cooler temperatures 21 – 24 °C (70 to 75° F) and easily digested food sources. As a result, they also tend to take over during the final stage of composting.

Macro-organisms

As mentioned earlier, larger organisms are involved in physically transforming organic material into compost. They are active during the later stages of composting – digging, chewing, sucking, digesting and mixing compostable materials. In addition to mixing materials, they break it into smaller pieces, and transform it into more digestible forms for microorganisms. Their excrement is also digested by bacteria, causing more nutrients to be released.

Micro- and macro-organisms are part of a complex food chain. This food chain consists of organisms classified as either first-, second-, or third-level consumers. The categories are based on what they eat and who eats them. First level consumers become the food for second level consumers, which in turn, are eaten by third level consumers. Soil ecologist Dr. Daniel L. Dindal gives an example of how the food chain works in *Ecology of Compost*. Mites and springtails eat fungi. Tiny feather-winged beetles feed on fungal spores. Nematodes ingest bacteria. Protozoa and rotifers present in water films feed on bacteria and plant particles. Predaceous mites and pseudoscorpions prey upon nematodes, fly larvae, other mites and collembolans. Free-living flatworms ingest gastropods, earthworms, nematodes and rotifers. Third-level consumers such as 1.5 m x 1.5m x 1.5m centipedes' rove beetles, ground beetles, and ants prey on second-level consumers.

Macro-organisms in compost piles

Ants: Ants feed on a variety of materials including fungi, seeds, sweets and other insects. They help the composting process by bringing fungi and other organisms into their nests. Ants can make compost richer in phosphorus and potassium by moving minerals around as they work.

Millipedes: Millipedes have wormlike segmented bodies, with each segment having two pairs of walking legs (except the front few segments). Millipedes help break down plant material by eating soft decaying vegetation. They will roll up in a ball when in danger.

Centipedes: Centipedes are flat, segmented worms with one pair of legs in each segment. They are third-level consumers that feed on soil invertebrates, especially insects and spiders.

Sow bugs: Sow bugs have a flat and oval body with distinct segments and ten pairs of legs. They are first-level consumers that feed on rotting woody materials and other decaying vegetation. Pill bugs look similar to sow bugs, but roll up in a ball when disturbed.

Springtails: Springtails are small insects distinguished by their ability to jump when disturbed. They rarely exceed one-quarter inch in length and vary in color from white to blue to black. Springtails are principally fungi feeders, although they also eat molds and chew on decomposing plants.

Flies: Flies are two-wing insects that feed on almost any kind of organic material. They also act as airborne carriers of bacteria, depositing it wherever they land. Although flies are not often a problem associated with compost piles, you can control their numbers by keeping a layer of dry leaves or grass clippings on top of the pile. Also, bury food scraps at least eight to twelve inches deep into the pile. Thermophilic temperatures kill fly larvae. Mites help to keep fly larvae reduced in numbers.

Beetles: Beetles are insects with two pairs of wings. Types commonly found in compost piles include the rove beetle, ground beetle, and feather-winged beetle. The feather-winged beetle feeds on fungal spores. Immature grubs feed on decaying vegetables. Adult rove and ground beetles prey on snails, slugs, and other small animals.

Snails and slugs: Snails and slugs are mollusks that travel in a creeping movement. Snails have a spiral shell with a distinct head and retractable foot. Slugs do not have a shell and are somewhat bullet shaped with antennae on their front section. They feed primarily on living plant material, but they will also attack plant debris. Look for them in finished compost before using it, as they could do damage to your garden if they move in.

Spiders: Spiders are eight-legged creatures and third-level consumers that feed on insects and small invertebrates. They can be very helpful for controlling garden pests.

Earthworms: Earthworms are the most important of the large physical decomposers in a compost pile. Earthworms ingest organic matter and digest it with the help of tiny stones in their gizzards. Their intestinal juices are rich in hormones, enzymes, and other fermenting substances that continue the breakdown process. The worms leave dark, fertile castings behind. A worm can produce its weight in castings each day. These castings are rich in plant nutrients such as nitrogen, calcium, magnesium, and phosphorus that might otherwise be unavailable to plants. Earthworms thrive on compost and contribute greatly to its quality. The presence of earthworms in either compost or soil is evidence of good microbial activity.

Factors Affecting the Preparation of Composting

There are certain key environmental factors which affect the speed of composting. The organisms that make compost need food (carbon and nitrogen), air, and water. When provided with a favorable balance, they will produce compost quickly. Other organism factors affecting the speed of composting include surface area/particle size, volume, and temperature.

Food

Organic material provides food for organisms in the form of carbon and nitrogen. As described earlier, bacteria use carbon for energy and protein to grow and reproduce. Carbon and nitrogen levels vary with each organic material. Carbon-rich materials tend to be dry and brown such as leaves, straw, and wood chips. Nitrogen materials tend to be wet and green such as fresh grass clippings and food waste. A tip for estimating an organic material's carbon/nitrogen content is to remember that fresh, juicy materials are usually higher in nitrogen and will decompose more quickly than older, drier, and woodier tissues that are high in carbon.

A C: N ratio ranging between 25:1 and 30:1 is the optimum combination for rapid decomposition. If ratio is more than 30:1 carbon, heat production drops and decomposition slows. It has been noticed that a pile of leaves or wood chips will sit for a year or more without much apparent decay. When there is too much nitrogen, your pile will likely release the excess as smelly ammonia gas. Too much nitrogen can also cause a rise in the pH level which is toxic to some microorganisms.

The C: N ratio does not need to be exact. Values in table 5 are calculated on a dry-weight basis. It is difficult to determine an exact C: N ratio without knowing the moisture content of the materials being used. Blending materials to achieve a satisfactory C: N ratio is part of the art of composting. A simple rule of thumb is to develop a volume-based recipe using from one-fourth to one-half high-nitrogen materials.

Table 5: Carbon and nitrogen ratios of some composting materials

S. No.	Material	C: N
1.	Corn stalks	50-100:1
2.	Fruit waste	35:1
3.	Grass clippings	12-25:1
4.	Hay, Green	25:1

5	Leaves, Ash	21-28:1
6.	Leave, pine	60-100:1
7.	Leaves, other	30-80:1
8.	Manure, horse and cow	20-25:1
9.	Paper	170-200:1
10.	Sawdust	200-500:1
11.	Seaweed	19:1
12.	Straw	40-100:2
13.	Vegetable waste	12-25:1
14.	Weeds	25:1
15.	Wood chips	500-700:1
16.	Black elder	21-28:1

Air

Proper aeration is a key environmental factor. Many microorganisms, including aerobic bacteria, need oxygen. They need oxygen to produce energy, grow quickly, and consume more materials. Aeration involves the replacement of oxygen deficient air in a compost pile with fresh air containing oxygen. Natural aeration occurs when air warmed by the composting process rises through the pile, bringing in fresh air from the surroundings. Aeration can also be affected by wind, moisture content, and porosity (spaces between particles in the compost pile). Composting reduces the pile's porosity and decreases air circulation. Porosity can be negatively affected if large quantities of finely sized materials such as pine needles, grass clippings, or sawdust are used. In addition, air circulation can be impeded if materials become water saturated.

Air movement in the pile can be improved with a few simple techniques. The easiest way to aerate a pile is to regularly turn it with a pitchfork or shovel. Turning will fluff up the pile and increase its porosity. Another option is to add coarse materials such as leaves, straw, or corn stalks. Other options include using a compost aeration tool (available from garden supply companies) or a ventilator stack. Stacks can be made out of perforated plastic pipes, chicken wire wrapped in a circle, or bundles of twigs. Ventilator stacks may be useful for large piles and should stick out the top or sides.

Moisture

Decomposer organisms need water to live. Microbial activity occurs most rapidly in thin water films on the surface of organic materials. Microorganisms can only utilize organic molecules that are dissolved in water. The optimum moisture content for a compost pile should range from 40 to 60 percent. If there is less than 40 percent moisture, bacteria slow down and may become dormant. If there is more than 60 percent, water will force air out of pile pore spaces, suffocating the aerobic bacteria. Anaerobic bacteria will take over, resulting in unpleasant odours.

The ideal percentage of moisture will depend on the organic material's structure. Straw and corn stalks will need more moisture than leaves, while food waste or grass clippings are not likely to need additional moisture. Since it is difficult to measure moisture, a general rule of thumb is to wet and mix materials so they are about as moist as a wrung-out sponge. Material should feel damp to the touch, with just a drop or two of liquid expelled when squeezed in your hand.

If a compost pile is too dry, it should be watered as the pile is being turned or with a trickling hose. Certain materials such as dead leaves, hay, straw, and sawdust should be gradually moistened until they glisten. These types of materials have a tendency to shed water or adsorb it only on the surface. If a pile is saturated with water, turn it so that materials are restacked. It may also help to add dry, carbon rich material.

Temperature

Temperature is another important factor in the composting process and is related to proper air and moisture levels. As the microorganisms work to decompose the compost, they give off heat which in turn increases pile temperatures. Temperatures between 32.2 and 60°C (90° and 140°F) indicate rapid decomposition. Lower temperatures signal a slowing in the composting process. High temperatures greater than 60°C (140° F) reduce the activity of most organisms.

Outside air temperatures can impact the decomposition process. Warmer outside temperatures in late spring, summer, and early fall stimulate bacteria and speed up decomposition. Low winter temperatures will slow or temporarily stop the composting process. As air temperatures warm up in the spring, microbial activity will resume. During winter months, compost piles can be covered with a tarp to help retain heat longer, but it is not necessary.

Novice composters and people interested in making fast compost may want to track temperatures. The most accurate readings will come from a compost thermometer or temperature probe. Compost thermometers are available from many garden supply companies.

Another method for monitoring temperature is to stick your fist into the pile. You can also place a metal pipe or iron bar in the middle of the pile, periodically pulling it out and feeling it. If the bar or the interior of the pile feels uncomfortably warm or hot during the first few weeks of composting, you will know everything is fine. If the temperature inside the pile is the same as the outside that is an indication that the composting process is slow. You can increase activity by adding nitrogen rich material and turning the pile.

Particle size

Particle size affects the rate of organic matter breakdown. The more surface area available, the easier it is for microorganisms to work, because activity occurs at the interface of particle surfaces and air. Microorganisms are able to digest more, generate more heat, and multiply faster with smaller pieces of material. Although it is not required, reducing materials into smaller pieces will definitely speed decomposition. Organic materials can be chopped, shredded, split, bruised, or punctured to increase their surface area. Don't powder materials, because they will compact and impede air movement in the pile.

For many yard trimmings, cutting materials with a knife, pruning shear, or machete is adequate. An easy way to shred leaves is to mow them before raking. It can collect them at the same time if mower has a bag attachment.

Another option is to use a lawn trimmer to shred leaves in a garbage can. Several different models of shredders and chippers are available for sale or rental to use in shredding woody materials and leaves. It is a good idea to wear safety goggles when doing any type of shredding or chopping activity. Hands should be kept out of the machine while it is in operation.

Kitchen scraps can be chopped up with a knife. Some ambitious people use meat grinders and blenders to make garbage soup from their food scraps and water. They pour the mixture into their heaps.

Volume

Volume is a factor in retaining compost pile heat. In order to become self insulating and retain heat, piles made in the Midwest should ideally be about one cubic yard. The one cubic yard size retains heat and moisture, but is not too large that the material will become unwieldy for turning. Homes located on lakes or in windy areas may want to consider slightly larger piles measuring 120 cm x 120 cm x 120 cm (4 feet x 4 feet x 4 feet). Smaller compost piles will still decompose material, but they may not heat up as well, and decomposition is likely to take longer.

Perparation, Requirement And Maintenance Of Vermi-Compost And Farm Yard Manure

PREPARATION OF ENRICHED FARM YARD MANURE

FYM is prepared in trenches of 6.0-7.5m (20-25 ft) long, 1.5-1.80 m (5-6 ft) broad and 0.9-1.05 m (3.0-3.5 ft) deep. Farm waste mixed with Earth should be spread under each animal in the evening for the absorbance of urine. The litter should be localized in the areas where urine generally drops and soaks into the ground and urine drains out in case of cemented floor. Each morning the urine soaked litter and dung should be well mixed and taken to the manure trench. A section of 0.9 m (3 ft) length of the trench from one end should be taken up for filling with daily collection of refuse from cattle shed. When the trench is filled to a height of 45-60 cm (1.6-2.0 ft) above ground level, the top of the heap is made dome shaped and plastered with cow dung mixed with soil. The manure becomes ready in about three months. By this time the next 0.9m (3 ft) length of the trench is being filled up. Generally two such trenches would be needed for three to four heads of cattle. It is possible to prepare 200-300 cu ft of manure (3-5 tons or 10-12 cartloads) per year per head of cattle. The farm yard manure should be reinforced by addition of superphosphate at 30-40 kg/trench before application to fields.

Use of superphosphate as chemical preservative will have following three advantages:

i. It will reduce loss of nitrogen as ammonia from farm yard manure.

ii. It will increase percentage of phosphorus in manure, thus making balanced one.

iii. Since tri-calcium phosphate produced with the application of Superphosphate to the farm yard manure is in organic form which is readily available to the plants, it will increase the efficiency of phosphorus utilization in acidic soils which tend to fix the available phosphorus (P_2O_5) of superphosphate into unavailable form.

VERMI-COMPOST

Vermi-culture and vermi-composting are two interlinked and inter dependent processes, which when conjoined, can be referred as vermi-technology. Vermi-culture can only be done on compostable or decomposable organic matter. Composting is the outcome of earthworm activities. So, both the processes can be brought about simultaneously. In other words, we can multiply earthworms for various uses and can obtain vermicast or vermi-compost at a faster pace, if we have a higher number of earth worms.

Requirements for Vermi-Composting

1. **Container:** Vermi-composting container can be of any shape or size and requirement depends upon quantity of waste to be composted and number of live earthworms we want to culture. On average, 2000 adult earthworms can be maintained in containers of 1 m² dimension. These with appropriate conditioning of composting material would convert approximately 200 kgs wastes every month. In a container of 2.23 x 2.23 meters, 10 kg earth worms can be maintained which have conversion rate of approximately one tonne per month. However, to have optimal conversion normally only upper 9-12 layers are composted. This should be softly scrapped off.

2. **Bedding material:** This is the lower most layer of earthworm feed substrate that is required to be vermicomposted. For this any biodegradable matter is used like banana stem peels, coir pith, coconut leaves, sugarcane trash, stems pf crops, grasses or husk. Waste or discarded cattle feed can also be used for bedding.

3. **Moisture content:** Moisture content during vermin-composting should be maintained between 30-40 per cent. If moisture is high, dry cow dung manure or leaf litter should be mixed with substrate.

4. **Temperature:** Requirement for optimal results is 20-30° C. However, survival of earthworms is even at lower temperatures and upto 48°C air temperature. Obviously with little provision of shade, temperature within worm feed substrate (material to be vermicomposted) can be reduced. For this it is desirable that substrate should not be tightly packed in containers.

5. **pH:** pH of substrate should be between 6.8 to 7.5.

6. **Cover of feed substrates:** This is required for reducing moisture loss and also save worms from extra movements (outside substrate) or from predators like ants. Moist gunny bag covers also help in conservation of moisture.

7. **Selection of right type of earthworm species:** Three types of earthworms can be used for vermin-culture. These are: (1) Epiges; (2) Endoges; and (3) Aneciques.

Maintenance of Vermi-composting Beds

Vermi-composting is a type of **composting** in which certain species of earthworms are used to enhance the process of organic waste conversion and produce a better end-product. It is a mesophilic process utilizing microorganisms and earthworms. Preparation of vermin-composting beds after necessary pre-tests and selection of suitable species is done. Steps for the described six schemes are as under:

» Available container is to be selected and cleaned from removing unwanted chemical or other material. At a bottom a 5 to 7.5 cm (2 to 3 inches) thick layer of any biodegradable matter is laid. A partly digested 5 to 7.5 cm (2 to 3 inches) layer of powdered cow dung is put over this layer. The whole material in bedding is sufficiently moistened (up to 40 % moisture) and then live earthworms are gently released over it. A box of 1 m x 1 m x 0.5 m high can hold around 1000 to 1500 earthworms which require about 30 to 40 kg of whole organic matter. On top, earthworm feed matter should be put in when previous layer disappears, i.e. is converted into vermin-compost. The vermin-compost should be periodically removed from the top. Finally on top a moist gunny bag should be kept.

» Dung and other feed materials are thoroughly mixed, watered and is subject to partial digestion for 14 to 21 days in layer of 30 cm thick. On this earthworms are released.

» In container bottom, a 2 cm thick layer of fine sand is laid. Over this, a 5 to 7.5 cm (2 to 3 inches) thick layer of saw dust is laid which is covered with a thin layer of garden soil. Thickness of each layer can be 5 to 7.5 cm (2 to 3 inches). Whole is watered and earthworms are released. The material is then covered with a moist gunny bag. This process for vermin composting is slow as saw dust takes time to partly decompose. However, can be useful for areas where saw dust is available.

» In case suitable container is not available, vermin-composting can be done on ground.

For this ground is levelled over this, soil free of stones, glass and any form of chemical contaminants is plastered and manageable sized platform is made. Depending upon availability of space and compostable organic waste, the size of platform is made. It could be 90 cm (3 feet) wide and 6-15 m (20-50 feet) long. Over the platform feed layer (9) partly digested feed is spread. After watering, live earthworms are introduced. Finally more layers of pre-tested waste-matter can be heaped. The whole is covered with a hessian cloth. If platforms are already under thatched roof shade, heaps can also be covered with broad leaves locally available. These however have to be periodically replaced as are decomposed and eaten up by worms.

» In another variation, heaps in semicircle can be made in same manner as in previous one. Composting can be done. Only advantage is that sufficiently moist dung does not spread if plastering with organic waste and cow dung has been done well.

» Rectangular or circular pits of dimension per requirement can be made and vermin-composting process can be taken up by any of the described methods. In this, however, packing or layering of the material has to be loose and not compared. To a certain extent common earthworm, *Metaphire posthuma* can also be used in this type of composting.

Collection of Vermin-compost

When vermin-compost and vermin-casting are ready for collection, top layer appear somewhat dark brown, granular as if used dry tea leaves have been spread over the layer. Watering should then stop 2-3 days and gently compost should be scrapped from the top layer or to a depth, it appears as vermi-compost. This should then be removed to a side and left undisturbed for 6 to 24 hours. If there are adults worms present then these would move down or away from the composted material. Some cocoon invariably goes along with the compost and would lead to natural dispersal of earthworms. From heaps, mature earthworms are easily removed by upturning the whole heap and leaving it for 6-12 hours. All mature earthworms go down and vermin-compost is removed from the top.

Reuse, Type, Treatment Of Grey Water

Grey water can be defined as any domestic wastewater produced, excluding sewage. It is the waste water from showers, baths, spas, hand basins, laundry tubs and washing machines. Water from dishwashers and kitchen sinks is often referred to as dark grey water, because it has a higher load of chemicals, fats and other organic matter. Water from toilets is called black water. The main difference between grey water and sewage (or black water) is the organic loading. Sewage has a much larger organic loading compared to grey water. Some people also categorize kitchen wastewater as black water because it has quite a high organic loading relative to other sources of wastewater such as bath water. People are now waking up to the benefits of grey water re-use, and the term wastewater is in many respects a misnomer. Maybe a more appropriate term for this water would be used Water.

REUSE OF WASTE WATER

With proper treatment grey water can be put to good use. These uses include water for laundry and toilet flushing, and also irrigation of plants. Treated grey water can be used to irrigate both food and non food producing plants. The nutrients in the grey water (such as phosphorus and nitrogen) provide an excellent food source for these plants.

Grey water may contain traces of dirt, food, grease, hair, and certain household cleaning products. While grey water may look dirty, it is a safe and even beneficial source of irrigation water in a yard. Keep in mind that if grey water is released into rivers, lakes, or estuaries, its nutrients become pollutants, but to plants, they are valuable fertilizer. Aside from the obvious benefits of saving water (and money on your water bill), reusing your grey water keeps it out of the sewer or septic system, thereby reducing the chance that it will pollute local water bodies. Reusing greywater for irrigation reconnects urban residents and our backyard gardens to the natural water cycle.

The easiest way to use grey water is to pipe it directly outside and use it to water ornamental plants or fruit trees. Grey water can also be used to irrigate vegetable plants as long as it doesn't touch edible parts of the plants. In any grey water system, it is essential to use "plant friendly" products, those without salts, boron, or chlorine bleach. The build-up of salts and boron in the soil can damage plants. While you're at it, watch out for your own health: "natural" body products often contain substances toxic to humans. It's estimated that just over half of household water used could be recycled as grey water, saving potentially hundreds of litres of water per day.

Principles in Reuse of Waste Water

It is believe that for residential grey water systems simple designs are best. With simple systems, the grey water is not send into an existing drip irrigation system, but must shape the landscape to allow water to infiltrate into the soil. Generally recommend simple, low-tech systems that use gravity whenever possible, instead of pumps. It is prefer irrigation systems that are designed to avoid clogging, rather than relying on filters and drip irrigation.

It is promote grey water reuse as a way to increase the productivity of sustainable backyard ecosystems that produce food, clean water, and shelter wildlife. Such systems recover valuable waste products–grey water, household compost, and human manure and reconnect their human inhabitants to ecological cycles. By modelling appropriate technologies for food production, water, and sanitation in the industrialized world, it is hope to replace the cultural misconception of wastewater with the possibility of a life-generating water culture.

It is believe more complex systems are best suited for multi-family, commercial, and industrial scale systems. These systems can treat and reuse large volumes of water, and play a role in water conservation in dense urban housing developments, food processing and manufacturing facilities, schools, universities, and public buildings. Because complex systems rely on pumps and filtration systems, they are often designed by an engineer, are expensive to install and may require regular maintenance.

Basic instructions for reuse of grey water

Grey water is different from fresh water and requires different guidelines for it to be reused.

- Don't store grey water (more than 24 hours). If store grey water the nutrients in it will start to break down, creating bad odours.

- Minimize contact with grey water. Grey water could potentially contain a pathogen if an infected person's faces got into the water, so your system should be designed for the water to soak into the ground and not be available for people or animals to drink.

- Infiltrate grey water into the ground, don't allow it to pool up or run off (knowing how well water drains into soil (or the soil percolation rate of soil) will help with proper design. Pooling grey water can provide mosquito breeding grounds, as well as a place for human contact with grey water.

- Keep system as simple as possible, avoid pumps, and avoid filters that need upkeep. Simple systems last longer, require less maintenance, require less energy and cost less money.

- Install a 3-way valve for easy switching between the grey water system and the sewer/septic.

- Match the amount of grey water your plants will receive with their irrigation needs.

The benefits of grey water reuse

Re-using water does not diminish our quality of life; however it can provide benefits on many levels.

Two major benefits of grey water use are:

- Reducing the need for fresh water. Saving on fresh water use can significantly reduce household water bills, but also has a broader community benefit in reducing demands on public water supply.

- Reducing the amount of wastewater entering sewers or on-site treatment systems. Again, this can benefit the individual household, but also the broader community.

Treatment of grey water for reuse

There are many ways by which to treat grey water so that it can be re-used. The various methods used must be safe from a health point of view and not harmful to the environment. These types of grey water systems rely on plants and natural microorganisms to treat the water to a very high standard so that it can be safely re-used. The main advantage with these types of systems is that they treat the grey water naturally, and also enhance the local environment because of the attractive plants used and the fauna attracted to them. There are other natural systems available to treat grey water. The type of system selected will depend on the specific application, and selection would be considered on a case by case basis.

Grey water treatment systems

These systems collect and treat (and some disinfect) the water to various levels of purity and hygiene. Several stages are involved in the treatment of water:

- » Filtration of solids (lint and hair).
- » Removal of pathogens and unwanted chemicals (such as salts and nutrients) using either micro-organisms or chemical treatment.
- » Disinfection by chlorination or UV light, though not all systems do this.

Treated water can be used in washing machines and toilets, as well as on the garden. If you don't have much garden to water, or if you don't need to water it in all seasons, this sort of system may be a more useful option as it can use the water elsewhere.

Cost of treatment system

Cost of treatment systems varied. Systems that treat grey water to 'Class A' level which is considered safe for watering plants intended for eating, but not for drinking or preparing food. There is also need to pay ongoing maintenance costs, to cover regular service call-outs and filter replacements. Installation costs tend to escalate if a lot of extra plumbing is required (if bathroom and laundry pipes are spread all around the house) or if pipes are in a concrete slab. The amount and location of water storage can also affect costs. Installing a system when building a new house (or doing major renovations) tends to be cheaper than retrofitting one. The cost of mains water in Australian cities is so low that it is unlikely to ever recoup the cost of a grey water treatment system. On the other hand, if consider that a garden makes up about 10% of the value of home, it might think the cost of a treatment system to keep the garden (and home value) growing during water restrictions is worthwhile.

Best option for recycling

If access to mains water, it's unlikely you'll want to bother with grey water recycling unless it is really want to do for the environment, or if don't have a rainwater tank. But if don't access to mains water, or produce a lot of grey water thanks to a large household, it could be worth investing in some sort of grey water recycling system. The system that best suits will depend on your situation:

Big garden (fruit or ornamentals)

The cheapest and simplest solution is to get a diverter and send water to the lawn or garden through sub-surface irrigation pipes. However, a higher level of treatment will be safer and give you more options.

Vegetable gardening

If want to water herbs and vegetables there is need to get a higher level of water treatment. Untreated grey water should definitely not be used on food that will be eaten raw.

Small or no garden

If produce a lot of grey water but don't have much garden, it requires a treatment system that allows the water to be used in toilet and/or washing machine.

Drought affected supply

In a drought-affected part of the country and rely on rainwater for household water supply, a grey water treatment system could help a lot. If plumbed into toilet and washing machine, it will save precious drinking water. It will also reduce the load on septic tanks or drought-stressed waterways (also consider a waste water treatment system that handles black water as well as grey water).

Installing a grey water treatment system will reduce how much water is used and reduce the amount of waste going into the sewerage system. While it won't necessarily save much money, it will get that warm fuzzy feeling for helping the environment.

Safety about the Treated or Untreated Grey Water

There are limits to do with untreated grey water because of the chemicals and bacteria in it, but treated grey water is somewhat safer to use.

- » Untreated grey water should only be used for sub-surface garden irrigation that is, through a network of pipes buried at least 1m below the ground – to reduce the risk of human or animal contact.

- » Pipes carrying untreated grey water must display relevant warning labels.

- » You can't store untreated grey water, because the bacteria and other pathogens could multiply to dangerous levels.

- » Use it immediately (or within 24 hours), and if it's raining, divert it to the sewer.

- » If someone in your family is sick with gastro or flu or another contagious disease, stop using the grey water.

- » Don't use grey water if you've been washing nappies or using bleaches or dyes.

- » Don›t water herbs, vegetables or pot plants with untreated grey water.

- » Your grey water shouldn›t escape from your property into a neighbouring one, into storm water systems or aquifers used for drinking water in fact it›s illegal.

Grey water is a complex substance and there are many things to consider if it is use safely and for maximum benefit.

Grey Water-Friendly Laundry Detergents

Choice tested washing machine run-off for chemicals that could harm garden plants and contaminate soil. Some laundry detergent products whose names imply they're environmentally friendly could in fact cause problems if used on your garden.

- » The components most likely to cause problems are phosphorus, salinity, sodium, and pH.

- » Small amounts of phosphorus can be useful for plants, and it's a major component of fertilizer. When it gets into waterways, however, it can cause excessive algal growth, leading to toxic algal blooms. The effect on your soil is varied depending on your soil type. Clay soils can deal with more phosphorus because the phosphorus binds to clay minerals and doesn't leach away. On sandy soils, excess phosphorus can leach into groundwater. Australian soils are typically low in phosphorus, and some native species can't tolerate high levels.

- » All laundry detergents contain salts, typically sodium salts such as sodium nitrate, sodium sulphate, sodium phosphate and sodium silicate. All laundry detergents are highly saline, and frequent long-term use would likely harm your garden, unless it was spread over a large area.

- » Sodium is particularly detrimental not only to plants, but soil. It affects the soil's permeability and causes a loss of structural stability.

- » Laundry detergents are highly alkaline (that is, has a high pH). A pH higher than 10 helps dissolve organic dirt, such as grease, oils and food scraps. Most biological systems prefer a pH between 6 and 9, so grey water with a high pH is likely to harm many plants and soil organisms.

- » Potential impacts are very much dose-dependent try reducing the amount of detergent you use, providing it still gets your clothes acceptably clean.

- » Water from front-loading washing machines tends to have a high concentration of detergent, so unless you're willing to use less than the recommended amount and compromise on the 'cleanliness' of your clothes, we wouldn't recommend using the wash water on your garden. The amount of rinse water may not be enough to make it worthwhile (about 30 to 50L per cycle enough to water about two square metres of garden).

The legal law

Different laws apply to grey water use, depending on area to area. Anyone can set up their own basic diversion system (check the safety warnings above), but if anyone considering installing a grey water treatment system, you'll need to:

- » Consult a licensed plumber for advice on the best system for your needs

- » Ask your local council if you're eligible for any green rebates

- » Consult your sewerage removal authority if you intend to redirect all or a major part of your used water

- » Check with your water supply authority and inform them of any changes to your plumbing.

Before you order any grey water diverter or treatment system, you need to check that it's accredited by your state health or environment department. It's unfortunate, and perhaps ironic, that although anyone can pipe their contaminant-loaded washing machine and bath water onto the garden, systems that provide a better level of treatment mean more bureaucratic dealings.

Important Tips for Grey Water Use in Gardening

By far the easiest way to use your grey water is on the garden. When you're calculating how much grey water you can put on your garden, reckon on about 20L per square meter per watering. The frequency of watering depends on the local climate, rainfall and the season, while the amount of water per square meter depends on soil type (20L is for loam soils).

Using these average figures though, the average person produces enough grey water to water 35 square meters of lawn or garden once a week.

So, to work out how many square meters of garden you can supply with grey water:

- » Either calculate the amount of grey water produced by your household each week, and divide it by 20
- » Or take the number of people in your household, and multiply by 35.

Tips

- » Keep an eye on the health of your plants. Grey water tends to be high in chemicals that alter the structure of the soil, and it also tends to be overused.
- » Bear in mind: sick looking plants could be suffering from overwatering, rather than the chemicals in the grey water.
- » Give your plants a break by using rainwater (if you have a rainwater tank) or tap water every six weeks.
- » Use compost to increase the organic content of your soil, improve its structure and help it survive the chemical onslaught.
- » If your grey water is untreated, don't use it to water edible plants or indoor plants.

Grey water from laundry to landscape

From the Washing Machine: Washing machines are typically the easiest source of grey water to reuse because grey water can be diverted without cutting into existing plumbing. Each machine has an internal pump that automatically pumps out the water- you can use that to your advantage to pump the grey water directly to your plants.

Laundry Drum: If you don't want to invest much money the system (maybe you are a renter), or have a lot of hard scape (concrete/patio) between your house and the area to irrigate, we recommend a laundry drum system. Wash water is pumped into a drum, a large barrel or temporary storage called a surge tank. At the bottom of the drum the water drains out into a hose that is moved around the yard to irrigate. This is the cheapest and easiest system to install, but requires constant moving of the hose for it to be effective at irrigating.

Laundry to landscape system, image credit: Clean water components

This system that gives flexibility in what plants to be irrigate and takes very little maintenance, recommend the laundry-to-landscape system. This system was invented by Art Ludwig. This grey water system doesn't alter the household plumbing: the washing machine drain hose is attached directly to a diverter valve that allows switching the flow of grey water between the sewer/septic and the grey water irrigation system. The grey water irrigation system directs water through 2.5 cm (1 inch) tubing with 1.25 cm (1/2 inch) outlets directing water to specific plants. This system is low cost, easy to install, and gives huge flexibility for irrigation. In most situations this is the number one place to start when choosing a grey water system.

Laundry-to-landscape grey water system

The laundry-to-landscape system is a simple system with easy distribution of grey water to multiple plants. It is relatively low cost, and easy to install. The washing machine's internal pump slightly pressurizes the grey water, so this system can irrigate plants across a flat yard.

The washer hose is connected to a 3-way valve that can divert grey water either to the sewer or the grey water system and piped outside with 1" rigid pipe, like PVC. This system requires 1" diameter tubing/pipe- larger diameter tubing will decrease the pressure and make it harder to spread out the water; and smaller diameter tubing will add too much strain on the pump. Outside 1" HDPE (high density polyethylene, black plastic tubing) is connected to the rigid pipe and "barbed" Tee fittings split the flow and allow grey water to spread out and water many plants. Advantages of laundry to landscape system are given below:

» Legal to install without a permit (in several states, like California) so long as state guidelines are followed.

» Grey water can travel slightly uphill (think small bumps in your yard) or longer distances across flat yards.

- » Accessible 3-way valve makes it easy to switch between the grey water system and the sewer/septic.
- » Very little maintenance.
- » Very little digging required.
- » Easy to spread out water and reach many plants (up to 20 with a top-loading machine, up to 8 with a front-loading machine, up to 10 with a top-efficient machine).
- » Flexible- Easy to change after installation

Disadvantages

- » Involves some very minor plumbing rerouting.
- » If poorly designed the system could add strain on washer pump and potentially shorten its life.
- » Some parts not readily available and must be ordered.

Basic materials

- » Kits available from Clean Water Components or Gray-2-Green
- » 1" brass three-way valve
- » 1" PVC pipe and fittings
- » "auto vent" or check valve used for air gap
- » If yard is higher than washer, use a swing check valve
- » 1" HDPE (high density polyethylene plastic) tubing and barbed fittings
- » Ball valves (1/2", used optionally)

1" barbed fittings and tubing are not available in most stores. You can mail order from Clean Water Components.com or try a large irrigation supply store. In urban areas you can often find the brand Blu-Lock or Eco-Lock in 1 inch. In areas with agriculture or vineyards, it's possible to find the 1" tubing in 100" sections at a local farm or irrigation store. In the SF Bay Area the Urban Farmer Store sells all the materials, in Los Angeles Aqua-Flo Irrigation carries all the parts.

Cost: $150-$300 in materials. Labor costs vary, but typically add a between $500-$2,000 to the materials depending on the size and complexity of the system.

Pump Longevity

To maximize the life of your pump avoid pumping grey water more than a few feet above the rim of the washer. If sending grey water long distances over flat ground the friction in the pipe will add some resistance on the pump. A safe rule of thumb distance is 15 m (50 feet) of horizontal pipe.

From the shower

Showers are great sources of grey water; they usually produce a lot of relatively clean water. To have a simple, effective shower system considers a gravity-based system (no pump). If yard is located uphill from the house, then you'll need to have a pumped system.

Branched drain

The branched drain system was also invented by Art Ludwig. Grey water in this system flows through standard 3.8 cm (1.5 inch size) drainage pipe, by gravity, always sloping downward at 2% slope, or 0.6 cm (1/4 inch) drop for every foot travelled horizontally, and the water is divided up into smaller and smaller quantities using a plumbing fitting that splits the flow. The final outlet of each branch flows into a mulched basin, usually to irrigate the root zone of trees or other large perennials. Branched drain systems are time consuming to install, but once finished require very little maintenance and work well for the long term.

From the sinks

Kitchen sinks are the source of a fair amount of water, usually very high in organic matter (food, grease, etc.). Kitchen sinks are not allowed under many grey water codes, but are allowed in some states, like Washington, Oregon, Arizona, and Montana. This water will clog many kinds of systems. To avoid clogging, recommend using a branched drain system with mulch basins, organic matter collects in the woodchips and decomposes. Since bathroom sinks don't typically generate much water, they can often combine flows with the shower water. Or, the sink water can be drained to a single large plant, or divided to irrigate two or three plants.

Constructed wetlands

Constructed wetlands are used to "ecologically dispose" of grey water. If produce more grey water than need for irrigation, a constructed wetland can help use up some extra grey water. Wetlands absorb nutrients and filter particles from grey water, enabling it to be stored for longer or sent through a properly designed drip irrigation system (though more filtration and pumping is also required). Grey water is also a good source of irrigation for beautiful, water loving wetland plants. If near a natural waterway and don't have anywhere else to direct grey water, a wetland can safely clean and soak-up grey water, protecting the creek. If an arid climate, or are trying to reduce fresh water use, don't recommend incorporating wetlands into grey water systems as they use up a lot of the water, which could otherwise be used for irrigation.

Pumped systems

If can't use gravity to transport the grey water (yard is sloped uphill, or it's flat and the plants are far away) there is need to pump grey water uphill. In basic pumped system grey water flows into a large (usually 190 L or 50 gallon) plastic barrel that is either buried or located at ground level. Inside the barrel an effluent pump pushes the water out through irrigation lines (no emitters) to the landscape. Pumps add cost, use electricity, and will break, so avoid this if possible.

Indoor grey water use

In most residential situations it is much simpler and more economical to utilize grey water outside, and not create a system that treats the water for indoor use. The exceptions are in houses that have high water use and minimal outdoor irrigation, and for larger buildings like apartments.

There are also very simple ways to reuse grey water inside that are not a "grey water system". Buckets can catch grey water and clear water, the water wasted while warming up a shower. These buckets can be used to bucket flush a toilet, or carried outside. There are also simple designs like Sink Positive, and more complicated systems like the Brac system. Earthships have an interesting system that reuse grey water inside with greenhouse wetlands.

Plants choice of grey water use

Low tech, simple grey water systems are best suited to specific, large plants. Use them to water trees, bushes, berry patches, shrubs, and large annuals. It's much more difficult to water lots of small plants that are spread out over a large area. (Like a lawn or flower bed).

Grey water friendly products

The products listed below do not contain high levels of salts or boron. The Oasis soaps and ECOS do not change the pH of the water, and in general any liquid detergents do not change the water's pH.

Laundry detergent

Oasis Laundry Detergent (liquid), ECOS liquid detergent, Vaska, Trader Joes Liquid Detergent, Puretergent and BioPac. Vaska also makes a commercial laundry detergent that is compatible with commercial washing machines. For Sinks use dishes and hands. Oasis All Purpose Cleaner, Dr. Bronner's any natural liquid or bar soap.

Body products

In general, the ingredients in personal care products are not harmful to plants including most shampoos and conditioners. If anybody wants to learn about what ingredients may be harmful to human health, visit the Cosmetic Database, a project of the Environmental Working Group and enter products to see how they rate for toxicity. Aubrey Organics is one company producing many types of shampoos and conditioners without toxic ingredients.

Plants choice for grey water irrigation

Whatever is water source, grow plants that produce food, provide habitat to wildlife, or create other beneficial uses like mulch, fertilizer, fuel, or building materials. In general, larger plants, such as trees, bushes, and perennials, are easier to irrigate with simple grey water systems than smaller plants. Turf grass, made up of hundreds of individual plants, is the most difficult to irrigate with grey water- don't recommend it. Remember it can safely irrigate any food plant so long as grey water doesn't touch the edible portion of the plant (no root crops).

Fruit trees

Most fruit trees thrive on grey water, and there are many delicious options! They can tolerate frequent watering and once established they can go long periods with no water.

Choosing a fruit tree: To start, use root stocks that are resistant to local diseases (ask at your local nursery or Cooperative Extension) and plant trees that are known to grow well in particular area. Trees of that area will also do better if it has good soil; adding compost may be helpful.

Drainage: Next, consider the drainage of the site. If drainage is poor, there will need to plant the tree on a mound and water it less to prevent diseases like crown rot. When planting trees ensure that the crown of the tree is above the mulch basin to prevent crown rot.

Salt: Fruit trees are generally salt sensitive and should not be irrigated with water from powdered detergents or other products containing salts. If your grey water source contains lots of salt (dishwasher detergent, for example, is high in salt), add salt-tolerant plants in the landscape, or irrigate frequently with rainwater to flush salts from the soil. Plants that thrive on recycled or reclaimed water (highly treated wastewater) are good choices for high-salt grey water. It's best to use plant-friendly products, those low in salts and free of boron, to ensure good quality irrigation water.

Other plants

Other perennials that thrive on grey water include edible shrubs and vines such as raspberries, thimbleberries, blackberries and their relatives, currants, gooseberries, filberts, rhubarb, elderberry, passion fruit, kiwi, hops, and grapes. Blueberries love acidic soil so you'd have to choose pH neutral soaps or use acidic mulch.

Irrigate at the drip line of your plants

Plant roots typically extend well past their drip line, the outer edge of the branched. Dig the mulch basin at or beyond the drip line and direct grey water into the basin.

Determine the water requirement of plants

Design simple grey water system to direct an appropriate amount of water to each plant; too much could over saturate the soil while too little could dry out the plants. Rough estimate of water for a typical fruit tree during the irrigation season (without rain to supplement) is given when design of system make sure to consult further resource such as The Water-Wise Home or the San Francisco Gray water Design Guidelines for Outdoor Irrigation, to get more detailed instruction on how to determine plant water requirements.

Approximate weekly irrigation needs for a medium sized fruit tree:

- » Cool climate: 30- 45 L (8-12 gallons)/week
- » Warm climate: 57-95 L (15-25 gallons)/week
- » Hot climate: 114-190 L (30-50 gallons)/week

The EPA (Environmental Protection Agency) has an on-line calculator to help estimate how much to irrigate in different parts of the country (based on climate).

Method to irrigate with grey water

With a simple grey water system, for example a laundry-to-landscape or branched drain system, grey water should be discharged onto mulch (either on the surface or subsurface depending on your state code requirements). Don't discharge grey water directly onto the bare ground, it can clog the soil by filling the small air gaps in its structure and then won't drain well. Mulch prevents this potential problem since it filters the particles, enabling grey water to soak into the soil below. Any type of mulch works, like wood chips, straw, or bark. More complex grey water systems filter grey water so it can be used in grey water-compatible drip irrigation tubing. In these systems the irrigation tubing is used to irrigate the same as in a conventional system and it can be covered with mulch.

With simple grey water systems remember to

Discharge onto mulch

» Have an air space between the pipe and the ground. This will prevent roots from growing back into the grey water pipe and clogging it. In states that require subsurface irrigation a "mulch shield", such as the irrigation valve box shown in the image, create this air space and prevent clogging.

Percolation test

In a well-functioning grey water system, grey water soaks into the ground without pooling or running off. In general, clayey soils drain very slowly, while sandy or gravelly soils drain quickly. A home percolation test is a simple way to measure how quickly your soil drains and to determine how much area you need to infiltrate the grey water. Professional percolation tests, required for septic leach fields, are expensive, and unnecessary for a small grey water system. Many grey water codes use a soil type chart to predict drainage rates instead of percolation rate data.

Grey water percolation tests should be conducted at the depth of discharge, ideally less than one foot.

Home soil percolation test

- » Dig a 15-30 cm (6"-12") deep hole in your future grey water infiltration zone.
- » Place a ruler (or stick marked in inches) in the bottom of the hole. The measuring device should reach the top of the hole.
- » Fill the hole with water several times to saturate the soil. This may take several hours or overnight in clayey soils.
- » Note the time. Fill the hole with water. When the hole is empty, note the time and calculate the time needed to drain the hole.
- » Convert this rate to minutes per inch (divide the minutes by inches- 120 min/5 inches is 24min/in)
- » Find out percolation rate on the chart below.

Soil percolation chart

Infiltration Rate (min/inch)	Area Needed (sq. ft/gal/day)	After filling the hole four times, the water level dropped 6 inches in 75 minutes. 75 divided by 6 is about 13 minutes/ inch.	Multiply the grey water flow (14 gallons per day) by the area needed (0.4). 14 X 0.4= 5.6, so need about 6 square feet of ground to absorb daily grey water flow.
0-30	0.4	13 min/inch is between 0 and 30, so we use this line.	Need 6 sq. feet for 14 gallons/day
40-45	0.7	-	-
46-60	1.0	-	If we were in this line we'd need 1.0 X 14 or 14 sq. feet.
61-120	2	-	-

Information from the california plumbing code

The California grey water code requires calculating infiltration area based on soil type rather than percolation rate. It is more accurate to do a percolation test than to rely on soil type but both provide important information. Regardless, if used their chart and assumed it had very clayey soil, then need 1.1 square foot per gallon of grey water per day.

Infiltration areas based on soil type

Type of soil	Sq. ft/ 100 gal/day	Gallons max. absorption/ sq. ft/ 24 hrs
Coarse sand or gravel	20	5.0
Fine sand	25	4.0
Sandy loam	40	2.5
Sandy clay	60	1.7
Clay with considerable sand or gravel	90	1.1
Clay with small amount of sand or gravel	120	0.8

So, if produced 14 gallons grey water per day then needs of 15.4 (round up to 16) square feet of infiltration area.

If 4 trees, then there is need an area of 16/4, or 4 square foot per tree. Most mulch basins have around 12 sq. feet of disposal area, so there's plenty of room.

Preparation of grey water systems

There are several manufactured grey water systems on the market in the US, Canada, and many more in Australia. All have the same basic concept: collect the grey water in a temporary storage tank, filter and maybe disinfect it then pump it out for toilet flushing or landscape irrigation.

In most single family homes with yards, using grey water directly for irrigation provides the greatest water savings for the least cost, embodied energy, energy use, and maintenance. In multi-family or commercial buildings with little irrigation need, reusing grey water to flush toilets can save significant amounts of water. Manufactured systems require regular maintenance, and a 6-month study in the UK found that several systems installed in private residences failed because their users didn't maintain them.

Preparation of grey water systems for irrigation

Aqua2Use: This system is housed in a small plastic box (about 60 cm x 60 cm x 30 cm or 2' x 2' x 1') with a series of filters inside and a low powered pump after the filters. The filters require manual cleaning. When they clog grey water can't flow through them and it instead flows to the sewer/septic. The filtered grey water from this system isn't clean enough for a drip irrigation system, it can be used with 1/4 inch outlets, or additional the system can include additional filters (not part of the Aqua2Use system) for smaller drip emitters. Users report problems with the pump and other technical difficulties, like the float switch getting hung up inside the box and burning out pump.

IrriGRAY: This system includes a tank, pump, filter, and grey water compatible drip irrigation equipment. The filter requires manual cleaning and when it clogs the pump cycles endlessly (until someone notices and shuts it off). The company, under the name Morrow Water Saver, is releasing several new products designed to automatize the system, including a self-cleaning filter. This system works well so long as there is someone to clean the filter regularly and flush the drip lines when they clog.

Nexus E-Water is designed for irrigation and toilet flushing.

Preparation of grey water systems for toilet flushing

In general, toilet-flushing grey water systems usually require frequent maintenance, manual cleaning of filters, and chemical disinfectant to prevent odours in the bathroom. They also tend to be relatively complicated, and it's critical that they be designed and installed properly. If you are considering such a system try to find people to talk to who've had the systems installed in their homes for at least a year, and be sure to find out the maintenance requirements of the system. Consider maintenance contact with the installer for any system that requires more than annual maintenance.

Nexus E-Water

Nexus E-Water is a home water and energy recycling system. It is NSF-350 certified, which means it will be easier to obtain permits for it in states like California. The system treats grey water for use in either toilets or for irrigation (no irrigation components are included in the system- have to design/purchase that portion separately). The system also harvests the heat energy in grey water using conventional heat pump technology. They expect to recycle 3 out of every 4 watts of energy needed to heat water in the home.

This grey water treatment technology cleans water in four steps (1) floatation, (2) media filtration, (3) carbon filtration, and (4) disinfection with UV. The key feature of our system is the central role of floatation in which use turbulence and aeration to float the bulk of the soap and large contaminants to the surface to be wicked away to sewer. The system does not require any chemical or biological balancing. The media filter is designed to be replaced once every 40,000 gallons e.g. once to twice a year. The UV bulb needs to be replaced once a year.

Costs: Around $6,000 for installation, with annual filter/UV replacement fees around $100.

Other systems

The **AQUS** and **Brac** systems use chlorine tablets to kill microbes and bad odours in grey water. Some users report problems installing and troubleshooting these systems, a chlorine gas odour in the bathroom, and problems with bacterial growth in the tank and pipes if chlorine use is discontinued. You can read about some problems if a house had with their system.

Water Legacy was a company with a system similar to Brac's that uses hydrogen peroxide and UV as disinfectant instead of chlorine. They were made to only be used with shower grey water, and have an inlet filter that requires regular cleaning. The company is no longer in business.

Performance of Toilet Flushing Systems

Kohler conducted a study on residential grey water systems and how they impact toilets over time. You can download a presentation of the study here. They installed four different types of systems that use shower water to flush toilets and tested them for one year. Studies and system users report that lower-cost systems ($3,000–$5,000) have maintenance issues, while the better-functioning ones cost a lot ($8,000–$9,000).

Commercial scale grey water systems

Laundromat grey water system in Chula Vista (Image credit ReWater.com)

Apartments, businesses, schools, and other situations with more water, more people, and more public exposure to the grey water will require a more complex and sophisticated system for reuse. These systems typically collect grey water in a large temporary tank, filter and disinfect it, and then pump it back to either flush toilets, or supply a drip irrigation system for the landscape.

There are many examples of platinum LEED buildings incorporating rainwater to flush toilets, and this system is referred to as a "grey water system." Technically this is not grey water, rather is a rainwater harvesting system to flush toilets. There is lack of consistency with terminology in these commercial scale situations.

At the commercial scale, it is often easier and more cost effective to treat all the building's waste water – grey water and black water – for reuse rather than separating out the grey water. This is especially true if the building is already constructed because the waste water is combined. In the United States, buildings such as the Solaris building in New York treat all the water on-site and then reuse it for toilet flushing. A membrane bioreactor (MBR) system is used to treat, store and reuse the waste water for toilet flushing, irrigation, and cooling systems. This reduces the fresh water taken from the city's water supply by over 75 per cent and decreases energy costs associated with pumping.

Living machines are another system used to treat combined waste water for reuse. The San Francisco Public Utilities Commission's headquarters in downtown SF uses a living machine to treat all wastewater in the building- on site. The building consumes 60 per cent less water than similarly sized building. Treated wastewater is used for toilet and urinal flushing.

Companies like Orenco Systems make on-site water treatment systems, the water can be reused for non-potable needs.

Industrial scale systems that treat water from clusters of buildings can include small waste water treatment plants, often incorporating wetlands and plants growing in greenhouses to treat the water for local reuse. Systems like Organica's greenhouse industrial park treatment plant in Shenzhen, China, treat water so it can be reused for toilet flushing or irrigation.

Grey water used in commercial building for toilet flushing

- » **Margot and Harold Schiff Residences**- Mercy Housing: Grey water from showers and sinks is filtered, disinfected, and used to flush toilets in all 96 units.

- » **Quayside Village gray water Demonstration Project:** A co-housing unit in British Columbia uses grey water for toilet flushing. They have gone through several types of filtering methods as there were maintenance issues.

Designers and installers of commercial scale systems

- » Rewater (CA)
- » Water Sprout (CA)
- » Hyphea Design Lab (CA)
- » Wahaso

- » Conservation Technology
- » 2020 Engineering (WA)
- » Sierra Watershed Progressive (CA)

Grey water systems in freezing climates

Evergreen Lodge recycles a million gallons of grey water a year, even in the snow. Year-round grey water reuse in cold climates can be a challenge. Many grey water users divert grey water to a sewer or septic system during the cold season but it's possible to reuse grey water even in the dead of winter.

Maintaining system in freezing weather requires additional planning and precautions to prevent system failures:

- » Pipes must drain completely of grey water. Any standing water in pipes could freeze and block the pipe, or potentially burst it. Gravity based systems must be carefully installed so all pipes slope downwards to avoid standing water. Make sure all water can drain out of the lines of any pumped systems (either drain it back to the tank or out into the landscape).
- » If you can't drain the lines in a pumped system create an automatic bypass tube at the beginning of the system. If there is freezing in the main line water can safely escape out the bypass, for example a tee with a tube running high up (high enough so grey water only flows out in case of a blockage).
- » Another option is a greenhouse irrigated by grey water to produce food and greenery all year long.
- » If you decide to shut off the system during the winter make sure to drain down any places containing standing water. You can install a drain-down valve at low points in the system (use a tee with a ball valve, keep the valve closed to use the system and open it to drain the line.)

Gravity, branched brain systems (at Evergreen Lodge)

Location: Yosemite, CA

Evergreen Lodge, located in the Sierra Nevada Mountains, recycles nearly a million gallons of grey water each year from 55 guest cabins, staff housing, and the commercial laundry facility. The 55 systems from the guest cabins are all simple, gravity fed branched drain systems and have operated problem-free for the past four years. All systems, both gravity-fed and pumped, were carefully designed and installed so no water was left standing in the pipes. The system designer and installer, Sierra Watershed Progressive, monitors and maintains the systems.

A diverter valve (not shown) redirects shower water from under the cabin to flow outside. The grey water pipe is painted brown and labelled. This system is under construction. The flow of shower water is divided multiple times with a "flow splitter." Each outlet ends in a large basin. Since the cabins are used by guests and the production of grey water fluctuates greatly (sometimes the cabins are empty, other times they are full), the basins were over-sized to provide surge capacity and enough area to infiltrate the potential large volumes of grey water. This basin will be filled with woodchips; they filter and absorb grey water, allowing it to slowly soak into the ground.

Grey water system in greenhouse

Location: Carrillo's, NM

Diverting grey water to an attached greenhouse reduces the risk of frozen pipes, creates a heat sink (warm grey water) to heat the house all winter long, and provides water to winter gardens. Indoor and greenhouse systems should be well drained and ventilated to avoid mould-promoting damp conditions.

Greenhouse grey water system

Pumice wick (detail)

Ampersand Sustainable Learning Center has no sewer connection or septic system. All grey water enters the greenhouse, and can be diverted to an outside vegetable bed. The water flows out a perforated 1" PVC pipe onto a pumice wick. The pumice releases water slowly into the surrounding soil. Excess water drains into the ground under the bed. Filter fabric covers the perforated pipe to allow easy maintenance while reducing human-grey water contact.

Grey water system for monitoring: Irrigation or wetland

Location: Missoula, MT

When you live over the aquifer that supplies your city's drinking water, returning grey water to the water table creates an intimate link with the water system. This demonstration system at MUD in Missoula, MT models year-round reuse of bathroom grey water in a cold climate. Its most visible component is a large, gravel-filled stock trough beside the back steps– filled with cattails and other beautiful wetland plants.

Goals

- » To reuse grey water from a bathtub for fruit tree irrigation during the summer and aquifer recharge during the winter.

- » To demonstrate effective grey water treatment and disposal strategies for floodplains and high-groundwater areas.

- » To study coliform, nitrogen, and phosphate removal in constructed wetlands- the system contains sampling ports before and after the wetland.

Site description: This single-family house has a bathroom at the rear (south side) of the house, and little fall between the drain and grade. A vented 30-gallon barrel under the back steps functions as a surge tank. Grey water passes through a woodchip biofilter that traps grease and food particles, and then enters a infiltrator inside a bathtub constructed wetland.

Cost: 1 1/2" 3-way valve $50, salvaged stock tank $0, salvaged pipe and fittings $25, filter fabric for frost-free sump $10, gravel $50. Total materials: $135.

Fixtures: Shower/ bath.

MUD's grey water system combines a number of elements to increase its flexibility, cold-weather capability, and ease of water monitoring.

Sewer diversion valve: Directly after draining from the tub, the water encounters a 3-way valve; here, system users can choose whether to direct the water to the sewer, or outside. Diversion to sewer allows flexibility in case of a long freeze, or when using non-biocompatible body products (like hair dye or bleach).

Surge tank: Grey water destined for irrigation flows through a surge tank, a 30-gallon barrel located under the steps that regulate flow into the wetland. Grey water is not stored, but flows out the bottom of the tank.

Constructed gravel wetland: Water directed outside can enter the wetland, where it will be treated by gravel-colonizing microorganisms and by plants. Pathogens, dirt, and soap are taken out of the water, rendering it safe for release into a sump (winter) or for irrigation.

Direct grey water irrigation: In summer, almost all the water can be diverted to bypass the wetland (it just needs enough to keep the plants alive) and enter subsurface distribution pipes. The current plan allows for the watering of the apple tree and a small flower bed, but can easily be altered to meet future needs.

Inline sampling valves: Both inflow and outflow pipes are equipped with valves to allow the sampling of water before and after treatment by the wetland. This monitoring will provide important data toward the development of grey water treatment strategies in this climate.

MUD's system is designed to comply with the grey water best practices incorporated into Arizona and New Mexico's grey water codes–the codes that Montana's soon-to-be-implemented grey water law is modelled on. These best practices include the option to divert grey water to the sewer, and subsurface infiltration into the top 9″ of soil, the "root zone" where most microorganisms that break down grey water reside. The system incorporates cold-climate strategies developed by John Todd in Vermont.

Ecological disposal and irrigation wetland

Location: Providence, RI

Goals: To divert all sink water from the sewer, model ecological wastewater treatment, and irrigate wetland and upland medicinal plants. Resident plans to reuse bathroom grey water after bathroom remodel (pipes are inaccessible).

Site description: This single-family house has two bathrooms and a kitchen sink. Grey water passes through a woodchip biofilter that traps grease and food particles, and then enters a infiltrator inside a bathtub constructed wetland.

Cost: 1 1/2″ 3-way valve $50, salvaged bathtub and pipe $25, fittings $15, plastic biofilter container $10. Total materials: $100. Concrete saw rental: $55.

Fixtures: Because residents grow and process large quantities of food and medicinal plants, the sink produces an estimated 25 gal/day of grey water. A 3-way valve diverts grey water to either the sewer or wetland.

Cold climate grey water wetland, Providence, RI

This system uses sink water to irrigate wetland plants and demonstrate biological wastewater treatment. Treated effluent irrigates medicinal herbs, but little water exits the wetland during the summer. Because the wetland is buried and located against the south side of the house, the wetland has yet to freeze after 2 winters of use.

Building constructed wetland with biofilter

An insulated woodchip biofilter removes grease and food particles. The biofilter's drain sleeves into a mesh-covered infiltrator for easy removal and maintenance.

Constructed wetlands

Ecological disposal wetland in Cuernavaca, MX.

Wetlands are nature's water purifiers. They remove nutrients filter sediments from floodwaters, and can be designed to remove nutrients and other pollutants from storm water and wastewater. Although grey water used for irrigation does not need treatment, constructed wetlands can be used to treat grey water destined for discharge into a local creek, river, pond, or estuary, or to create backyard wildlife habitat. (Note: It is not legal or recommended to discharge grey water into a waterway, but since this problematic practice occurs all over the world, constructed wetlands can help mitigate the damage.) They can also be incorporated into more complex systems to treat water for toilet flushing.

Tips Treating Grey Water in a Wetland

Ecological disposal wetland from an office with no sewer connection

> » Grey water contains nitrate and phosphate, which are plant nutrients. Discharging nutrients onto soil is beneficial and improves plant growth. Discharging nutrients into surface water (ponds, lakes, rivers, or streams) constitutes pollution, because in excess, nutrients cause algae blooms. After the algae dies, microbial decomposition removes oxygen from the water. Since residential grey water can contain nutrients, the water should never be discharged into streams, ponds, natural wetlands, or shallow aquifers that are hydrologically connected to surface waters.

> » A residential grey water system should not contain toxins or dangerous pollutants. Avoiding toxic cleaning products is the best way to keep household toxins out of your local ecosystem. Wetlands can remove or render harmless some industrial pollutants,

but require constant monitoring and special design. Don't rely on simple backyard wetlands to treat household hazardous waste!

» Wetlands transpire large amounts of water. If your grey water goal is to irrigate a garden or orchard, or return water to the local aquifer, a wetland may be counterproductive.

Backyard grey water wetlands are great if

» The site is located close (within 50 ft.) to a stream or pond, or within a floodplain–a wetland can reduce nitrate and phosphate levels in grey water, thereby preventing accidental nutrient pollution of the waterway.

» A household produces more grey water than it requires for irrigation. The wetland plants can use up the excess water and create wildlife habitat in the process.

» In cold climates, irrigation water is only required for part of the year. During winter months, frost-protected wetlands can treat water for release into the aquifer below the frost line.

» Wet land plants are being grown for beauty or habitat value. These plants love water and grey water is a way to water them.

Wetlands can be incorporated into other situations, but they do add a level of complexity to the system, take up more space, and reduce the amount of water available for irrigation.

Design and installation considerations

» Constructed wetlands can be divided into two types. In subsurface-flow or reed bed systems, plants grow in gravel or soil. In surface-flow or pond systems, plants float in the water. For backyard grey water wetlands, we recommend reed bed systems in bathtubs filled with gravel, mulched with woodchips, and planted with cattails, bulrushes, and other wetland plants. In these systems grey water never surfaces above the mulch, which keeps people and animals from contacting the grey water, and is also is a requirement for many codes. Pond systems, which have standing water, are never allowed under state codes. If you want a pond system the grey water must first be filtered by a reed bed wetland, this prevents the unsightly, smelly, and potentially unhealthy pond of raw grey water.

» For bathtub and stock tank wetlands from shower or washing machine systems, we use a surge tank with a 1" outlet to allow water to flow slowly into the wetland.

For large, in-ground wetlands, an infiltration zone filled with 2" to 4" cobble can replace the surge tank.

- » Use a woodchip biofilter to reduce flow rate and remove food particles.

- » The wetland inflow pipe must be screened; otherwise, plant roots will clog the pipe. We build simple infiltrators from milk crates covered in 1/4" hardware cloth. Sew hardware cloth together with wire–cattail roots are tenacious!

- » Place 2" to 4" cobble around the wetland outlet, and screen the outflow with 1/4" hardware cloth. The outflow hose should be at least 1" in diameter to avoid clogging (the bathtub's overflow drain works well–attach a bathtub drain assembly to the overflow hole).

- » Plant a mix of wetland plants with rhizomatic roots to maximize wildlife benefit and treatment at a range of temperatures. Use native species, but ask a local naturalist where to harvest them sustainability or buy them from a nursery. For cold climate systems, include sedges (*Carex spp.*), which take up nutrients down to 4 degrees C. When choosing a sedge species make sure it has rhizomatic roots, not fibrous matting roots, as they can clog the wetland.

- » Weed out grasses and other plants with fibrous roots as they will clog the pore space in the gravel. Once a year, cut back the tops of wetland plants and use them to mulch your garden. After several years, you may need to thin the wetland plants–donate them to a friend's constructed wetland or local restoration project!

5 Types, Significance, Choice Of Cover Crops

There is growing interest in the use of short-season summer annual legumes or grasses as cover crops and green manures in vegetable production systems. Cover crops can provide a significant source of nitrogen (N) for subsequent crops; reduce erosion, runoff, and potential pollution of surface waters; capture soil N that might otherwise be lost to leaching; add organic matter to the soil; improve soil physical properties; impact insect and disease life cycles; and suppress nematode populations and weed growth. There can be potential drawbacks, such as cooler soils in the spring, and the additional cost of seeding the cover crop. These factors must be considered depending on the particular cash crops and cover crops being grown.

Land does not have to be taken out of production in order to incorporate cover crops into cropping systems. Cover crops are usually grown in the off-season to provide benefits to the subsequent cash crop. In North Carolina, summer cover crops can be planted in the production window immediately following spring harvest and prior to fall planting of vegetable crops. The best use of cover crops maximizes the benefits described above without reducing the yield or quality of commercial cash crops.

There is growing interest in the use of short-season summer annual legumes or grasses as cover crops and green manures in vegetable production systems. Cover crops can provide a significant source of nitrogen (N) for subsequent crops; reduce erosion, runoff, and potential pollution of surface waters; capture soil N that might otherwise be lost to leaching; add organic matter to the soil; improve soil physical properties; impact insect and disease life cycles; and suppress nematode populations and weed growth. There can be potential drawbacks, such as cooler soils in the spring, and the additional cost of seeding the cover crop. These factors must be considered depending on the particular cash crops and cover crops being grown.

A cover crop is a type of plant grown to suppress weeds, help build and improve soil, and control diseases and pests. Cover crops are also called "green manure" and "living mulches." They are called "green manure" because they provide nutrients to the soil like manures and "living mulches," are the cover crops prevent soil erosion. Once grown, cover crops are usually mowed and then tilled into the soil. Rye is a plant commonly used as a cover crop.

SIGNIFICANCE OF USE OF COVER CROPS

Recent environmental and ecological awareness has started resurgence in cover crop use. Although cover crops have been used for centuries but today's modern farmer has grown up in a generation which has replaced the use of cover crops with widespread use of fertilizers and herbicides. Now as some farmers are trying to use cover crops, they lack the information and experience to be successful. University and private researchers are going back to the basics and are experimenting with beneficial plants to determine how they can best be used in modern farming practices to supplement or replace purchased inputs. Cover crops have an important role in successful sustainable farming systems.

Benefits of Cover Crops

The main purpose of a cover crop is to benefit the soil and/or other crops, but is not intended to be harvested for feed/Food or sale. Some of the primary benefits from cover crops include:

1. **Soil quality improvements:** Soil tilth is improved whenever a plant establishes roots and grows into compacted areas. Water infiltration is improved as well. When a field lays fallow for a period of time, the surface tends to seal and water will run off. Cover crops protect the soil surface and reduce sealing. Also, beneficial organisms in the soil, such as earthworms, thrive when fresh plant material is decomposing.

Cover crops add organic matter to the soil. Organic matter levels tend to improve with the addition of cover crops. Organic matter provides benefits to the soil and the subsequent crop in many different ways. Organic matter improves the physical condition of the soil by improving soil tilth, stability of soil aggregates, water infiltration, air diffusion, and by reducing soil crusting. The addition of organic matter can also increase the populations of soil microbes and earthworms, which in turn, contribute to efficient nutrient cycling and improvements in soil structure. Finally, organic matter additions can also increase nutrient retention in the root zone.

2. **Erosion control:** Cover crops reduce wind and water erosion on all types of soils. By having the soil held in place by cover crops during the fall, winter, and early spring, loss of soil from erosion is greatly reduced. Scientists have estimated that the United States has lost 30% of its topsoil in the past 200 years due to agricultural practices that do not return organic matter to the soil and that leave bare fallow soils for a significant portion of the year. Erosion has long-term costs such as loss of agricultural productivity and aquatic habitat as well as sedimentation of rivers, reservoirs, and estuaries. There are also short-term costs to farmers. The USDA estimates that farmers are losing approximately $40.00 worth of fertilizer per acre per year in runoff from farm fields. Cover crops can help reduce soil erosion by keeping the soil covered during high rainfall periods when it would normally be bare. Farmers also report improved soil structure, stability and permeability, decreased crusting, and increased water infiltration.

3. **Fertility improvements:** Legumes can add substantial amounts of available nitrogen to the soil. Non-legumes can be used to take up excess nitrogen from previous crops and recycle the nitrogen as well as available phosphorus and potassium to the following crop. This is very important after manure application, because cover crops can reduce leaching of nutrients.

Cover crops enhance the nitrogen cycling in the plant-soil system. Grass or non-leguminous cover crops can help keep N in the plant-soil system by utilizing residual N that would otherwise be lost to leaching. The nitrogen is assimilated into the tissue (biomass) of the cover crop and then, as the cover crop decomposes, it is released to subsequent crops. Leguminous cover crops, such as cowpea, soybean, and velvetbean, can also "fix" significant amounts of nitrogen for use by subsequent crops. Through a symbiotic association with the legume, *Rhizobium* bacteria convert atmospheric nitrogen into a form that the legume can use for its own growth. When grass or legume cover crops are killed or incorporated, soil microorganisms decompose their residue. In a process called mineralization, the N in the plant tissue is converted by soil microbes into a form (nitrate) that subsequent plants can use. Nitrogen in the above ground biomass of the cover crops varies considerably within species, but legumes generally contribute anywhere from 60-200 lb of N per acre. This nitrogen is mineralized over an extended period of time, with an average of 50% of the total N contained in the cover crop available to subsequent crops.

The ratio of carbon to nitrogen (C: N) in the cover crop biomass affects how much and when the nitrogen contained in the cover crop will be available for subsequent crop uptake. The microbial population that decomposes the cover crops are made up of 10 parts carbon to 1 part nitrogen. When a cover crop is incorporated, the population of soil microorganisms increases in response to the added food source. As the population of these microbes increase, this 10:1 carbon to nitrogen ratio must be maintained. Consequently, as the microbial populations increase, N contained in the cover crop and the soil may be immobilized or "tied up" as part of the physical structure of the microbes. As a result, the cover crop N may not be available for uptake by the following crop. When the microbes die, the N is "mineralized" and becomes available for subsequent crop use.

For example, a mature sorghum-sudangrass cover crop may have a relatively high C:N ratio of 50:1. This ratio only supplies about 20% of the N needed by the microorganisms when they utilize this carbon source. As the soil microbial population increases in response to the incorporation of the grass cover crop into the soil, maintenance of a microbial C:N ratio of 10:1 requires the immobilization of all the N contained in the sorghum-sudangrass biomass as well as additional available soil N. If a vegetable crop follows the sorghum-sudangrass, additional N would be required to insure adequate crop growth.

On the other hand, legumes have relatively low C:N ratios and therefore little or no immobilization can be expected from their incorporation into soil. In fact, legume N can be quickly mineralized, perhaps even before the subsequent crop has a high demand for it. Approximately 65 to 70% of the carbon in cowpea and sunnhemp residue (C:N = 32 and 24, respectively) is mineralized in the first two weeks after soil incorporation. If the N is mineralized quickly and available N exceeds the crop need, this N can be subject to leaching losses.

Achieving synchrony of N release from decomposing legume residue and crop N demand is expected to increase the overall efficiency of use. Timing of cover crop kill can affect this synchrony. Generally, the more mature the cover crop, the higher the C:N ratio and the slower the decomposition. Better utilization of N is important to prevent N pollution of surface and groundwater through runoff and leaching.

Cover crops the enhance nitrogen cycling by reducing nitrate leaching losses. Nitrogen is the most difficult nutrient to manage in agricultural systems. It is necessary in large quantities for adequate crop growth and yield, but it is also extremely mobile in soil. Planting a non-leguminous cover crop after a spring vegetable harvest can "trap" leftover, residual soil nitrogen through cover crop uptake and reduce the potential for leaching losses through the summer months. Additionally, the cover crop utilizes excess water in the soil which also helps limit leaching losses.

Leaving cover crop residues on the soil surface, rather than incorporating them, have advantages and disadvantages with regard to efficiency of utilization of cover crop N. Maintenance of surface residues can result in increased losses of cover crop N to the atmosphere via a process called denitrification. On the other hand, immobilization and slower residue decomposition can result in reduced leaching losses. Also, a higher concentration of crop residues and organic matter near the soil surface can increase the diversity of microorganisms and fauna at the surface, which can result in greater recycling of N in the soil ecosystem.

4. **Suppress weeds: Cover crops can reduce weeds in subsequent crops.** While the cover crop is growing, it will suppress the germination and growth of early spring weeds through competition and shading. When killed and left on the surface as mulch, cover crops continue to suppress weeds, primarily by blocking out light. Cover crops can also suppress weeds chemically - Some plants release chemicals, either while they are growing or while they are decomposing, which prevent the germination or growth of other plants (allelopathy). Researchers have effectively used cover crops of wheat, barley, oats, rye, sorghum, and sudangrass to suppress weeds. Weed suppression has also been reported from residues and leachates of crimson clover, hairy vetch, and other legumes.

5. **Cover crops impact on insects: Cover crops can also impact insect populations either negatively or positively.** Cover crops attract beneficial and pest insects into cropping systems. Both can disperse to cash crops when the cover crop matures or dies. Prior to the arrival of important insect pests, beneficial insects attracted into an area by a cover crop may reach sufficient population densities to maintain pest populations in adjacent crops below economic threshold levels. Research in Georgia reported high densities of bigeyed bugs, lady bugs, and other beneficial insects in vetches and clovers. It has also been reported that assassin bugs have destroyed Colorado potato beetle feeding on eggplant planted into strip-tilled crimson clover. Beneficial insects, such as lady beetles or ground beetles, may be encouraged by planting cover crops.

6. **Cover crops impact on diseases:** Pathogens can either be enhanced or inhibited by cover cropping systems. The impact of the cover crop on the pathogen will depend upon the nature and life cycle requirements of the pathogen. For example, if the pathogen survives best on surface residue and the cover crop is left on the surface as mulch, then the level of disease may increase. On the other hand, increases in soil organic matter content can enhance biological control of soil-borne plant pathogens through direct antagonism and by competition for available energy, water and nutrients. In experiments in Georgia, *Sclerotium rolfsii* was reduced in cover cropped no-till systems.

Organisms that cause disease can by affected by decreases in temperature, increases in moisture, reductions in soil compaction and bulk density, and changes in nutrient dynamics. Whether or not the cover crop is taxonomically related to the subsequent crop will also influence whether or not disease cycles are interrupted or prolonged. Some cover crops have also been shown to reduce nematode populations, including velvetbean, sorghum-sudangrass, and sunnhemp.

7. **Cover crops impact the economics of farming operations:** Cover crops affect on the economics of farming operations differently and depending on the cover crop and the main crop. Growers need to account for cover crop seed and planting costs - Quantifiable savings, on the other hand, can include reduced fertilizer and herbicide applications, and reduced costs of pest and disease control. Growers will have to determine how they want to account for less apparent long-term savings such as, reduced soil erosion, increased organic matter content, improved soil physical properties, reduced nitrate leaching, and enhanced nutrient cycling.

Choice of Cover Crops

The desired purpose of the cover crop will help determine the most appropriate species. If the purpose of a cover is to provide readily available, biologically-fixed N for subsequent crops, then the grower should choose a legume like cowpea, which fixes nitrogen and has a low C:N ratio. If the cover crop will be managed as a surface mulch for weed suppression, the grower should choose a high C:N, heavy biomass producer with demonstrated weed suppression characteristics, such as sorghum-sudangrass.

Planting time of cover crops

While adopt the cover crops, farmers should know the few points. First, they must determine how their farming system can accommodate cover crops. Ideally, as soon as one type of crop is utilized (harvested, killed, or incorporated) the next is planted. Look for open periods in each field's rotation where a cover crop can be planted.

Type of Cover Crops

Selection of a cover crop depends on when it can be planted and the goal for its use. There are many cover crop species. Legumes cover crops such as clovers, hairy vetch, field pea, annual medic, alfalfa and soybean fix atmospheric nitrogen into a form plants and microorganisms can use. Non-legume species such as rye, oats, wheat, forage turnips, oilseed radish and sudangrass recycle existing soil nitrogen and other nutrients and can reduce leaching losses. A combination of 2 or more types of cover crops may be beneficial for quick establishment and improved nutrient utilization.

Seeding

Seed cost, method of seeding, and method of killing the cover crop should also be considered. Often, farmers are able to grow their own cover crop seed and save on seed costs. Larger seeds, such as hairy vetch and winter peas can be planted through the normal hopper of a drill. Clovers may require a small-seeders attachment. Broadcasting cover crop seed can also be done with a fertilizer cart followed by light incorporation. If a legume is planted that has not been grown before, the proper inoculants should be added when seeding.

Managing growth

There are some drawbacks to using cover crops that farmers should be aware of. If the cover crop does not naturally winter kill, then a farmer should have a method for killing the crop before it competes with the next cash crop. This can be done mechanically or by using herbicides. Cover crops can act as weeds if not controlled. Timing is important when killing a spring growing cover crop. Cover crops should be allowed to grow as long as possible in the spring to add additional nutrients to the soil and suppress weeds, but they can also use up soil moisture and hurt the following cash crop if dry conditions exist. More management time is required to determine how and when it would be best to control the cover crop. Hard seed that did not germinate may also require control later in the season. Like any other crop, moisture is needed to establish a cover crop. This is especially critical during July and August.

Failures may occur due to late establishment as well. Additional problems may occur if the cover crop harbors armyworm, grubs or cutworms that can attack the cash crop.

Return on investment

The economics of using cover crops can be calculated by the savings of purchased nutrients and herbicides versus the additional cost of using cover crops. There are also other factors that are not easily credited to cover crop use, such as improved soil tilth, enhancing soil biology, and improving organic matter content of the soil. Overall, as the cost of fertilizer and herbicides continue to increase, the benefits of using cover crops in a sustainable farming system will become more attractive to modern farmers. More research is needed to quantify the effects of cover crops so that research based recommendations can be given to producers.

Non-legume cover crops

Buckwheat *(Fagopyrum esculentum):* Buckwheat is a very rapidly growing, broadleaf summer annual which can flower in 4 to 6 weeks. It reaches 75cm (2 $^1/_2$ ft) in height and is single-stemmed with many lateral branches. It has a deep tap root and fibrous, superficial roots. It can be grown to maturity between spring and fall vegetable crops, suppressing weed growth and recycling nutrients during that period. Buckwheat flowers are very attractive to insects, and some growers use this cover as a means to attract beneficial insects into cropping systems. Buckwheat is an effective phosphorous scavenger. It is succulent, easy to incorporate, and decomposes rapidly. The main disadvantage to buckwheat is that it sets seed quickly and may, if allowed to go to seed, be a weed problem in subsequent crops.

The optimal time to incorporate buckwheat is a week after flowering, before seed is set. The seeding rate is 35 to 100 kg/ha (75 to 225 lb/ha); higher rates are used when broadcasting. Seed should be drilled 1.5 cm (½ inch) deep, or broadcast and incorporated with a light dusking. Buckwheat can be planted anytime in the spring, summer or fall, but is frost-sensitive.

Sorghum-sudangrass *(Sorghum bicolor X S. sudanense):* Sorghum-sudangrass is a cross between forage or grain sorghum and sudangrass. It is a warm-season annual grass that grows well in hot, dry conditions and produces a large amount of biomass. Often reaching 180 cm (6 ft) in height, it can be mowed to enhance biomass production. Sorghum-sudangrass is very effective at suppressing weeds and has been shown to have allelopathic properties. The roots of sorghum-sudangrass are good foragers for nutrients (especially nitrogen) and help control erosion. Research on nematode suppression by sorghum-sudangrass is not conclusive. Some studies have shown that nematode populations have been higher in vegetables following sorghum-sudangrass, while other studies have shown that sudangrass suppresses nematode levels.

Sorghum-sudangrass does well when planted in mixtures, providing effective support for viney legumes like velvetbean. It can be planted 1.5 to 3.5 cm ($^1/_2$ to 1 $^1/_2$ inches) deep from late spring through midsummer at a rate of 50 kg/ha (112 lb/acre). If frost-killed, the residue can provide no-till mulch for early planted spring crops like potatoes.

The main advantages of sorghum-sudangrass include its ability to scavenge nitrogen, grow quickly for erosion control, suppress weeds, and suppress some nematodes. A possible disadvantage is nitrogen tie-up and a resulting limit of available N for subsequent crops.
German (foxtail) millet *(Setaria italica):* German or foxtail millet is an annual warm season

grass that matures quickly in the hot summer months. It is one of the oldest of cultivated crops. Although German millet has a fairly low water requirement, it doesn't recover easily after a drought because of its shallow root system. Grain formation requires 75 to 90 days. German millet forms slender, erect, and leafy stems that can vary in height from 60 to 150 cm (2 to 5 ft). The seed can be planted from mid-May through August at a rate of 25 to 37 kg/ha (62 to 75 lb/ha). A small seeded crop, German millet requires a relatively fine, firm seedbed for adequate germination. In order to avoid early competition from germinating weed seed, German millet should be sown in a stale seed bed or closely drilled in the row. Coarse sands should be avoided.

Pearl Millet *(Pennisetum glaucum)*: Pearl millet is a tall summer annual bunchgrass that grows 120 to 360 cm (4 to 12 ft.) tall. It is also often referred to as cattail millet because its long dense spike-like inflorescences resemble cattails. The mature panicle is brown. Though it performs best in sandy loam soils, pearl millet is well adapted to sandy and/or infertile soils; Pearl millet can be planted from late April through July at a rate of 6 to 15 kg/ha (12 to 37 lb/acre). Pearl millet matures in 60 to 70 days. In studies in North Carolina, pearl millet was not as readily killed by mechanical methods (mowing and undercutting) as German or Japanese millet.

Japanese Millet *(Enchinochloa frumentacea)*: Japanese millet is an annual grass that grows 60 to 120 cm (2 to 4 ft) tall. It resembles, and may have originated from barnyardgrass. The inflorescence is a brown to purple panicle made up of 5 to 15 sessile erect branches. Japanese millet is commonly grown as a late-season green forage. If weather conditions are favorable, it grows rapidly and will mature seed in as little as 45 days. Japanese millet can be planted from April-July at a rate of 22 to 27 kg/ha (50 to 62 lb/ha). It performs poorly on sandy soils.

Legumes

Cowpea *(Vigna unguiculata)*: Other common names for this plant include blackeye, crowder, and southern pea. Cowpea is a fast growing, summer cover crop that is adapted to a wide range of soil conditions. Having a taproot that can obtain moisture from deep in the soil profile, it does well in droughty conditions. Vigorous cowpea varieties compete well against weeds. A high nitrogen producer, cowpea yields average 35 to 45 q/ha (3000 to 4000 lb/ha) of dry biomass containing 3 to 4% nitrogen. Maximum biomass is achieved in 60 to 90 days. Residues are succulent and decompose readily when incorporated into the soil. Cowpea can be planted in the spring after all danger of frost through late summer. Cowpea seed can be drilled in rows 15 to 20 cm (6 to 8 inches) apart at 45 kg/ha (100 lb/ha) or broadcast at 80 to 137 kg/ha (175 to 300 lb/hae). However, higher seeding rates are necessary in late summer if soil moisture is likely to be limiting. Recommended cultivars include Ironclay and Redripper. Ironclay matures later than Redripper. Plants normally grow up to 60 cm (2 ft) tall, but some cultivars can climb when planted in mixtures with other species. Good mixture options are sorghum-sudangrass and German foxtail millet. When killed mechanically, cowpeas can have considerable regrowth after mowing and undercutting in some years.

Soybean *(Glycine max):* Soybean is one of the best economic choices for a summer legume cover crop. It is an erect, bushy plant that grows 60 to 120 cm (2 to 4 ft) tall, establishes quickly, and competes well with weeds. When grown as a green-manure crop, late maturing varieties usually give the highest biomass yield and fix the most nitrogen. While most of the roots are in the top 20 cm (8 inches) of the soil, some roots can penetrate up to 15 cm (6 ft) deep. Soybean will withstand short periods of drought if they are well-established. Soybean will grow on nearly all types of soils, but are most productive on loam soils. Soybean planted as cover crops should be broadcast or closely drilled at 67 to 112 kg/ha (150 to 250 lb/ha). Some new viney forage types (e.g. quailhaven soybeans) are available or are being developed that have the potential to produce more biomass than traditional soybean varieties.

Velvetbean *(Mucuna deeringiana):* Velvetbean is a vigorously growing, warm-season annual legume native to the tropics but well adapted to southern conditions. It performs well in sandy and infertile soils. Most cultivars are viney and some can attain a stem length of 9m (30 ft). The leaves of velvetbean are trifoliate with large ovate leaflets. Pods are hairy, up to 6 inches long, and contain 3 to 6 seeds. Velvetbean is an excellent green manure crop, producing high amounts of biomass that decomposes readily to provide N for subsequent crops. Velvetbean should be seeded into warm soils at 27 to 37 kg/ha (62 to 87 lb/ha) in 100 cm (40 inch) rows. Velvetbean seed should not be drilled because the very large seed can be damaged in conventional drills. When grown for seed, velvetbean should be sown in a mixture with an upright crop like sorghum-sudangrass or corn. Velvetbean vines will climb stems of these grasses and flowers produced will get necessary air circulation.

Sunnhemp *(Crotalaria juncea):* Sunnhemp is a tall, herbaceous, warm-season annual legume that has been used extensively for soil improvement and green manuring in the tropics. The erect fibrous stems are competitive with weeds. It grows rapidly and can reach a height of 270 cm (9 ft) in 60 days. It can tolerate poor, sandy, droughty soils but requires good drainage. Sunnhemp tolerates moderate alkalinity and a soil pH below 5 reduces growth. Sunnhemp should be broadcast at 40 kg/ha (90 lb/hae) or seeded in 100 to 105 cm (3.5 foot) rows at 6 to 7 kg/ha (12 to 17 lb/ha). Higher seeding rates will lead to prolonged succulence of the stems and are recommended if the crop will only be grown for 28 to 35 days. Lower rates are recommended if grown for seed production. Sunnhemp becomes fibrous with age, but the plants will remain succulent for about 8 weeks after seeding. It can be integrated into cropping systems by sowing it in late summer after corn. It will produce high biomass yields and N in the months before frost. Residues left on the soil surface over the winter months will facilitate no-till crop production the following spring. Seed is not currently readily available, but if demand were to increase, seed availability would most likely respond. While forage of some *Crotalaria* species is toxic to animals, sunnhemp forage is not. Sunnhemp should also not to be confused with the weed species, and the growing season in North Carolina is not long enough for sunnhemp to produce viable seed.

Carbon Farming: Green And Brown Manuring

CARBON FARMING

Carbon farming is the process of changing agricultural practices or land use to increase the amount of carbon stored in the soil and vegetation (sequestration) and to reduce greenhouse gas emissions from livestock, soil or vegetation (avoidance). Carbon farming potentially offers landholders financial incentives to reduce carbon pollution, but should always aim to achieve multiple economic and environmental co-benefits. The Department of Primary Industries and Regional Development can provide scientific assessments of the technical feasibility and risks, but anyone contemplating participating in carbon farming should seek appropriate legal and technical advice.

Impact of Carbon Farming

Carbon farming is the process of managing soil, vegetation, water and animals to increase carbon storage and reduce greenhouse gas emissions. The main gases of interest to agriculture are carbon dioxide (CO_2), nitrous oxide (N_2O) and methane (CH_4). Agriculture is a significant emitter of greenhouse gases and also has a significant opportunity to reduce emissions and sequester carbon. We have described the carbon farming management options below for farm managers wanting to make decisions about adopting carbon farming technologies and participating in carbon farming funding schemes.

Soil

» Carbon farming: applying biochar to increase soil organic carbon

Carbon farming: green and brown manuring

- » Carbon farming: claying as a method of increasing soil carbon content
- » Carbon farming: green and brown manuring
- » Reducing nitrous oxide emissions from agricultural soils
- » Liming to increase carbon sequestration
- » Composting to avoid methane production
- » Soil organic carbon and carbon sequestration

Cattle

- » Reducing methane emissions from cattle using feed additives (ERF approved)
- » Cattle breeding for lower greenhouse gas emissions
- » Managing pastures and stocking rates to reduce methane emissions from cattle

Sheep

- » Sheep genetics in methane reduction
- » Managing sheep pastures to reduce methane production
- » Nutrition and feed additives to reduce methane emissions

Fuels

- » Liquid bio-fuel as a replacement for fossil fuel

Plantings

- » Permanent environmental plantings to earn carbon credits

Regional natural resource management groups: Need to manage the natural resources of different regions of the particular areas or country. Green or brown manuring is the practice of returning plant material to a soil to increase soil organic matter, improve soil fertility and reduce weeds. We provide this information to support land manager decisions about investing in carbon farming.

Green Manuring

Green manuring incorporates green plant residue into soil with cultivation, commonly with an offset-disc plough. Green manuring aims to kill weeds and control seed set while building soil organic matter and nitrogen status. More than one tillage pass may be required for a successful kill, and cultivation may lead to losses of soil organic matter and cause soil structure damage. In India, legumes like sunhemp, dhanicha are used as green manuring crops. However, in Western Australia, pulses are preferred over cereal crops for green manuring because they add more nitrogen to the soil. Green manuring pulse crops carry the risk of increased nitrous oxide (N_2O) emissions, which may discount increases in soil carbon sequestration. In some locations it may be possible to grow a summer crop for example, broad-leafed plants such as sunflower and safflower, or grasses such as sorghum and millet for green manuring, especially on sand plain soils and in higher rainfall areas. However, high water use over summer and high carbon to nitrogen (C:N) ratios, which may tie up nitrogen, can depress subsequent yields.

Brown Manuring

Brown manuring is a 'no-till' version of green manuring, using a non-selective herbicide to desiccate the crop (and weeds) at flowering instead of using cultivation. A follow-up treatment may be required to control survivors. The plant residues are left standing, helping to retain surface cover and soil structure. Soil organic matter is increased.

A variation on brown manuring is mulching, where the crop or pasture is mowed, slashed or cut with a knife roller and the residue is left lying on the soil surface. This mulch reduces soil moisture loss through evaporation. Mulched residues break down more rapidly than for normal brown manuring because of the increased contact with soil and smaller pieces.

Benefits of Green and Brown Manuring

This activity is best suited to cropping soils with loamy to clay surface textures. Clay soils can sequester more carbon as soil organic carbon (SOC). Green or brown manuring solely for carbon credits is not viable because there isn't an approved methodology for generating carbon credits from this activity, but green or brown manuring could be one component of a farming system.

Other benefits

» Improved soil fertility (largely observed in leguminous green manures) achieved by building SOC and nutrient status, and increasing buffering capacity to moderate changes in pH

» Reduced weed burdens, particularly when herbicides are not an option or effective, or a break is required

» Improved soil structure and providing a protective cover for the soil surface: this increases water infiltration and retention, reduces wind and water erosion risk, and reduces the impact of extreme temperatures.

Threats of green and brown manuring for carbon storage

» There is limited data that clearly quantifies the change in SOC linked solely to manuring practices. Good data is needed to calculate the potential carbon sequestration value.

» More-specialised machinery, such as stubble rollers and mulchers, may need to be purchased.

» There will be a revenue loss from not cropping the paddock in the year of manuring: this may be offset in future years if improved soil quality increases the performance of subsequent crops.

» Green manuring can result in increased methane production through anaerobic decay.

» Green manuring over summer (an opportunistic activity) will reduce soil moisture for the following crop: this loss may be offset by the increased SOC and nutrient level.

» These practices will likely to be part of rotational cropping management for the paddock, making it difficult to isolate the component(s) of the farming system affecting SOC change.

Maintaining SOC in rain-fed Australian farming systems after manuring: The content of SOC in rain-fed Australian farming systems is influenced by soil type, climate and management:

Soil type: clay/loam soils can support higher concentrations of organic carbon than sandy soils

Climate: higher rainfall supports more biomass leading to higher SOC (areas receiving less than 400 millimetres per year have limited ability to generate SOC; high temperatures will lead to faster rates of biomass and SOC decomposition, with the release of carbon dioxide, methane and nitrous oxide

Management: soil disturbance through tillage will encourage SOC loss. Minimum tillage and no-till minimise SOC loss.

Brown Manure Crop Management (Pea Crop)

1. **Seed procurement**

    » Ideally all seed would be purchased from a reliable seller at a price that is mutually acceptable.

    » Develop an ongoing agreement if possible.

    » If producing own seed the following should be considered:

    » Choose a paddock with a low weed burden

    » Choose a paddock geographically isolated from other pea crops or pea stubble.

    » If running controlled traffic farming (CTF) tracks, consider blocking the seed runs on wheel tracks to reduce disease spread from crop damage.

2. **Time of sowing**

    » Sowing a brown manure crop should not compromise the timing of cash crop sowing.

    » Sowing into warm soils before weeds have germinated will maximise biomass production, competition with weeds and ultimately Nitrogen fixation.

    » If sowing field peas in March or early April, choose varieties with the highest tolerance to Bacterial Blight, provided they are late maturing varieties that produce large amounts of biomass.

    » When comparing trial results for biomass production, ensure that the biomass production is being measured at spray out timing, not simply peak biomass for the crop grown through to harvest.

    » Any extra growth past mid-September is wasted, as Wild Oat flowering time generally dictates sprayout timing.

    » In areas with high Field Pea intensity, consider delaying sowing until May to avoid Ascochyta spore showers.

3. **Weed management in brown manure crops**

    » Aim to minimise Wild Oat seed set in cereal crops prior to Brown Manure crops

    » Ryegrass populations can be successfully reduced with the combination of a Brown Manure crop followed by a canola crop.

    » In situations where grass weed densities are high enough to reduce the establishment of a Brown Manure crop, pre-emergent herbicides may be beneficial, such as: Trifluralin and Metribuzin

    » In many cases, sowing field peas into warm soil allows them to out compete the majority of weeds without the addition of herbicides.

4. **Disease management in brown manure crops**

    » Bacterial Blight is the most important disease of field pea Brown Manure crops, although Ascochyta black spot should also be considered.

    » Bacterial Blight is most effectively managed by reducing physical damage on pea crops, especially when this coincides with frosty conditions.

    » Avoid driving, riding or walking through crops once established.

    » Where possible, do not allow animals to travel through pea crops (sheep, cattle, kangaroos).

    » Severity of Ascochyta black spot is highly dependent on the amount of spore release prior to emergence.

    » Dry conditions from spray out through to sowing will increase disease risk. If wet conditions are experienced between October and March, the majority of Ascochyta spores will have been depleted prior to sowing.

    » If choosing alternative species, the carryover of other diseases must be considered, most importantly:

    - Sclerotinia Stem Rot of Canola (Lupins, Faba Beans and Clover are alternative hosts)

    - Take-all and Crown Rot of Wheat (carried by Oats, Triticale and many other annual grasses)

    » Consider that if a pea crop is sown in March and dies in August from Bacterial

Blight it is likely to have already produced 3-4 t/ha of biomass and fixed more than 60 kg/ha of Nitrogen. From that point it could be treated as a long fallow.

- » While Lupins and Faba Beans may not be susceptible to Bacterial Blight, the carryover of Sclerotinia may have a much larger impact on the following canola crop.
- » Disease risk for following cash crops should be more heavily weighted than disease risk in manure crops.

5. **Spraying out brown manure crops**

- » Must be sprayed out by mid September.
- » Flowering time for Wild Oats or Ryegrass will usually dictate the time of spraying.
- » Consider adding an insecticide to reduce insect pressure in the following crop.

Refer to the accompanying document "Brown Manure Crops – Sprayout Decisions" for a decision matrix of product choice and timing.

6. **Brown manure stubble management**

- » There are a number of operations that can be used to manage Brown Manure Stubbles.
- » Rolling stubbles
- » Can provide some weed control through smothering of weeds under stubble mulch.
- » Improves soil contact to increase breakdown
- » Can improve the coverage of weeds in a "double knock" situation.
- » Cultivation
  - Can range from knife points through to offset discs
  - Any machinery passing through pea stubbles will cause significant shattering when stubbles are dry. The affect is much less when stubble is wet or dewy.
  - Can provide weed control, especially important if weeds escape the knockdown process (eg. Glyphosate resistant weeds)
  - Timing of cultivation needs to be considered in the context of the spraying and sowing systems.

Carbon farming: green and brown manuring

7. **Sowing canola into brown manure stubble**
   - When > 150 mm rain falls between September sprayout and April sowing, stubble breakdown is generally sufficient for most sowing equipment to sow through Brown Manure stubbles.
     - ◊ Operations such as rolling and cultivation will further increase breakdown.
     - If < 100 mm rain falls September to April, stubble breakdown will be much less.
     - Under these conditions, blockages will occur with a range of tyned sowing equipment.
     - Fitting coulters in front of tynes will improve trash flow, provided:
       - ◊ Coulters are directly in line with tynes
       - ◊ Stubble is dry, so coulters cut through the stubble
       - ◊ There is sufficient distance between coulters and tynes for stubble to flow through.
   - When stubbles have not broken down over Summer, small falls of rain, or even heavy dew is enough to create blockages. This can be managed by:
     - Manipulating stubbles during summer or early autumn, as discussed above.
     - Commencing canola sowing earlier, with the understanding that sowing may only be possible during daylight hours.
     - Sowing canola into Brown Manure stubbles in dry conditions (when these are likely to be the only paddocks with moisture for establishment) and switching to canola on cereal stubbles if rain occurs.
   - Refer to the document "Sowing into Brown Manure Stubble" to aid decision making through summer and autumn.

8. **Managing canola after brown manure crops**
   - Early crop growth is driven by soil temperature
     - Sow early in April to improve crop vigour
     - Extra moisture under Brown Manure stubbles will often present additional canola sowing opportunities.

- » Nitrogen from Brown Manure crops may not mineralise until the following spring, so a small amount of Nitrogen (either Pre-sowing or At-sowing) will promote growth and development until mineralisation occurs.

- » Use the full suite of knockdown, pre-emergent and post-emergent herbicides available to control weeds in the canola crop to ensure that costs are minimised in the following wheat crop.

Brown manure legumes may be one way for growers who have removed livestock from their operations to lessen the risk of total cropping in a drier climate. A crop production system involving brown manure legumes can be as profitable as continuous cropping and, even if slightly less profitable, has much lower production and financial risk due to lower input and operating costs. Brown manure cropping involves growing a grain legume crop with minimal fertiliser and herbicide inputs to achieve maximum dry-matter production before the major weed species have set viable seed. The grain legume crop is sprayed with a knockdown herbicide before seed-set to kill the crop and weeds, ideally no later than the start of the crop's pod development to also conserve soil moisture. A second knockdown herbicide is generally applied to achieve a 'double knock'. This is different to green manure where the crop and weeds are cultivated.

Brown manuring with legumes should be considered, especially by growers in southern New South Wales, as diminishing growing-season rainfall is putting downward pressure on yields. To counter this, most growers are increasing the quantity and cost of inputs, particularly herbicides and nitrogen. This adds even more to production and financial risks. Consequently, growers are putting more effort into crop rotations, especially crop sequences where wheat follows crops such as lucerne or oilseeds. However, over the past, relatively dry decade, the benefits of lucerne to subsequent crop production have been challenged due to pastures failing to establish under cereal crops in dry springs, as well as poor crop performance following lucerne where soil moisture recharge has not occurred prior to cropping. It has reached the point that, for many larger growers, canola remains the only viable cash break crop. However, this requires increasing quantities of artificial nitrogen to maintain yields and grain protein; weed control has also become more problematic.

Legumes option

By contrast, brown manuring with legumes, even if not immediately as profitable as continuous cropping, significantly lowers seasonal risk due to the lowered costs. While vetch is a common brown manure crop, early-sown field peas may be more competitive against weeds and potentially produce more dry matter. Higher dry-matter production should lead to higher nitrogen accumulation, and more stubble cover provides shading to reduce evaporation and reduce sunlight available to germinate weeds.

Brown manure legume crops provide three major benefits over long fallows:

- » Competition for weeds (reducing knockdown herbicide use during the growing season);
- » Accumulation of soil nitrogen; and
- » The maintenance of ground cover during the growing season and over the summer preceding the next crop.

The major disadvantage of brown manure crops compared with long fallowing is the cost of the grain legume seed ($30 to $35 per hectare), plus the cost of sowing, although this is low in the overall scheme of things.

Crop sequences

Grain legume crops such as lupins have traditionally been followed by wheat, which responds well in terms of yield and grain protein. However, in dry springs many of these wheat crops 'blow up', due to high early dry-matter production depleting soil moisture. This results in reduced wheat yields and high protein grain, but also high screenings. Given the desire to establish canola early with stored soil moisture and adequate nitrogen to optimise yield potential, canola is now being grown after brown manure crops. This enables almost complete prevention of annual ryegrass and wild oat seed-set in two successive years, depleting the seedbank to the extent that control measures may not be necessary in the following two cereal crops. The two-year broadleaf crop sequence of brown manure legume followed by canola is also predicted to provide control of crown rot. Reduction of take-all levels under high-disease-pressure weather conditions should also be adequate to allow early (mid-April) sowing of the first wheat crop with little root disease risk. The ability to sow early with confidence (subject to variety) should lead to higher wheat yield potential.

A common crop sequence now being adopted is a brown manure legume, followed by canola, wheat and feed barley. While field peas have generally been the first brown manure crop grown, vetch is being adopted in the second sequence to minimise the disease experienced with a shorter break between field pea crops.

Economics

An economic analysis of two farming systems was done in southern NSW with a 450-millimetre annual rainfall.

The two farming systems analysed were:

- » Continuous cropping of wheat and canola only; and
- » Continuous cropping, but including brown manure field peas grown on 25 per cent of the arable area.

The economic analysis is based on two similar-sized properties in southern NSW, with two family labour units performing most of the operations. Farm data from properties that have adopted brown manure peas have shown 25 to 30 per cent yield increases for both canola and wheat crops grown in the two years following brown manure field pea crops. The analysis assumes a 20 per cent increase in yield above average in the first two crops following brown manure peas. Wheat prices have been adjusted to reflect protein levels. It shows the estimated capital required for each of the farming systems. The difference in plant investment is due to a larger header and bins being required for the continuous cropping system.

The working capital requirement of the continuous cropping system is higher than the brown manure field peas system due to the higher inputs (herbicides, fungicides and artificial nitrogen). The amount of working capital required is a measure of the system's degree of risk because there is virtually a guarantee that costs of continuous cropping will be higher, but there is no guarantee that gross income will be higher. This results in the potential for a greater loss to occur in that year if seasonal conditions are unfavourable, leading to the potential for this additional working capital to become long-term debt.

The brown manure system is considered to be robust and low risk in drier seasons, as there is less potential to spend money on crop inputs in the pursuit of higher yields. The annual trading results measured by earnings before interest and taxes (EBIT) and three key financial ratios are observed.

EBIT is a measure of profitability after allowances for plant replacement and family labour. Based on the assumptions used, predicted EBIT from continuous cropping is slightly higher than that from the brown manure legume system. There is little difference between the financial ratios, except that while the gross income and EBIT from continuous cropping is higher, it has the lower EBIT margin, due to its higher costs relative to income. This lower EBIT margin suggests a higher degree of risk with this system.

The results of the comparison are sensitive to the price of nitrogen fertiliser. A $100 per tonne increase in the price of urea increases costs in continuous cropping by $19,200 a year compared with $5200 in the brown manure legume system. This would bring the respective EBITs within $2000 of each other.

GREEN MANURE

History

In agriculture "green manure" refers to crops which have already been uprooted (and have often already been stuffed under the soil). The then dying plants are of a type of cover crop often grown primarily to add nutrients and organic matter to the soil. Typically, a green manure crop is grown for a specific period of time, and then plowed under and incorporated into the soil while green or shortly after flowering. Green manure crops are commonly associated with organic farming, and are considered essential for annual cropping systems that wish to be sustainable.

The value of green manure was recognized by farmers in Ancient Greece, who ploughed broad bean plants into the soil. Chinese agricultural texts dating back hundreds of years refer to the importance of grasses and weeds in providing nutrients for farm soil. It was also known to early North American colonists arriving from Europe. Common colonial green manure crops were rye, buckwheat and oats. Traditionally, the incorporation of green manure into the soil is known as the fallow cycle of crop rotation, which was used to allow soils to recover.

By green manuring is meant the turning under of a green crop, for the enrichment of the soil. Some German writers have limited this definition to the turning under of legume crops, but there seems no warrant for this restriction. Strictly speaking, the green crop must have been produced for the purpose of being turned under.

This definition is, however, a very narrow one. Any plant material whether sod, rye, weeds, clover, cowpeas, or crop residues, when turned under accomplishes the purpose of green manuring, namely, to add organic matter to the soil. This is equally true whether this plant material is green or dry. It is not possible to distinguish sharply green manuring from cover cropping, though each term has a basic meaning of its own. The term "cover crop" was first used by Prof. L. H. Bailey to designate a crop especially planted to cover the ground in winter and to serve as a protection to the roots of trees. When such a cover crop is rye, or clover, and is turned under in spring, it becomes a green-manure crop. Of late in the United States the term "cover crop" has tended to displace the term green-manure crop, so that the two terms have become, to a great degree, synonymous. A "catch crop" may also serve as a green-manure crop, and indeed a great deal of green manuring is done with catch crops. It seems necessary, therefore, to point out that, while definitions may be made, the practices carried on under these terms run into or overlap one another and any attempt at sharp separation must be artificial. With this understanding, the following definitions may be given:

Green manuring is the practice of enriching the soil by turning under un-decomposed plant material (except crop residues), either in place or brought from a distance.

Types of green manuring: There are two types of green manuring:

1. **Green manuring *in-situ*:** When green manure crops are grown in the field itself either as a pure crop or as intercrop with the main crop and buried in the same field, it is known as Green manuring *In-situ*. E.g.: Sannhemp, Dhaicha, Pillipesara, Shervi, Urd, Mung, Cowpea, Berseem, Senji, etc.

These crops are sown as:

   i. Main crop,

   ii. Inter row sown crop,

   iii. On bare fallow, depending upon the soil and climatic conditions of the region.

Upland rice is commonly grown during rainy season in Himalayan hills. Low or no application of fertilizer nutrients and heavy weed infestation are the major constraints leading to low productivity of rice in this region. Thus, the present investigation was conducted to see the effect of in-situ green manure in supplementing nutrients and saving of labour due to reduction in manual weeding from 2 to 1. Different combinations of chemical fertilizer with green manure or FYM were evaluated. In-situ green manuring with sunhemp improved the overall productivity of upland rice. Highest yield attributes, yield and sustainability index was recorded with 100 per cent NPK + green manure, which were at par with 75 per cent NPK + green manure. However, these values were lowest under 100 per cent NPK application. Higher productivity and saving of labour under green manuring resulted in higher net return and benefit: cost ratio. Soil available nutrients (NPK) and net balance of NPK were also higher under 100 per cent NPK + green manure. The study clearly suggests that in-situ green manure is a viable low-cost technology for enhancing productivity and profitability of upland rice in the fragile ecosystem of Himalayan hills.

Rice needs adequate nitrogen, particularly under rainfed lowland situation, to ensure its growth for optimum grain yield. Traditional practice of applying entire dose of N fertilizer during sowing causes its low recovery. Nonetheless, it does not sustain over subsequent growth of rice, mainly due to uncontrolled waterlogged situation. Green manuring along with chemical N application was reported as advantageous ensuring sustained availability of adequate N over the period. A field study on green manuring rice under rainfed lowland situation was conducted during wet seasons in 2001 to 2004 in the Central Rice Research Institute, Cuttack, India. The photosensitive, tall and long-duration rice variety, Durga was grown along with green gram (*Phaseolus radiatus*) for green manuring under three mixed cropping stands, that is, additive system, replacement system, and broadcast-sown stands. Green gram, also known as mung bean, is a widely grown pulse crop. As a leguminous crop, it is capable of contributing considerable amount of green matter and N into the soil. Thus it can be used for green manuring to facilitate N nutrition to the rice grown in its association. Rice was sown at 15 × 20 cm spacing (plant × row), except in broadcast-sown stands where rice and green gram seeds were mixed thoroughly before sowing and broadcasted evenly without maintaining any plant/row arrangement. In the additive system, one row of green gram was intercropped after every two consecutive rice rows, when usual row spacing was changed therein. Thus the adjacent rice rows were 10 cm apart from green gram row. In the replacement system, one rice row after every two consecutive rice rows was replaced with one row of green gram without changing usual row spacing, that is, the adjacent rice rows were 20 cm apart from green gram row. These green-manured mixed stands were compared with two pure stands of rice, that is, rice grown with usual dose of 40 kg N/ha and rice grown without N. Similar management practices were followed for both the crops altogether and no additional fertilizers were applied in green gram. Growth dynamics of green gram as evidenced from its plant height and dry matter recorded at regular interval showed no variation over the years. It was incorporated into the soil at its flowering stage accumulating 7.00 to 25.60 kg N/ha over the years under different stands. Rice growth due to green manuring in the replacement system remained consistently better than other stands in all the years. As a consequence, rice under the replacement system significantly out-yielded other stands producing 2.41 to 2.52 t/ha over the years, which was at par with that treated with fertilizer N alone (2.48 and 2.60 t/ha), while grain yield (2.12 to 2.20 t/ha) under the additive system was significantly higher than broadcast-sown crop (1.77 and 1.84 t/ha). In addition, N utilization pattern revealed that crops under the replacement system derived higher Agronomic efficiency (43.50 to 53.50 kg grain/kg N applied), N uptake (27.75 to 29.30 kg/ha), and N recovery (56.25 to 66.00%, respectively) over the years as compared with that in other cropping geometry.

A field experiment entitled "*In-situ* green manure incorporation effects on nutrient dynamics of *kharif* maize (*Zea mays* L.)" was conducted during *kharif*, 2012 on clay soil of the Agricultural College Farm, Bapatla, under rainfed condition. The treatments consisted of 3 different green manures (dhaincha, sunnhemp and pillipesara) and 3 different ages of their incorporation (60, 45 and 30 days) with an additional control (where no green manuring was done). The experiment was conducted in a randomized block design (RBD) with factorial concept and replicated thrice.

Age of incorporation of green manures only had a significant influence in increasing plant height and drymatter production at 30 DAS of maize. Similar trends were also observed for plant height and drymatter production in maize at remaining stages.

Quantity of green manure biomass production was maximum (13.8 t/ha) in dhaincha followed by sunnhemp and pillipesara and all these were found to be significantly superior to one and another. Age of incorporation of green manures also had a significant effect on biomass production. The 60 days aged greenmanure recorded maximum biomass (18.1 t/ha) which was found to be significantly superior to 45 and 30 days aged green manures. A similar trend was also observed in respect of drymatter production with regard to these green manures. Quantity of green manure biomass production was maximum (13.8 t /ha) in dhaincha followed by sunnhemp and pillipesara and all these were found to be significantly superior to one and another. Age of incorporation of green manures also had a significant effect on biomass production. The 60 days aged green manure recorded maximum biomass (18.1 t/ha) which was found to be significantly superior to 45 and 30 days aged green manures. A similar trend was also observed in respect of drymatter production with regard to these green manures.

Days to 50 per cent tasseling and silking of maize were significantly reduced due to age of incorporation of green manures whereas, no significant difference was observed due to incorporation of different green manures. Number of cobs per plant due to *in-situ* incorporation of green manures was also not significant. No significant difference in respect of cob length, number of grains per cob, cob weight, test weight and shelling percentage was observed due to *in-situ* incorporation of different green manures but significant increase was observed due to age of incorporation of green manures. However, the interaction was found to be non significant. All green manure treated plots recorded a significant increase over control.

Age of incorporation had a significant influence on grain yield of maize. Maximum grain yield (78.71 q/ ha) was recorded when incorporated at 60 days which was significantly superior to 45 and 30 days incorporation of green manures. However, grain yield obtained from 45 and 30 days aged incorporated green manure plots did not attain the level of significance with each other. The minimum grain yield (57.61 q/ ha) was observed in control where no green manuring was done. All the green manure treated plots recorded significantly higher grain yield of maize over control. The percentage increase in grain yield due to 60, 45 and 30 days age of incorporation over control was 36.6, 22.0 and 14.8 per cent, respectively. All the three green manure crops remained at a par with one and another. However, the interaction between green manures and their age of incorporation was remained statistically at a par. Similar results were observed for stover yield.

Nitrogen content and uptake at harvest were found to be higher in dhaincha incorporated plot which was significantly superior to sunnhemp and pillipesara incorporated plots, however, no significant difference was observed between sunnhemp and pillipesara incorporated plots. Age of incorporation of green manures also had a significant effect on nitrogen content and uptake of maize at harvest. It was found maximum where green manures were incorporated at 60 days which was significantly superior to 45 and 30 days incorporated green manures. Whereas, N content and uptake at 45 and 30 days incorporated green manures was remained on a par with each other. All the green manure treated plots recorded a higher N content and uptake in maize at harvest and found significantly superior to control. Nitrogen use efficiency (NUE) of different green manures was highest in dhaincha followed by sunnhemp and pillipesara.

Age of incorporation of green manures also had an effect on the NUE. The maximum (18 kg/ kg N) NUE was observed with 60 days age of incorporation followed by 45 and 30 days age of incorporation. Overall, there was an increase in microbial population due to incorporation of different green manures and their ages of incorporation. But, the increase in microbial population was more or less was affected due to their ages of incorporation. The pH of the soil 30 DAS as well as at harvest and bulk density at harvest of maize did not vary significantly either due to incorporation of different green manures or their ages of incorporation. Maximum organic carbon content (0.73 %) in soil 30 DAS of maize was recorded with dhaincha incoporated plot which was significantly superior to sunnhemp and pillipesara. Organic carbon contents of dhaincha and sunnhemp as well as sunnhemp and pillipesara incorporated plots were statistically remained on a par with each other. Organic carbon content of 60 days (0.75 %) incorporated green manures was found to be significantly superior to 45 and 30 days incorporated green manure plots. However, organic carbon content in 45 and 30 days incorporated green manure plots did not differ significantly with each other. All the green manure treated plots recorded significantly higher organic carbon content over control. A similar trend was observed after harvest of maize which was decreased.

Undecomposed green manure biomass portion at 30 DAS of maize was found maximum (3.69 t/ha) with dhaincha incorporated plot followed by sunnhemp and pillipesara, and all these were found to be significantly superior to one and another. Due to age of incorporation of green manures, maximum undecomposed portion was observed with 60 days age of incorporation followed by 45 and 30 days age of incorporation, but all these were significantly differed with one and another.

The soil available N at 30 DAS of maize did not differ significantly due to incorporation of different green manures. Soil available N at 60 days age of incorporation was found to be significantly superior to 30 days age of incorporation. However, 60 and 45 days age of incorporation as well as 45 and 30 days age of incorporation didn't differ significantly with each other. The entire green manure treated plots recorded significantly higher soil available N over control. A similar trend was observed for available soil N at harvest of maize crop.

Similar trend of response was observed in respect of P and K at 30 DAS and at harvest of maize crop as that was observed in respect of soil available N except soil available K at harvest which was not significant. Among all the treatments tried, highest return per rupee investment (N. 4.01) was obtained where dhaincha was incorporated at 45 days followed by dhaincha incorporated at 30 days and sunnhemp incorporated at 30 days.

Overall, it can be concluded that green manuring with dhaincha either at 60 days or 45 days is more beneficial than green manuring with either sunnhemp or pillipesara for *kharif* maize in realizing higher grain yields, improvement of soil physical properties and soil fertility status.

2. **Green leaf manuring:** It refers to turning into the soil green leaves and tender green twigs collected from shrubs and trees grown on bunds, waste lands and nearby forest area. E.g.: Glyricidia, wild Dhaicha, Karanj.

Green and Brown Manures in Mediterranean-Type Environments

Dryland wheat production systems

This research on the use of green and brown manures in wheat cropping systems in Mediterranean-type environments in the light of contemporary pressures on cropping systems including changing climates, increasing costs and declining profit margins.

Green and brown manuring have been demonstrated to have benefits in terms of weed control, delaying the development of resistance to herbicides, reducing populations of disease organisms, altering soil water, soil quality and biology, erosion control and contributing to the nutrition of subsequent crops. However, few researchers have attempted to measure more than one of these variables, which presents difficulties in both interpreting the causes behind results of field trials and in estimating the total benefit of manuring, and hence its consequences for profitability.

Well-designed experiments have been reported on component mechanisms (such as weed numbers or N_2 fixation). However, these experiments are often not taken through to maturity of the crop following the manuring treatment. As a result, there is limited yield and grian quality data on which to base sound analyses of profitability. A few reports are available which present the impact of manuring on wheat yield and profitability in specific areas and systems but the results vary widely. For such reports to be of value, further research is needed into the factors inducing the changes in response (climate, soil type, or the specifics of the farming system at the time the treatments are imposed) and the mechanisms by which these act. Thus, research is needed into both the mechanisms and yield benefits that flow from the individual responses to manuring.

Two further limitations to determining the economic benefit of manuring emerge. Firstly, impacts are primarily reported for a single year after a single manuring treatment. However, if measurements are made over a number of years, effects can often still be detected. More studies aimed to assess the longer-term impacts of manuring on soil health, disease prevalence, and weed populations are required. Secondly, there has been very little effort to explore the whole-farm impact of using manures. These impacts could include effects on other farm enterprises as well as business-level impacts such as potential changes in labor requirements, cash flow, and risk.

The incorporation of manuring into wheat production systems may have multiple on-farm and off-farm benefits. However, there is a substantial research requirement before these approaches could be recommended. The highest priority is a sound demonstration of short- to medium-term economic benefits to growers. Without this, adoption can be expected to be poor.

Ideal characteristics of a good manuring

- » Yield a large quantity of green material within a short period.
- » Be quick growing especially in the beginning, so as to suppress weeds.
- » Be succulent and have more leafy growth than woody growth, so that its decomposition will be rapid.
- » Preferably is a legume, so that atmospheric 'N' will be fixed.
- » Have deep and fibrous root system so that it will absorb nutrients from lower zone and add them to the surface soil and also improve soil structure.
- » Be able to grow even on poor soils.

Ideal stage of green manuring

A green manuring crop may be turned in at the flowering stage or just before the flowering. The majority of the G.M. crops require 42 to 56 days after sowing at which there is maximum green matter production and most succulent.

Disadvantages of green manuring

i. Under rain fed conditions, the germination and growth of succeeding crop may be affected due to depletion of moisture for the growth and decomposition of G.M.

ii. ii) G.M. crop inclusive of decomposition period occupies the field least 75-80 days which means a loss of one crop.

iii. Incidence of pests and diseases may increases if the G.M. is not kept free from them. Application of phosphatic fertilizers to G.M. crops (leguminous) helps to increase the yield, for rapid growth of *Rhizobia* and increase the 'P' availability to succeeding crop.

A cover crop is one planted for the purpose of covering and protecting the ground during winter.

A catch crop is a rapidly growing crop, following a main crop during the same season and occupying the ground for a few weeks only.

A shade crop is one used in hot regions to shade the ground during summer and thus prevent beating from rains, excessive heating of the soil, or injury to trees from reflected heat.

Functions of Green Manure Crops

Green manures usually perform multiple functions that include soil improvement and soil protection:

- » Leguminous green manures such as clover and vetch contain nitrogen fixing symbiotic bacteria in root nodules that fix atmospheric nitrogen in a form that plants can use. This performs the vital function of fertilization. If desired, animal manures may also be added.

- » Depending on the species of cover crop grown, the amount of nitrogen released into the soil lies between 45-250 kg (100-500 pounds)/ha. With green manure use, the amount of nitrogen that is available to the succeeding crop is usually in the range of 40-60 per cent of the total amount of nitrogen that is contained within the green manure crop.

Average biomass yields and nitrogen yields of some green manure crops

Cover crops	Biomass yield (t/ha)	N (kg/ha)
Sweet clover	4.37	120
Berseem clover	2.75	70
Crimson clover	3.50	100
Hairy vetch	4.37	110

- » Green manure acts mainly as soil acidifying matter to decrease the alkalinity / pH of alkali soils by generating Humic acid and Acetic acid

- » Incorporation of cover crops into the soil allows the nutrients held within the green manure to be released and made available to the succeeding crops. This results immediately from an increase in abundance of soil microorganisms from the degradation of plant material that aid in the decomposition of this fresh material. This additional decomposition also allows for the re-incorporation of nutrients that are found in the soil in a particular form such as nitrogen (N), potassium (K), phosphorus (P), calcium (Ca), magnesium(Mg), and sulfur (S).

- » Microbial activity from incorporation of cover crops into the soil leads to the formation of mycelium and viscous materials, which benefit the health of the soil by increasing its soil structure (i.e. by aggregation).

- » The increased percentage of organic matter (biomass) improves water infiltration and retention, aeration, and other soil characteristics. The soil is more easily turned or tilled than non-aggregated soil. Further aeration of the soil results from the ability of the root systems of many green manure crops to efficiently penetrate compact soils. The amount of humus found in the soil also increases with higher rates of decomposition, which is beneficial for the growth of the crop succeeding the green manure crop. Non-leguminous crops are primarily used to increase biomass.

- » The root systems of some varieties of green manure grow deep in the soil and bring up nutrient resources unavailable to shallower-rooted crops.

- » Common cover crop functions of weed suppression. Non-leguminous crops are primarily used (e.g. Buckwheat). The deep rooting properties of many green manure crops make them efficient at suppressing weeds.

- » Some green manure crops, when allowed to flower, provide forage for pollinating insects. Green manure crops also often provide habitat for predatory beneficial insects, which allow for a reduction in the application of insecticides where cover crops are planted.

- » Some green manure crops (Winter wheat and winter rye) can also be used for grazing.

- » Erosion control is often also taken into account when selecting which green manure cover crop to plant.

- » Some green crops reduce plant insect pests and diseases. Verticillium wilt is especially reduced in potato plants.

Incorporation of green manures into a farming system can drastically reduce, if not eliminate, the need for additional products such as supplemental fertilizers and pesticides.

Limitations to consider in the use of green manure are time, energy, and resources (monetary and natural) required successfully growing and utilizing these cover crops. Consequently, it is important to choose green manure crops based on the growing region and annual precipitation amounts to ensure efficient growth and use of the cover crop(s).

Nutrient creation

Green manure is broken down into plant nutrient components by heterotrophic bacteria that consume organic matter. Warmth and moisture contribute to this process, similar to creating compost fertilizer. The plant matter releases large amounts of carbon dioxide and weak acids that react with insoluble soil minerals to release beneficial nutrients. Soils that are high in calcium minerals, for example, can be given green manure to generate a higher phosphate content in the soil, which in turn acts as a fertilizer.

The ratio of carbon to nitrogen in a plant is a crucial factor to consider, since it will impact the nutrient content of the soil and may starve a crop of nitrogen, if the incorrect plants are used to make green manure. The ratio of carbon to nitrogen will differ from species to species, and depending upon the age of the plant. The ratio is referred to as C:N. The value of N is always one, whereas the value of carbon or carbohydrates is expressed in a value of about 10 up to 90; the ratio must be less than 30:1 to prevent the manure bacteria from depleting existing nitrogen in the soil. *Rhizobium* are soil organisms that interact with green manure to retain atmospheric nitrogen in the soil. Legumes, such as beans, alfalfa, clover and lupines, have root systems rich in rhizobium, often making them the preferred source of green manure material.

Green Manure Crops

Winter green manure crops are oats and rye.

Other green manure crops:

- » Alfalfa, which sends roots deep to bring nutrients to the surface.
- » Buckwheat in temperate regions
- » Cowpea
- » Clover (e.g. annual sweet clover)
- » Fava beans
- » Fenugreek
- » Lupin
- » Millet
- » Mustard
- » Sesbania
- » Sorghum

- » Soybean
- » Sudangrass
- » Sunn hemp, a tropical legume
- » Velvet bean (*Mucuna pruriens*), common in the southern US during the early part of the 20th century, before being replaced by soybeans, popular today in most tropical countries, especially in Central America, where it is the main green manure used in slash/mulch farming practices
- » Vetch (*Vicia sativa, Vicia villosa*)
- » Tyfon, a *Brassica* known for a strong tap root that breaks up heavy soils.
- » Ferns of the genus *Azolla* have been used as a green manure in Southeast Asia.

Effect of Green Manure/ Leguminous Cover Crops

Biomass production and nitrogen accumulation

Green manuring involves soil incorporation of any field or forage crop whereas cover crop is any crop grown to provide soil cover and to improve physical, chemical or biological properties of soil regardless of whether it is later incorporated. In conservation agriculture the residues of cover crops are always left on the surface and are incorporated biologically rather than with tillage implements. Among the cover crops, legume species are considered the most beneficial to provide nitrogen which is the most limiting source of nutrients in tropical soils. These green manure/leguminous cover crops are known to benefit the succeeding crops by way of symbiotic nitrogen fixation and mobilization of lesser available forms of plant nutrients, improves soil structure and decrease leaching loss of nutrients. This makes it possible to achieve significant responses in the yield of crops such as corn and cotton (nitrogen demanding) that are sown after legumes. The phytomass build-up by cowpea, dhaincha and sunhemp on sandy loam soil at Ludhiana was 17.1 to 31.7, 16.4 to 29.8 and 18.7 t/ ha, respectively at 42 to 49 days age. The maximum plant height, fresh shoot and root biomass in sunhemp at all the three growth periods but the grain yield and total biomass production at harvest stage (60 DAS) were maximum in case of Pakistani Janter (*Sesbania aculeate*). Conducted study on green manure legumes at Palanpur and observed maximum phytomass and biomass production with sunhemp (17.8 and 3.8 t/ ha, respectively) while the minimum was recorded in S*esbania* (15.4 and 2.3 t/ ha, respectively) at 55 days of growth. In Zimbabwe, observed that mucuna (*Mucuna pruriens*) recorded a peak biomass level of about 4 t/ha at 112 days after planting similar in that regard to cowpea (*Vigna unguiculata*) but inferior to sunhemp (*Crotalaria juncea*), 12 t/ha and lablab (*Lablab purpureus*) 9 t/ha. The research trials conducted at Punjab Agricultural University, Ludhiana confirmed that sunhemp recorded more plant height, higher phytomass and biomass at 45 and 60 days after sowing (DAS) as compared to dhaincha. Results of two year field experiment conducted at Dharwad, revealed that sunhemp green manure recorded significantly higher green matter yield and dry matter yield (11.38 and 2.02 t/ha, respectively) than cowpea (8.58 and 1.35 t/ha, respectively) and dhaincha (7.55 and 1.25 t/ha, respectively). The results of trial confirmed that sunhemp and cowpea as the best cover crops with high biomass and weed suppression while mucuna was the least. Sunhemp consistently yielded higher cover biomass averaging 11.2 t/ha over the two seasons while mucuna had a consistently lowest average biomass yield of 4.1 t/ha.

The higher biomass yield and nitrogen accumulation was recorded in sunhemp (4.6 and 78 kg N/ha, respectively) than in dhaincha (2.9 and 57 kg N/ha, respectively at 60 DAS. Likewise, green biomass potential, dry matter accumulation and nitrogen contribution of *Crotalaria juncea* has been reported to be higher than traditionally grown *Sesbania aculeate* and highest nitrogen accumulation was obtained with *Crotalaria juncea* (92-121 kg/ha) followed by *Sesbania rostrata* (34-67 kg/ ha).

The nitrogen addition at seven weeks age from green manures crops *viz.*, dhaincha, cowpea and sunhemp was 77.5, 83.3 and 94.3 kg/ha, respectively at research trial conducted at PAU, Ludhiana. Field investigation at New Delhi found that *Sesbania* accumulated significantly higher amount of biomass (4.75 t/ha dry weight) and nitrogen (131 kg/ha) as compared to dual purpose legumes i.e. cowpea and green gram, which added 3.26 and 2.79 t/ha of stover, and 49, 47 kg N/ha, respectively into the soil.

At Dharwad reported that the increase in phytomass and biomass of cowpea and sunhemp over green gram was 41.4 and 25.9; 29.9 and 24.9 per cent, respectively. The sunhemp accumulated significantly higher N (50.4 kg/ha) followed by cowpea (44.3 kg/ha) while the lowest N accumulation of 25.6 kg/ ha was recorded with green gram. Similarly the findings of another study at Dharwad found that dhaincha green manure showed significantly higher N content on pooled basis (3.04 %, respectively) when compared to sunhemp (2.98 %) and cowpea (2.79 %) but sunhemp recorded significantly N accumulation (60.1 kg/ha) than dhaincha (37.9 kg/ha) and cowpea (37.6 kg/ha).

However at Dehradun noticed that all the legumes accumulated almost equal quantity of biomass at 30 days but at 45 days whereas N content varied in different crops and decreased with advancement in age. N accumulation was similar with *dhaincha* and sunhemp and was comparatively more than with cowpea at both the stages, due to relatively higher N content and biomass accumulation in the former crops.

The green biomass potential, dry matter accumulation and N contribution of *Crotalaria juncea* has been reported to be higher than traditionally grown *Sesbania aculeate* in mollisols of crop research station at Pantnagar. The 60 days old crop accumulates about 170 kg N, 20 kg P and 130 kg K/ha. The N content of 45 days old crop was higher as compared to 60 days old crop. The N accumulation by *Crotalaria* at 45 days stage was 106 and 112 kg/ha in 2001-02 and 2002-03, respectively while 60 days old *Crotalaria* accumulated 162 and 177 kgN/ha in 2001-02 and 2002-03, respectively.

Effect of green manure/cover crops on maize

Tillage is more effective for killing cover crops but tilled cover crops break down rapidly once these are incorporated into the soil and may lead to nutrient losses through leaching. However, in no-till production system these leguminous cover crops can be killed by chemical (herbicides) and mechanical methods (mowing, rolling, undercutting). The maximum N generated by legumes was during mid-bloom or when one half of the inflorescence is fully expanded and the other half of blooms are in bud. Therefore, cover crops in conservation tillage system are terminated during early reproductive growth with herbicides or mechanically.

Three methods for mechanically killing cover crops were evaluated in a study in Eastern North Carolina: undercutting, mowing and rolling. Mowing was accomplished with a flail mower, which leaves the finely chopped residue evenly distributed over the bed. Rolling was accomplished using the same flail mower but with the mowing tines disengaged. The undercutting was performed with a single blade that severed cover crop roots approximately two inches deep and an attached roller laid the undercut cover crops flat. The effectiveness of kill of cowpea and *Sesbania* by mow was 98 and 100 per cent, respectively, which was highest followed by undercut and least in roll method.

In field experiment in South Africa, the cover crops i.e. oat (*Avena sativa*), grazing vetch (*Vicia dasycarpa*), faba bean (*Vicia faba*), lupin (*Lupinus angustifolius*) and forage pea (*Pisum sativum*) were killed at flowering stage by rolling them and applying glyphosate at a rate of 5 l/ha and subsequent sowing of maize was done under no-till system.

From Nigeria, reported that maize crop succeeding legumes recorded higher plant height and dry matter than that succeeded by fallow or maize by itself in rotation. This may be attributed to better availability of symbiotically fixed N by the legume to the succeeding crop. Also reported from Tanzania that number of grains cob^{-1}, cob weight and grain yield of maize significantly increased where sole legumes preceded it in rotation. The highest grain yield (4.56 t/ha) was recorded from plots where sole *Crotalaria* preceded it.

Conducted an experiment on loamy sandy soils and reported that the plant height and dry matter production of maize was significantly higher with the incorporation of different green manure crops like cowpea, clusterbean and *Sesbania* than that of fallow irrespective of level of N applied. Similarly, reported that when green manure crops like cowpea and *Sesbania aculeata* of 45-50 days old were incorporated fifteen days prior to maize resulted in significant increase in maize growth parameters and supplemented fertilizer N by 60 kg/ha.

Similarly, from black clay soils of Dharwad, observed that sunhemp green manuring to maize recorded significantly higher plant height, leaf area index, dry matter accumulation than that of cowpea green manuring. However found from silty clay loam soils of Dehradun that growth attributes of maize were significantly superior with live mulching of sunhemp + *Leucaena* to that of no-mulching.

From red sandy soil of Raichur that green manuring with subabul loppings in maize with 100 per cent recommended dose N (RDN) application recorded maximum number of grains/cob, grain weight/cob and test weight than those with either of them. From black clay soils of Dharwad, showed that sunhemp green manuring in maize increased cob length, cob girth, number of grains per cob of maize grown with sunhemp green manuring than that of cowpea green manuring.

At Ludhiana on loamy sand soil, demonstrated that maize grain yield was significantly increased with the incorporation of 20 t/ha phytomass of cowpea and this yield was equal to the yield observed with supplementation of 120 kg Nha^{-1}. Similarly, reported increase in maize grain yields by 1.2 t/ha with sunhemp green manuring.

A field investigation at Dehradun found that the beneficial effect of live mulching with sunhemp or *leucaena* was similar (12.3-14.7 %), while their combined application increased the maize yield by 19.1 per cent over no-mulching in maize sown under rainfed conditions. The highest maize grain yields (9.13 t/ ha in 2010-11 and 9.38 t/ha in 2011-12) were found in the treatment, which received green manure of *S. aculeata*, which were significantly higher than the other treatments. The lowest yields (7.54 t/ ha in 2010-11 and 6.78 t/ha in 2011-12) were recorded in the control. The straw yields were also followed the same trends.

Four legume cover species, Sunhemp (*Crotolaria juncea*), Mucuna (*Mucuna pruriens*), Lablab (*Dolichos lablab* var Highworth) and Cowpea (*Vigna unguiculata* var Agrinawa) were evaluated for their effect on subsequent maize yield in the Eastern Cape Province of South Africa. The results revealed that subsequent maize yield was significantly higher in the sunhemp plots, 64.2 per cent more than weedy fallow plot. Mucuna, lablab and cowpea had maize grain yield increases of 16.6, 33 and 43.2 per cent, respectively.

Growth and yield of maize was more after *Sesbania* followed by green gram and cowpea at New Delhi and there was saving of N to the extent of 57-67 kg/ha with *Sesbania*, and 37-49 kg/ha with cowpea and green gram. Whereas, sunhemp green manuring in maize recorded significantly higher grain and stover yield (55.1 and 95.8 q/ha, respectively) over cowpea green manuring (50.3 and 91.7 q/ ha, respectively) and also resulted in significantly higher N P K (225.5, 30.1, 208.6 kg/ha) uptake as compared to cowpea (211.7, 27.9, 193.5 kg/ha) at Dharwad.

Similarly, noticed significant increase in N uptake by maize with the application of fertilizer N up to 90 kg/ha + green manuring of cowpea and up to 120 kgN/ha following fodder cowpea or fallow. Sunhemp and *Leucaena* mulching increased N uptake of maize by 20.4 and 25.0 per cent and of wheat by 22.7 and 26.8 per cent%, respectively and their combined application resulted in an increase of 33.9 and 44.8 per cent over no-mulching at Dehradun. The enhanced moisture and nutrient contribution due to the added mulch material lead to increased biomass production, and hence higher N uptake.

The incorporation of *Sesbania aculeata* or sunhemp before maize planting saved the fertilizer N up to 45-60 kg/ ha on acid lateratic soil of Kharagpur. Similarly, in a study on N economy through green manuring in maize on cultivators fields of Ropar and Patiala districts reported that green manuring with cowpea or *Sesbania aculeata* (45 to 50 days old) prior to sowing of maize substituted for nearly 60 kg/ha of fertilizer N needed by the succeeding maize.

Cover crop residue left on the soil surface in conservation tillage systems decompose and add to the soil organic matter content. These carbon inputs improve soil aggregation, which maintains soil structure resulted in enhances water retention and reduces erosion. Cover crop roots also improve soil porosity and increase the infiltration rate thus reducing runoff from the field. An increase in soil moisture observed with increased amount of cover crop residue in a conservation tillage system as compared to conventional tillage.

Field studies at Ropar and Patiala districts of Punjab revealed that organic carbon content of soil increased from 0.55 (N_{60}) to 0.58 per cent with green manure+N_{60} kg/ha and available nutrient status of the soil increased from 154, 17.9, 178 to 167, 19.7 & 189 kg/ha for N, P and K, respectively. *Sesbania* green manuring and mungbean residue incorporation in rice-wheat cropping system increased soil organic carbon over summer fallow by 0.11 to 0.14 per cent, N by 0.01 per cent and available P by 5.0 to 5.5 kg/ha.

The sunhemp had the highest residual total N (0.165%), which was comparable to mucuna (0.154%) whereas, lablab (0.067%) and cowpea (0.097%) did not differ but had higher N than the weedy fallow (0.031%). There were also significant differences ($p<0.05$) in soil organic carbon and total phosphorus among the plots with the different cover crop species. Only the mucuna plot had significantly higher soil organic carbon than the weedy fallow plot. In terms of total phosphorus, only cowpea had significantly higher levels than all the other treatments.

The highest organic matter (1.36%) and total N (0.068%) in G_1 (*S. aculeata*) followed by G_2 (*Mimosa invisa*) and G_3 (*V. radiata*). The lowest organic matter (0.65%) and total N (0.033%) were found in the control treatment. The increase of organic matter and total N contents in green manure treated plots might be due to addition of more biomass. The control plots showed declining trends of soil organic matter and total N status after two years. The *S. aculeata* showed the lowest bulk density followed by *Mimosa* and *V. radiata* and it was highest in control. The significantly highest soil moisture content was found in G_1 (*S. aculeata*) followed by G_2 (*M. invisa*) and G_3 (*V. radiata*). The lowest moisture content was observed in the control.

A field experiment at Dharwad reported that sunhemp green manure added on an average 57.6, 9.7, 35.6 kg total N P K as compared to cowpea green manure, which added 47.4, 6.5 and 25.4 N, P AND K kg/ha, respectively. The infiltration rate (cm/hr) was also significantly higher in sunhemp (3.67 cm/hr) as compared to cowpea (3.23 cm/hr) while bulk density remained non-significant.

In a study at New Delhi found that mean soil organic carbon content after cowpea (0.59 %), green gram (0.66 %) and *Sesbania* (0.66 %) was significantly more than after fallow (0.51 %). Similarly, the mean $KMnO_4$-N content showed maximum improvement after *Sesbania* (271 kg/ha), followed by green gram (256 kg/ha) and cowpea (252 kg/ha), all of which were significantly superior to fallow (210 kg/ha). An increase of 28.5 & 18.5 per cent in the soil organic carbon and infiltration rate, respectively in rainfed maize when combined application of *Leucaena* and sunhemp was done in rainfed maize for three years where as the bulk density decreases from 1.44 to 1.36 g/cc.

The treatment of in-situ green manuring with sunhemp has recorded the highest soil moisture content both at 15cm (16.37%) and 30cm (16.13%) soil depth almost at all growth stages of maize in south-eastern dry zone of Karnataka. While, the same treatment recorded significantly lower soil temperature at all growth stages, significant and numerically higher infiltration rate (9.68 cm/hr) and improved available soil nutrients (N, P and K) and organic carbon content, which resulted in higher maize yield of 5269 kg/ha. Higher net income of Rs 30098/ha and B: C of 2.53 was also realized in the same treatment.

Interaction effect of leguminous cover crops, no-tillage and nitrogen on maize

Cover crops are receiving increased attention for their ability to enhance their multi-functionality of cropping systems, particularly in no-till (NT) farming. These cover crops have the ability to "Jump start" no-till perhaps eliminating any yield decrease especially in initial years. No-tillage, N and cover crops are known to play an important role in conserving or increasing soil organic carbon and soil total N but the effects of their interactions are less known.

a. Cover crops and no-tillage

In an experiment in Rachuonyo in Kenya, *Lablab perpureus* as cover crop together with conservation tillage gave maize grain yields of 2.6 t/ha against 1.8 t/ha in control. A field experiment consisting of conventional tillage, no-tillage with varying percentage of surface residue coverage (0, 33, 66 and 100 per cent) and no-tillage + 33% residue + cover crop of either *Vicia* sp. or *Phaseolus vulgaris* L. at Patzcuaro watershed in Central Mexico, revealed that no-tillage with moderate amount of crop residue (33%) and planted to leguminous cover crop rapidly improved the soil quality characteristics.

Maize sown with no-till conditions after pigeon pea and sorghum cover crops gave significantly higher grain yields in direct seeding mulch based cropping (DMC-14) than the bare fallow (BF) treatment in Brazilian Cerrados. The N uptake was significantly lower in the bare fallow treatment as compared to the pigeon pea, pearl millet with Congo signal grass and sorghum cover crop treatments. The DMC-14 field, cover crops increased N uptake of zero-N fertilized maize with 25 to 66 kg N/ha as compared to BF, which resulted in an increase in grain yield of 0.4 to 2.4 Mg/ha.

In a field experiment at Ohio, no-till sown corn after cover crop of clover gave significantly higher grain yield as compared to corn sown without clover. The yield and economics of corn sown after clover under no-till system and corn sown with conservation tillage (without clover) was at par but benefits of soil quality improvement i.e. soil tilth and active carbon was there in no-till system with clover as cover crop.

Field experiment consisting of minimum tillage with 3 Mg/ha crop residue mulch of the previous crop (MTR), minimum tillage without residue mulch (MT) and conventional tillage without residue mulch (CT) was conducted for maize-wheat sequence under rainfed conditions of *kandi* region of Punjab. Soil organic matter content, water retention, infiltration of water and aggregation of the surface soil was improved in the MTR. Pooled grain yield in the MTR treatment remained below the CT treatment during the first two years but was subsequently became greater than the CT indicating the necessity of using residue mulch in conjunction with minimum tillage in order to improve soil quality and sustain/improve crop production.

Evaluated the effects of green manures (vetch, oat and none) either tilled into the soil (CT) or cut and left on the surface as a mulch (NT), on maize yield and soil quality on Andisols located in Central Mexico. Vetch cover crop produced higher maize yield and significantly higher N uptake than oat cover crop. Yield of both grain and stover were higher in CT than in NT. Soil under NT had significantly higher soil organic carbon than under CT whereas, the results were non-significant for the green manures. Significantly higher soil total N concentrations were found under NT than under CT, under both vetch and oat. Soil under NT had more rapid infiltration and had significantly lower penetration resistance than under CT.

b. Cover crops and nitrogen

A field study was conducted consisting of 3 green manure treatments (*Sesbania cannabina*, *Sesbania rostrata* and no green manure) and 5 N levels ($N_0, N_{30}, N_{60}, N_{90}, N_{120}$) revealed that green manuring with *Sesbania cannabina*, *Sesbania rostrata* with 90 kg N/ha to maize crop increased grain yield by 10.69 and 10.49 q/ha over control, respectively.

At Lincon NE, observed significant interaction of cover crop and N levels for maize grain yields. With no nitrogenous fertilizer, there was positive response in terms of grain yield to cover crop and was negative with 60 kg N/ha and response to N was considerably greater without than winter cover crop. The grain yield was positively co-related with soil organic carbon and total soil N and negatively with bulk density.

An incorporation of sunhemp as green manure along with application of 100 per cent RDN recorded significantly higher grain yield (6.31 t/ha) and stover (10.61 t/ha) over lower rates of N application on black soils. Similarly, observed maize response to fertilizer N at all the rates they applied (0 to 90 kg N/ha) even with sunhemp or *Leucaena* mulching at Selaki, Dehradun. The legumes cowpea and sunhemp reduced fertilizers needs of subsequent corn crop by 36 kg/ha in Zimbabwe.

A field experiment consisted legume mulching, viz. in situ grown sunhemp and *Leucaena* pruning, along with varying N levels, viz. 0, 30, 60 and 90 kg N/ha (to maize), and 0, 40 and 80 kg N/ ha (to wheat) at Dehradun revealed an improvement in organic C and total N status of soil, and a decrease in bulk density associated with an increase in infiltration rate due to legume mulching and N fertilization at the end of 4 cropping cycles. In a field experiment consisting of living mulch (crown vetch cv. Penngift and Pennmulch; flatpea, cv. Lathco; birdsfoot trefoil cv. Empire and Steadfast; hairy vetch) and N rates (0-225 kg N/ha) in corn at central Pennsylvania revealed that average N fertilizer equivalency of Penngift, Pennmulch, Empire, Steadfast and Lathco was 71, 45, 44, 13 and 50 kg/ha, respectively at 0 kg N rates. Their N fertilizer efficiency decrease to zero with increasing N fertilizer rates. Bulk density, soil organic carbon, infiltration rate was not significantly improved even 10 years of Penngift living mulching.

c. **Cover crops, no-tillage and nitrogen**

In order to evaluate the long term effect of tillage, N and cover crops on soil organic carbon (SOC) and soil total nitrogen (STN), an experiment consisting of two tillage methods :conventional tillage (CT) and no-tillage (NT), four N fertilization rates (N_0, N_1, N_2 and N_3) and four soil cover crop (CC) types (C- no cover crop; NL – non-legume CC; LNL – low nitrogen supply legume CC, and HNL – high N supply legume CC) was conducted in Italy. In the NT system the soil organic carbon and soil total N in the top 30 cm soil depth increased by 0.61 and 0.04 Mg/ha/yr, respectively while under the CT system it decreased by a rate of 0.06 and 0.04 Mg/ha/yr, respectively. The N_1, N_2 and N_3 increased the soil organic carbon content in the 0-30 cm soil layer at a rate of 0.14, 0.45 and 0.49 Mg/ha/yr, respectively Only the higher N fertilization levels (N_2 and N_3) increased soil total N content, at a rate of 0.03 and 0.05 Mg/ha/yr. NL, LNL and HNL cover crops increased soil organic carbon content by 0.17, 0.41 and 0.43 Mg/ha/yr and -0.01, +0.01 and +0.02 Mg/ha/yr, respectively.

In a field experiment at Kanas State, study the effect of sunhemp (SH), late maturing soybean (LMS), four N rates on soil properties under no-till conditions. The result showed that averaged across N rates, the soil organic carbon was 1.3 times greater under SH and 1.2 times under LMS as compared to control plot. At same depth, average across cover crops treatments the soil organic carbon increased systematically with increase in N application and it was 1.3 times from 0 to 100 kg N/ha. Cover crops also increase mean weight diameter of aggregates by 80 per cent in 0-7.5 cm depth. Soil temperature was greater under cover crops by 35 per cent than control and reduces soil temperature by 4^0C at 5cm depth and 1^0c at 30 cm depth.

The effect of leguminous cover crops and their time of chopping on no-till maize productivity

The experiment consists of nine treatment combinations (3 cover crop and 3 chopping dates) and additional control treatment in which no cover crop was sown. These ten treatments were laid out in a RBD with four replications. Leguminous cover crops i.e. sunhemp, cowpea and dhaincha were sown after harvest of wheat in three sowing dates (ten days interval) under no till conditions with recommended package of practices. The 25, 35 and 45 days old cover crops were chopped in situ and left uniformly on the surface followed by planting of maize under no till conditions as per the treatments.

Leguminous cover crops

The highest phytomass was observed with sunhemp, which was statistically similar with cowpea but significantly better than dhaincha cover crop. Sunhemp cover crop recorded the highest biomass yield, which was significantly superior to cowpea and dhaincha during both the years.

Phytomass and biomass of leguminous cover crops increased significantly with the age and it was highest under 45 days chopping treatment. Among the chopping time of 25 and 45 days, the biomass yield of sunhemp remained statistically at par with cowpea but chopping time of 35 days resulted in significantly higher biomass yield of sunhemp over the cowpea, which further out yields dhaincha cover crop for both the years.

Maximum N content was recorded with dhaincha that was statistically superior to cowpea and sunhemp cover crop. The N content recorded in cowpea was numerically higher than sunhemp but the differences were non-significant in both the years. The N content decreases with increase in chopping time and differs significantly among chopping time of 45 and 25 DAS. Significantly higher N accumulation was recorded in sunhemp as compared to cowpea and dhaincha cover crop but it was statistically similar in dhaincha and cowpea crop during both the years. Likewise, significant increase in N accumulation was found with the increase in chopping time during both the years. The cover crops accumulate about equal amount of N at chopping time of 45 DAS.

Effect of cover crops on maize and soil properties

The leguminous cover crops had no significant effect on emergence count of the maize crop during both the years. The maximum plant height, LAI and DMA at 30, 60 DAS and at harvest stage of maize was recorded under sunhemp cover crop, which was significantly higher over cowpea and dhaincha, however, the differences were statistically same in dhaincha and cowpea cover crops.

Among the three cover crops, the days to 50 per cent tasseling and silking of maize crop was significantly lower under sunhemp cover crop as compared to cowpea and dhaincha cover crops during both the years. No significant differences in maturity days were recorded though the maturity days were taken less under sunhemp.

Maize cob length, 1000-grain weight and grain weight per cob increased significantly under sunhemp as compared to cowpea and dhaincha, however, differences were statistically similar under cowpea and dhaincha cover crop during both the years.

The maximum maize grain and stover yield was obtained with sunhemp cover crop, which was significantly better than cowpea and dhaincha. However, the differences were non-significant in dhaincha and cowpea cover crops. The average increment in maize grain yield under sunhemp was 8.7 and 11.2 per cent over cowpea and dhaincha cover crops, respectively during the study period.

No significant changes in HI and shelling percentage of maize were observed among cover crops, though the maximum was recorded in sunhemp during both the years.

The periodic and total N uptake of maize was higher under sunhemp, which was significantly higher than cowpea and dhaincha cover crops during both the years. The protein, oil, starch and total sugar content of maize grain did not differ significantly under various cover crops treatments. No significant differences were observed among the cover crops pertaining to the soil NPK though N and K were highest under sunhemp, whereas, highest P content recorded after dhaincha during both the years. No significant differences pertaining to SOC and infiltration rate were found among the various leguminous cover crops but maximum was observed under sunhemp cover crop during both the years.

No significant variation in soil bulk density was noticed among different cover crops, it was lower under sunhemp cover crop during the year of 2013 and 2014. The minimum consumptive use in maize crop was observed under sunhemp > cowpea > dhaincha, whereas, the average WUE under sunhemp cover crop was 10.6 and 14.6 per cent higher compared to cowpea and dhaincha cover crop, respectively during the study period. The soil temperature did not vary with the use various cover crops during maize season, though the lowest was recorded under sunhemp cover crop during both the years.

Effect of chopping time on maize and soil properties

The chopping time had no significant effect on emergence count of the maize crop during both the years. The plant height, LAI and DMA of maize increased significantly with the successive delay in chopping time at 30, 60 DAS and at harvest. The days to 50 per cent tasseling and silking decreased significantly with delay in chopping time. However, the differences in days were non-significant with respect to physiological maturity.

The cob length, thousand grain weight and grains weight per cob of maize improved significantly with the chopping time from 25 to 45 DAS. The maize grain and stover yield increased significantly with delayed chopping time. Cover crops chopped at 45 days resulted in 25.1 and 13.2 per cent higher grain yields as compared to 25 and 35 days chopping treatment, respectively. HI and shelling percentage of maize was achieved less with chopping time of 45 days, which was statistically same with 35 days but significantly better than 25 days chopping time during both the years.

Periodic and total N uptake by maize (grain and straw) increased significantly with varying chopping treatment and it was found less than 45 days chopping treatment. The highest protein content was obtained with maximum delay in chopping time of 45 days, which was statistically superior to minimum delay in chopping time of 25 days. No significant differences in oil content, starch and total sugar content was observed with delay in chopping time during both the years.

The NPK content, SOC and infiltration rate increased with different chopping time but the differences were non-significant. In general, bulk density of soil decreases with progressive delay in chopping time and this reduction was more prominent at 0-5 cm soil depth. The consumptive use in maize decreased with delay in chopping time, whereas, WUE was increased with the maximum delay in chopping time of 45 days which was 15.8 and

30.6 per cent higher as compared to lower chopping time of 35 and 25 DAS. The chopping time delayed up to 45 days reduced the soil temperature by 1.0 and 0.9 °C at 1430 hrs during the growing season of maize in 2013 and 2014, respectively.

Interaction effect of cover crop and chopping time

The sunhemp chopped at 35 days resulted in substantial increase in plant height, LAI and DMA of maize at 30, 60 DAS and at harvest as compared to 35 days chopping time of cowpea and dhaincha but remained at par under all cover crops chopped at 45 days.

The days to 50 per cent tasseling and silking of maize under the sunhemp chopped at 35 days remained statistically similar with all the three cover crops chopped at 45 DAS but significant reduction was observed under both cowpea and dhaincha chopped at 35 DAS.

The interactive effect of leguminous cover crops and their chopping time resulted in highest maize cob length, thousand grain weight and weight of grains per cob, grain and stover yield under sunhemp chopped at 45 days, which was statistically at par with 35 days chopping of sunhemp and 45 days chopping of cowpea and dhaincha but significantly better than all other treatment combinations.

No significant differences in periodic and total N uptake by maize was observed among the three cover crops at chopping period of 25 and 45 days, whereas, sunhemp chopped at 35 days resulted in significantly higher N uptake as compared to cowpea and dhaincha chopped at 35 DAS.

Comparison of control *vs* rest of treatments

Leguminous cover crops resulted in significant increase in maize plant height, LAI and DMA at 30, 60 DAS and at harvest as compared to control during both the years. Significant decrease in days to 50 per cent tasseling and silking in maize crop was observed with in-situ chopping of leguminous cover crops as compared to control plot. Application of leguminous cover crops led to significant improvement in cob length, thousand grain weight and grains weight per cob of maize crop as compared to control treatment. The treatment combination of leguminous cover crops and chopping time increased the grain yield significantly by 14.6 and 16.1 per cent over the control during the year 2013 and 2014, respectively. Similar trend was observed for stover yield HI and shelling percentage of maize increases with the use of leguminous cover crops but the differences were statistically similar as compared to control. Treatment combination of cover crops and chopping time resulted in significant increase in periodic and total N uptake of maize over the control treatment during the year 2013 and 2014. No significant differences with respect to grain protein, oil, starch and sugar content of maize was observed between cover crops treatment combinations and control treatment during both the years.

The NPK content of soil improved with cover crops but differences were statistically similar to control. The treatment combinations of cover crops and chopping time resulted in 5.2 and 8.8 per cent increase in infiltration rate and SOC over the control during the year of 2013 and 2014, respectively. Numerically decrease in soil bulk density was observed with leguminous cover crops over the control. The CU decreased and WUE increased with treatment combinations of cover crop and chopping time as compared to control treatment though the differences were not huge. The average soil temperature (0-5 cm depth) was reduced by 1°C with in-situ chopping of cover crops as compared to control during both the years.

It can be concluded that the use of leguminous cover crops resulted in significant increase in maize grain yield over the control and among the cover crops sunhemp proved superior over cowpea and dhaincha cover crops. The cover crops chopped at 45 DAS resulted in significant higher grain yield as compared to 25 and 35 days hopping time. Treatment combination of sunhemp along with chopping period of 35 days resulted in statistically similar maize grain yields as compared to sunhemp, cowpea and dhaincha cover crops chopped at 45 DAS. The soil properties like SOC, NPK status, infiltration rate and bulk density also improves under leguminous cover crops and with delay in chopping days although results obtained were statistically similar. The cover crops resulted in 20.1 per cent more WUE than control treatment and among cover crop, the sunhemp recorded 10.6 and 14.6 per cent average increase in WUE of maize as compared to cowpea and dhaincha cover crop, respectively.

Effect of chopping of leguminous cover crops and nitrogen levels on productivity of no- till maize

The experiment comprised of 16 treatment combinations and was laid out in a SPD with three replications. After manual harvest of wheat, the direct sowing of leguminous cover crops was done in main plots. These cover crops were in-situ chopped at 45 days after sowing for subsequent no-till sowing of maize crop. Therefore main plot consists of maize sown in chopped cover crops (sunhemp, cowpea and dhaincha) and control (no cover crop) treatment and subplots comprised of four nitrogen levels (0, 50, 75 and 100 per cent of recommended dose) to be applied to the maize crop. The total rainfall received during maize crop season was 476.0 and 394.0 mm during the year 2013 and 2014, respectively.

Leguminous Cover Crops

The highest phytomass was obtained from cowpea followed by sunhemp and dhaincha cover crop. The biomass yield of cover crops was highest under sunhemp > cowpea > dhaincha cover crop during both the years. The highest nitrogen content was observed in above ground biomass of dhaincha followed by cowpea and sunhemp cover crop during the study period. However, the average N accumulation in sunhemp, dhaincha and cowpea was 162.4, 155.7 and 154.7 kg N/ha, respectively during both the years.

Effect of leguminous cover crops on maize and soil properties

No significant differences in emergence count of maize were observed under cover crops and control during both the years. The highest plant height, LAI and DMA of maize at harvest was obtained under sunhemp cover crop, which was significantly higher than cowpea, dhaincha and control during the study period. The differences were statistically similar under cowpea and dhaincha cover crops.

Under sunhemp cover crop recorded significant reduction in days taken for 50 % tasseling and silking as compared to cowpea, dhaincha and control. No significant differences in days were observed between cowpea and dhaincha but took significantly lesser days as compared control. There was no significant differences regarding days taken for physiological maturity of maize was observed among cover crops though maturity days were minimum under sunhemp < cowpea < dhaincha < control treatment.

Maximum cob length, grain weight per cob, thousand grain weight, grain and stover yield was observed under sunhemp cover crop which was significantly higher than cowpea, dhaincha and control. The differences were non-significant under cowpea and dhaincha but significantly higher than control in both the years. The increase in grain yield with sunhemp over cowpea and dhaincha was averaged to the tune of 10.4 and 12.6 per cent respectively during both the years. HI and shelling percentage was affected by various leguminous cover crops but proved to be superior to control.

A significantly higher N uptake was recorded under sunhemp cover crop as compared to cowpea, dhaincha and control but no noteworthy differences in N uptake by maize were observed between cowpea and dhaincha cover crops during both the years. The protein content of maize grain was notably influenced by leguminous cover crops and control treatment during the study period.

Oil, starch and total sugar content of maize grain were not significantly affected by leguminous cover crops during both the years. Among the cover crops, sunhemp recorded highest infiltration rate and SOC as compared to control but the treatments effect were non-significant during both the years

NPK status of soil improved after the addition of cover crops although the increase was statistically non-significant as compared to control treatment. No significant differences in soil bulk density were observed among different cover crops during the study period. Minimum CU was recorded under sunhemp cover crop followed by cowpea, dhaincha but it was higher under control. However, the maize crop under sunhemp cover crop recorded 37.9, 14.3 and 12.2 per cent higher WUE as compared to control, dhaincha and cowpea cover crop treatments. Cover crops resulted in lowering the soil temperature as compared to control and the effect was more pronounced under sunhemp cover crop when temperature was recorded at 1430 hrs.

It can be concluded that the sunhemp cover crop recorded significantly higher grain yield of maize as compared to control (no cover crop), dhaincha and cowpea treatment. Likewise, sunhemp cover crop recorded 37.9, 14.3 and 12.2 per cent higher WUE as compared to control, dhaincha and cowpea cover crop treatments as averaged for both the years. The addition of leguminous cover crop resulted in slight improvement in soil properties like SOC, NPK status, bulk density and infiltration rate as compared to control. The grain yield of maize improved significantly with subsequent increase in nitrogen level up to RN_{75}. Hence, usage of leguminous cover crops resulted in saving of 25 per cent of the recommended N dose. Maize can be sown after sunhemp, dhiancha and cowpea cover crops by chopping at 35, 45 and 45 DAS, respectively with the application of 75 per cent nitrogen of the recommended for obtaining higher grain yield, improving the soil properties, saving of nitrogen 25 per cent of the recommended and conservation irrigation water.

Climate Change And Global Warming Impact On Agriculture

The agricultural sector represents 35 per cent of India's Gross National Product (GNP) and as such plays a crucial role in the country's development. Food grain production quadrupled during the post-independence era; this growth is projected to continue.

The impact of climate change on agriculture could result in problems with food security and may threaten the livelihood activities upon which much of the population depends. Climate change can affect crop yields (both positively and negatively), as well as the types of crops that can be grown in certain areas, by impacting agricultural inputs such as water for irrigation, amounts of solar radiation that affect plant growth, as well as the prevalence of pests.

The Indian Agricultural Research Institute (IARI) examined the vulnerability of agricultural production to climate change, with the objective of determining differences in climate change impacts on agriculture by region and by crop.

Agriculture and Climate Change are Related: Causes and Effects

Agriculture and climate change are deeply intertwined. The effects of global warming on food supply are dire, whilst world population is increasing. It's time to change the way agriculture affects the environment, and vice versa.

The relationship between agriculture and climate change is problematic to say the least, and it is putting food safety at risk. Using the "which came first, the chicken or the egg?" question as an analogy, it is difficult to understand exactly when this conflict began. Over time, has the effect of global warming on agriculture and food supply been to decrease crop production or has intensive agriculture contributed to climate change by causing average global temperatures to increase?

- » The world population is increasing
- » The effect of climate change on crop production: how is climate related to agriculture?
- » How does agriculture contribute to climate change?
- » Agriculture and climate change: is agroecology the answer?
- » How does agriculture affect the environment? Eating habits matter, especially in Europe

World Population

Population increase is a determining factor that must be immediately taken into consideration if we wish to gain a clearer picture of this dichotomy. The world population is in fact rapidly increasing and according to the United Nations Department of Economic and Social Affairs (UN/DESA) it could increase to 9.7 billion people by 2050, compared to today's 7.8 billion December, 2019. At the same time, crop yields, mainly grain and corn, could decrease by 50 per cent over the next 31 years because of altered climatic conditions. A risk we must avoid and prevent, especially at this moment in history in which the number of people affected by famine is slightly decreasing. There are nearly 795 million people who regularly still don't have enough food to eat, The State of Food Insecurity in the World 2015 report by the International Fund for Agricultural Development (IFAD) and World Food Programme (WFP) calculates. This number was 1 billion in 1990-1992.

Definition of Climate Change

Climate change is the direct consequence of global warming. Here's everything you need to know about the causes and effects of one of the biggest threats facing our time. The discovery and acknowledgement of global warming dates back to the 19th century when Svante Arrhenius, Swedish chemist and physician who was awarded the Nobel Prize in Chemistry in 1903, outlined his theory according to which carbon dioxide has an impact on climate patterns, causing climate change. Since then, the awareness that humanity influences climate and causes anthropogenic effects (climate change) has become increasingly popular. In the first half of the 20th century scientists believed (or hoped) that oceans maintained CO_2 levels in the atmosphere unvaried by absorbing most of human-related emissions.

Climate change occurs when changes in Earth's climate system result in new weather patterns that remain in place for an extended period of time. Climate change is any significant long-term change in the expected patterns of average weather of a region (or the whole Earth) over a significant period of time. Climate change is about abnormal variations to the climate, and the effects of these variations on other parts of the Earth. Climate change is a problem that is affecting people and the environment. Human-induced climate change has, e.g., the potential to alter the prevalence and severity of extreme weathers such as heat waves, cold waves, storms, floods and droughts.

The first measurements CO_2 on Mauna Loa volcano, Hawaii

In 1957, however, this assumption was called into question by scientists Roger Revelle and Hans Suess who demonstrated that despite oceans absorb extra CO_2, they do it more slowly than expected and this would lead to an increase in the average global temperature. Their research was confirmed in the 1960's and 1970's when a group of chemists started conducting accurate measurements of greenhouse gases from the Mauna Loa Observatory (MLO), Hawaii. They asserted that carbon dioxide concentration in the atmosphere was gradually rising.

In particular, Charles David Keeling of the Scripps Institution of Oceanography of San Diego, California, made a long-term analysis that became the well-known Keeling Curve, which traces carbon dioxide concentration in the atmosphere month after month, year after year. The fluctuation of concentrations led researchers to define the graphic as the reproduction of the Earth's breathing. These new figures led the issue to be included in the agenda of some of the most important international scientific meetings in order to start studying the issue in the 1980's technologies started to become more advanced and accurate.

Gases cause the greenhouse effect

Scientists discovered that not only carbon dioxide (CO_2) causes global warming, but also a number of gases including methane (CH_4), nitrous oxide (N_2O), ozone (O_3) and, indirectly, water vapour (H_2O). All of them contribute to creating the so-called "greenhouse effect". In addition, there are other gases such as CFCs (chlorofluorocarbons) that have been regulated by Montreal Protocol in 1987 as they are responsible for the depletion of the ozone layer. Estimates published in 1985 showed that these other gases together have an impact on global warming just as much as CO_2, making the issue more serious and complicated than previously expected. The greenhouse effect, however, is a natural phenomenon that can be described as the capacity of the atmosphere of absorbing and retaining sun rays' humidity and warmth. This is why the presence of these gases is crucial for life on Earth. Without greenhouse gases, and thus Earth's atmosphere, the average temperature would be 18 ºC, while the greenhouse effect keeps the average temperature around 14-15ºC.

Global warming and its causes

What's behind global warming is an increase in the global average temperature due to high concentrations of CO_2 and other gases that no longer derive from nature alone but are also linked to human activities. The global increase in carbon dioxide is mainly caused by the fossil fuels humanity is relentlessly burning to produce energy (responsible for 75.2 per cent of greenhouse gas emissions), which is used to meet the electricity and heating consumptions (32.6 %) and for the transport industry (14.2 %). The increase in methane and nitrous oxide is mainly linked to the farming industry (16.1%): this means that agricultural and climate change are intimately related.

Scientists have determined that the major factors causing the current climate change are greenhouse gases, land use changes, and aerosols and soot. Global warming is the long-term rise in the average temperature of the Earth's climate system. It is a major aspect of current climate change, and has been demonstrated by direct temperature measurements and by measurements of various effects of the warming. It is a major aspect of current climate change, and has been demonstrated by direct temperature measurements and by measurements of various effects of the warming. The term commonly refers to the mainly human-caused increase in global surface temperatures and its projected continuation. Global warming is projected to have a number of effects on the oceans. Ongoing effects include rising sea levels due to thermal expansion and melting of glaciers and ice sheets, and warming of the ocean surface, leading to increased temperature stratification.

Deforestation contributes to the increase in CO_2 in the atmosphere and climate change. Deforestation, too, contributes to the increase in carbon dioxide in the atmosphere: forests, especially tropical forests, absorb and retain CO_2. This is why their destruction, along with preventing absorbing CO_2, releases more carbon dioxide, which was previously "naturally stored". Since the early 1990's deforestation led to a 15-25 per cent increase in CO_2. By adding the emissions from agriculture and the consequent deforestation it is 21 per cent of the total CO_2 emitted into the atmosphere between 2000 and 2010, equal to 44 billion tonnes.

Indian, self sufficiency and sustainability in food grain production is under threat due to the climatic variability and changes that had occurred in the last decade. In spite of technological advances such as improved crop varieties, production and protection measures and irrigation facilities, weather and climate are playing key role in Indian agriculture. Several climatic models predict that global warming in future may reduce over large area of semi-arid grasslands in North America and Asia. It is predicted that there will be a 17 per cent increase in desert land at global level due to climate change expected from a doubled concentration of atmospheric CO_2.

It is interesting to mention here the scientists, common man and politicians all over the world has started thinking that global warming will have a major impact on agro-ecosystems that is why United Nations Environment Program (UNEP) along with World Meteorological Organization (WMO) established the Inter-Governmental Panel On Climate Change (IPCC) in 1988 to periodically assess the state of global environment and to advice various UN agencies on climate change.

Climate change can be defined as a statically significant variation in either the mean state of the climate or in its variability persisting for an extended period (typically decades or longer). Intergovernmental Panel on Climate Change (IPCC) defined climate change as "Any change in climate over time, whether due to natural variability or as a result of human activity". United Nations Framework Convention on Climate Change (UNFCCC) also defined climate change as "A change of climate, which is attributed directly or indirectly to human activities that alter composition of the global atmosphere, which are in addition to natural climate variability observed over comparable time period".

The effects of climate change have reached such an extent that irreversible changes in the functioning of the planet are feared. Some of the main effects of climate change with specific reference to agriculture and food production especially during the last decade are:

» increased occurrence of storms and floods

» increased incidence and severity of droughts and forest fires

» increased frequency of diseases and insect pest attack and vanishing habitats of plants and animals.

Carbon dioxide (CO_2) is the major anthropogenic green house gases (GHGs). Its annual emissions have been increased between 1970 and 2004 by about 80 per cent, from 21 to 38 gigatonnes (94.4 to 170.8 ton), and represented 77 per cent of total anthropogenic GHGs emissions in 2004. The rate of growth of CO_2-equivlent emissions was much higher during the last decade of 1995-2004 [0.92 Gt (4.14 ton) CO_2- equivalent per year] than last two and a half decade of 1970-1994 [043 Gt (1.93 ton) CO_2-equivlent per year]. The largest increase in GHG emissions between 1970 and 2004 has come from energy supply, transport and industry, while residential and commercial buildings, forestry (including deforestation) and agriculture sectors have been growing at a lower rate.

Change in climate is mainly a result of increased production of CO_2, methane (CH_4), Nitrous oxide (N_2O), ozone, water vapors, Chlorofluorocarbons (CFCs), which resulted in increase in atmospheric temperature, disturbance in quantity and distribution of rainfall, melting of glaciers, rise in sea level etc. the concentrations of CO_2, CH_4, N_2O and CFCs between 1000-1750 AD were 280 ppm, 700 ppb, 270 ppb and 0 ppt, respectively (Table 2), but in 2005, these values increased to 379 ppm, 1774 ppb, 319 ppb and 5.03 ppt, respectively. These increases in concentration of green house gases have resulted in warming of the atmosphere by 0.74 °C during 1906-2005. Eleven out of the twelve years (1995-2006) rank amongst the twelve warmest years since 1850. The rate of warming has been much higher in the recent decades and the minimum temperature at night has been increasing at twice the rate of day time maximum temperature.

Causes of Climate Change and Major Contributors

The change in climate of the world has mainly brought out by rapid industrialization, deforestation, increased agricultural operations, combustion of fossil fuels, increased number of vehicles, etc. and the driving force behind these factors is ever increasing human population requiring more food and space to live. It is resulted in global warming. This is happened due to the increase in concentration of GHGs in the atmosphere, and leads to a phenomenon widely known as 'Greenhouse Effect'. Amongst various sources of GHGs, agriculture is considered a major contributor primarily through the emission of MH_4 and N_2O. As per Indian Network for Climate Change Assessment (INCCA) Report (2010), the net GHGs emissions were 1727.7 million tons (Mt) of CO_2 equivalent from India in 2007. The major cause to climate change has been ascribed to the increased levels of GHGs like CO_2, MH_4, N_2O, and CFCs. Beyond their natural levels due to the uncontrolled human activities such as burning of fossil fuels, increased use of refrigerants, and enhanced agricultural activities. These GHGs are nearly transparent to the visible and near infra-red wavelengths of sunlight but they absorb and re-emit downward a large fraction of the longer infra-red radiation emitted by earth. As a result of this heat trapping, the atmosphere radiates large amounts of long wavelength energy downward to the earth's surface and long wavelength radiant energy received on earth is increased.

Table 1. Sources of GHGs emission in Indian agriculture

Source	CH_4 (Million ton)	N_2O (Million ton)	CO_2 equivalent (Million ton)
Enteric fermentation	10.10	-	212.09
Manure management	0.12	-	2.44
Rice cultivation	3.37	-	84.24
Agricultural soil	-	0.22	64.7
Crop residue burning	0.25	0.01	8.21
Total	13.84	0.23	371.68

Source: INCCA (2010)

Agriculture sector emitted 371.68 million tons of CO_2 -equivalent, of which 13.76 million tons is and 0.15 million tons is N_2O. The major sources in the agricultural sector are enteric fermentation (57.06%), rice cultivation (22.66%), agricultural soils (17.41%), livestock manure management (0.66%) and burning of crop residues on the fields (2.21%). The crop production sector (manure management, rice cultivation, soil and field burning of crop residues), thus contributes CO_2 42.94 per cent, CH_4 27.02 per cent and N_2O 100 per cent to the total emissions from agriculture.

Table 2. Relative increase in GHGs influenced by anthropogenic activities

| Gas | Year | | Atmospheric lifetime (years) | Anthropogenic source |
	1750	2005		
CO_2	280ppm	379ppm	Variable	Fossil fuels combustion, land use conversion, cement production
CH_4	715ppb	1774ppb	12.2	Fossil fuels, rice stubbles, water dumps, livestock
N_2O	270ppb	319ppb	120	Fertilizer, industrial processes combustion
CFC's	0	503ppt	102	Liquid coolants, foams

Source: Kaur and Hundal (2008)

Carbon Dioxide

It is a colorless, odorless non-flammable gas. CO_2 is the most prominent greenhouse gas in earth's atmosphere. Every year humans add over 30 billion tons of CO_2 in the atmosphere by these processes. The rapid increase in atmospheric concentrations of carbon dioxide over years is linked with combustion of fossil fuels, conversion of forested land to agricultural use and changes occurring in various carbon pools and fluxes (Table 2). There has been growing concern in recent years that these high levels of greenhouse gases may not only lead to changes in the earth's climate system, but may also alter ecological balances through effects on vegetation. Terrestrial ecosystems act as both source and sink and large uncertainty exists in understanding the current carbon status and its spatial and temporal variability. CO_2 is increasing at the rate of 1.5 ppm per year. Half of the green house effect is expected due to CO_2. The world's countries contribute different amounts of heat-trapping gases to the atmosphere. The table below shows data compiled by the Energy Information Agency (Department of Energy), which estimates carbon dioxide emissions from all sources of fossil fuel burning and consumption (Table 3). Here we list the 20 countries with the highest carbon dioxide emissions (Energy Information Agency 2008).

Table 3. Global scenario of CO_2 emission

Country	Total emission (Million meteric ton of CO_2	Per capita emission (Ton/per capita)
China	6534	4.91
United States	5833	19.18
Russia	1729	12.29
India	1495	1.31
Japan	1214	9.54
Germany	829	10.06
Canada	574	17.27
United Kingdom	572	9.38
South Korea	542	11.21
Iran	511	7.76
South Arabia	466	16.56
Italy	455	7.82
South Africa	451	9.25

Mexico	445	4.04
Australia	437	20.82
Indonesia	434	1.83
Brazil	428	2.18
France	415	6.48
Spain	359	8.86
Ukraine	350	7.61

Source: Energy Information Agency (Department of Energy) 2008

In general developed countries and major emerging economy nations lead in total carbon dioxide emissions. Developed nations typically have high carbon dioxide emissions per capita, while some developing countries lead in the growth rate of carbon dioxide emissions. Obviously, these uneven contributions to the climate problem are at the core of the challenges the world community faces in finding effective and equitable solutions.

Methane

Methane is a colourless, odourless, flammable gas. It is formed when plants decay and where there is very little air. It is often called swamp gas because it is abundant around water and swamps. Methane is the second most important greenhouse gas after carbon dioxide and contributes about15% to the global warming. Rice cultivation (Table 2) has been accredited as one of the major source of anthropogenic methane. With the intensification of rice cultivation to meet the growing global food demand, CH_4 emission from this important ecosystem is anticipated to increase. It was observed that all India monthly average atmospheric concentration of methane ranges from 1693 to 1785 ppb. A systematic seasonal pattern was observed in methane concentration, which was mostly influenced by rice growth characteristics. It was found that January to June is associated with relatively lower concentration of methane (1699-1708 ppb) in India, which characteristically increases from July to September (1747-1785 ppb) with further gradual decline from October to December (1768-1704 ppb). The satellite based spatial variability of methane is in accordance with field based methane emission measurements. The spatial distribution of methane over Indian region is associated with agricultural practices particularly rice cultivation. It was observed that Indo-Gangetic plain including North-eastern region, parts of Chhattisgarh, Orissa and Andhra Pradesh showed higher methane concentration (> 1730 ppb) as compared to hilly regions of Jammu and Kashmir (< 1710 ppb).

Source

» Rice fields are one of the most important sources of atmospheric CH_4 with a global emission of CH_4 estimated between 60 and 150 Tg/year (Tg= 1million ton). Flooding of rice fields stops the influx of atmospheric oxygen into the soil and decomposition of organic matter becomes anaerobic.

» Methane is also found in the digestive track of ruminants and in the guts of various insects of which termites are the most important. Methane has two times greater capacity for global warming than that of CO_2. About 20 per cent of the global warming is caused by methane.

The total output of methane into the atmosphere from all sources in the world is estimated to be 535 Tg/year. India's contribution to global methane emission from all sources is 18.5 Tg/year. Agriculture largely rice and ruminant animals are the major (68%) is the major contributor to its emission.

Nitrous Oxide

Nitrous oxide (N_2O) is another colourless greenhouse gas however, it has a sweet odour. Nitrogen oxides play a central role in tropospheric chemistry. An improved knowledge of the global tropospheric distribution of NOx ($NO+NO_2$) is important for climate change studies. NOx and volatile organic compounds are emitted in large quantities due to human activities such as vehicles and industry (Table 2). The knowledge of the ozone distribution and its budgets is strongly limited by a severe lack of observations of NO and NO_2 in the troposphere. The technique used to retrieve total slant columns of atmospheric trace species from Satellite measurements is the Differential Optical Absorption Spectroscopy (DOAS). The DOAS technique allows the determination of concentrations of atmospheric species, which leave their absorption fingerprints in the spectra. Spatial distribution of tropospheric NO_2 concentration was analyzed over India. It was observed that high concentration of NO_2 distribution is associated with coal-mine and thermal power locations as well as major metropolitan cities of India.

Source

» This gas is released naturally from oceans and by bacteria in soils

» Through nitrogen based fertilizers

» Disposing of human and animal waste in sewage treatment plants

- » Automobile exhaust
- » N_2O gas has risen by more than 15per cent since 1750.
- » Nitrogen based fertilizer use has doubled in the past 15 years.
- » Oxides of nitrogen (gaseous form) have 10-1000 time greater effect on global warming than that of CO_2.

Chlorofluorocarbons (CFC's)

Fluorocarbon is a general term for any group of synthetic organic compounds that contain fluorine and carbon. The major source of CFCs is liquid coolants and foams (Table 2). CFC's are emitted into the atmosphere; they break down molecules in the Earth's ozone layer. CFC's have 10,000 times greater potential for global warming than that of CO_2. CFC's are used for refrigeration, aerosol propellant and for insulation. These are responsible for 15 per cent of the green house effect. The substitutes for CFC's are hydro fluorocarbons (HFC's). CFC's do not breakdown the ozone molecule, but they do trap heat in the atmosphere, making it a greenhouse gas, aiding in global warming.

Emission of GHG from various food products from crop and animal

Basically four stages of life cycle of food products i.e., production, processing, transportation and preparation are contributed in the emission of GHGs. Food products from animal determined the CH_4 emission, while food products from crop determined the emission of CH_4 (from rice cultivation) and N_2O (from all crops). Emission of CO_2 occurred during farm operations, production of farm inputs, transport, processing and preparation of food. Production of food products varied considerably in GHG emission (Table 4). For example, emission of GHG from production of ordinary rice was about 10.2 and 43.3 times higher than production of wheat and vegetables, respectively. Higher emission in rice was because of CH_4 emission under anaerobic soil condition, whereas, wheat, vegetables and other crops are grown in aerobic soil conditions and there is no CH_4 emission. Potato and other root vegetables have high productivity, resulting in low emission of GHG per unit food product. Production of food (meat and milk) from animal emitted larger amount of GHG compared to food from crops because of emission of methane by ruminants. The nature of GHG also varied for different food items. The food products from animal such as mutton, poultry meat, dairy products and fish dominated the CH_4 emission. On the other hand, the food products from crop contributed to N_2O emission except rice, which contributed to CH_4 as well as N_2O emission. Application of synthetic nitrogen fertilizers in agriculture was responsible for a major part of the N₂O emission. The GWP of food items was larger on dry weight basis than that with fresh weight basis However, as the foods are generally consumed fresh, the results lower GWP.

Table 4. Emission of GHGs due to production of various food products from crop and animals.

Crop/animal	GHG emission (g/kg)			
	CH_4	N_2O	CO_2	GWP (CO_2 eq.)
Wheat	0.0	0.3	45.0	119.5
Rice	43.0	0.2	75.0	1221.3
Cauliflower	0.0	0.1	13.3	28.2
Brinjal	0.0	0.1	12.5	31.1
oilseed	0.0	1.3	50.0	422.5
Mutton	482.5	0.0	0.0	12062.7
Egg	0.0	2.0	1.0	588.4
Milk	29.2	0.0	0.0	729.2
Fish	25.0	0.3	18.8	718.3
Apple	0.0	1.0	41.7	331.4
Banana	0.0	0.2	10.0	71.6
Spice	0.0	2.5	100.0	845.0

GHGs: Greenhouse gases

Effect of Climate Change oOn Ecosystem

Plant and animal species

The rapid pace of changes may be too quick for many organisms to adjust to changing habitats. About 80 per cent of the existing forests are undergoing a change in the type of vegetation. Many species of plants and animals might not be able to cope with climate change and could, therefore, face "Extinction".

Sea level

Sea levels have risen between 4-10 inches since 1990. By 2100, there will be 60 cm rise in the sea level and it will continue to rise 60-90 cm per century, for 1000 years. It increases the salinity of freshwater throughout the world and cause coastal lands to be washed under the ocean. Warmer water and increased humidity will encourage tropical cyclones. Changing wave patterns could produce more tidal waves and strong beach erosion. Increased water vapour in the atmosphere, glaciers and polar ice caps appear to be melting, floods and droughts are becoming more severe. About 50 million acres of Asian region are already subjected to seasonal floods.

Temperature

Change in atmospheric temperature has been reported by many workers. An increase in air temperature over last 100 years (1850-1899 to 2001-2005) of 0.76°C was recorded in India. In many parts of Northern India, there is an increase in minimum temperature by about 1°C in *rabi* season. On long term trends of surface temperature in India from a period of 1900-1982 from 73 weather stations distributed over the country and found warming trend of 0.04°C/decade. Indian mean annual temperature has shown significant warming trend of 0.05 °C/decade during 1901-2003, the recent period 1971-2003 has a relatively accelerated warming rate of 0.22 °C/decade. Analysis of meteorological data of Punjab revealed that the maximum temperature has decreased from normal at Ballowal Saunkri and at Bathinda, however, for other locations no trend could be established. The k*harif* maximum temperature decreased at a rate of 0.04 °C/ year at Ballowal Saunkri and at Bathinda. The annual and seasonal minimum temperature has increased at the rate of 0.07°C/year over past three decades at Ludhiana. At Patiala the annual and k*harif* minimum temperature has increased at the rate of 0.02 °C/year and at Bathinda the annual, k*harif* and *rabi* minimum temperatures has increased at the rate of 0.03, 0.02 and 0.05°C/year, respectively. However no trend of change in minimum temperature was observed at Ballowal Saunkri and Amritsar.

Rainfall

Summer monsoon rainfall during 1901-2000 has shown significant decreasing trends in the sub-divisions of NE India, viz. Nagaland, Mizoram, Manipur and Tripura (-12.5mm/decade), Orissa (-11.0 mm/decade) and East Madhaya Pradesh (-14.0 mm/decade). Significant increasing rainfall trends in Konkan and Goa (27.9 mm/decade) and coastal Karnatka (28.4 mm/decade) along the west-cost and in Haryana, Chandigarh and Delhi (13.6 mm/decade) and Punjab (18.6 mm/decade) were noticed in North India. Similarly, the winter monsoon rainfall has shown significant increasing trend in the sub-divisions of Marathwada (5.4 mm/decade), Telengana and North interior Karnataka (4.5 mm/decade), in central India and also in Gujarat (1.2 mm/decade). An overall increase in rainfall over a period of 1970-1998 at different locations in Punjab, but during the period from 1999 to 2005 below normal rainfall was received at all the five locations (Ballowal Saumkri, Amritsar, Ludhiana, Patiala and Bathinda) during the year 1999, 2002, 2004 and 2005. In the year of 2000 below normal rainfall was recorded at Ludhiana and Patiala and during 2001 below normal rainfall was recorded at all the four locations except at Ludhiana. This resulted in arresting the increasing trend of rainfall at different locations in the state. Hence, no significant trend in increases/decreases of rainfall was noted at all locations except at Ballowal Saunkri where a significant decreasing trend was noticed.

Shift in monthly rainfall

Any shift in monthly rainfall has a direct bearing on agriculture as it influences various agricultural operations mainly the time of sowing and subsequent crop growth therefore necessitating shift in time of sowing and cropping patterns to match the modified rainfall regime. A study from Karnataka covering the period 1991-2000 indicated that shift in rainfall peaks by 14-21days. Similarly, observed the trends of decreasing pattern in pre-monsoon rainfall in some parts of Chhattisgarh region in May and June proving detrimental to pre-sowing operations of rice.

Projected Climate Change in India

Projected change in seasonal temperature and rainfall is being presented in Table 5.

Table 5. Climate change projected for India

Year	Season	Temperature change (°C)		Rainfall change (%)	
		Lowest	Highest	Lowest	Highest
2020	Annual	1.00	1.41	2.16	5.97
	Rabi	1.08	1.54	-1.95	4.36
	Kharif	0.87	1.17	1.81	5.10
2050	Annual	2.23	2.87	5.36	9.34
	Rabi	2.54	3.18	-9.22	3.82
	Kharif	1.81	2.37	7.18	10.52
2080	Annual	3.53	5.55	7.48	9.90
	Rabi	4.14	6.31	-24.83	-4.50
	Kharif	2.91	4.62	10.10	15.18

Source: Lal (2001).

Projected impacts of climate change on Indian Agriculture

» Productivity of most cereals would decrease due to increase in temperature. Reports indicate a loss of 10-40 per cent in crop production by 2100 AD. Greater loss expected in *rabi*. Increased droughts and floods are likely to increase production variability

» The potential effect of climate change on agriculture is the shifts in the sowing time and length of growing seasons geographically

» Increased temperature would increase fertilizer requirement for the same production targets. However, increase in CO_2 concentration can lower pH, thereby, directly affecting both nutrient availability and microbial activity

» The effect of temperature rise will lead to an increase in soil biological activity as well as physical and chemical processes.

» The average atmospheric temperatures are expected to increase more near the poles than at the equator.

» Increased temperature resulting from global warming is likely to reduce the profit from cultivation and will compel farmers of lower latitudes to option for maize and sorghum, which are better adapted to higher temperature

» In mid-latitudes, crop models indicated that warming of less than a few °C and the associated increase in CO_2 concentrations will lead to positive responses and generally negative responses with greater warming.

The history of CO_2 concentrations in the atmosphere

CO_2 concentration in the atmosphere passed from 280 ppm (parts per million, i.e. the ratio between the molecules of greenhouse gases and air. For example, living in a world with 350 ppm means there are 350 molecules of greenhouse gases in 1 million molecules of air) prior the industrial revolution to 400 ppm in 2017. It's a record level as CO_2 emissions in the atmosphere reached 41 billion tonnes, yearly. This exponential growth began just a few decades ago. The United Nations Intergovernmental Panel on Climate Change (IPCC), in fact, has chosen 1750 as the reference year to start studying climate change. To fully understand what living in a world with 400 ppm means it only takes to consider that natural CO_2 levels in the past 650,000 years were comprised between 180 to 300 ppm. Methane passed from 715 ppb (parts per billion) in 1750 to 1,880 today. Nitrous oxide passed from 270 ppb to 328 ppb.

This has already caused an increase in the global average temperature of at least 1 degree compared to pre-industrial levels, according to the World Meteorological Organization (WMO). Even if the concentrations of all greenhouse gases will be kept at 2000 levels, the temperature is expected to increase by 0.1 degrees every decade as oceans struggle to absorb gases. As a consequence, CO_2 concentration in the atmosphere would be twice pre-industrial levels by 2050 (about 550 ppm). Actually, current emissions will make us achieve 550 ppm in 2035.

In order to curb such a catastrophic evolution of the phenomenon, we should at least achieve the minimum goal of not exceeding 450 ppm by 2050, compared to 400 ppm suggested by many scientists. This goal would allow keeping the average temperature rise within 2 degrees, in accordance with the Paris Agreement on climate achieved in Paris in 2015. It should be noted that the international agreement urges the parties to do everything in their power to try not to exceed 1.5 degrees. In order to succeed, global emissions should peak by the end of the decade and decrease by 5 per cent each year, until reaching a total drop of 80 per cent by 2050 compared to current emission levels. Realistically, countries' pledges made so far to reduce emissions lead experts to predict a temperature rise of over 3 degrees.

Effect of elevated co_2 concentration on agriculture

Elevated CO_2 concentrations often stimulate plant growth, because photosynthesis is stimulated by elevated CO_2, at least in C_3 plants, secondly in almost all species, stomatal closure is induced in response to the increased availability of CO_2. Lower stomatal conductance can result in reduced evapotranspiration, which in turn can result in comparably higher soil moisture at any given plant biomass, or to the maintenance of higher plant biomass at any given level of soil H_2O.

Effect of elevated CO_2 concentration on crops

Carbon dioxide is vital for photosynthesis and hence for plant growth. An increase in atmospheric CO_2 concentration affects agricultural production by climate change and change in photosynthesis and transpiration rate. The direct effect of increased concentrations of CO_2 are generally beneficial to vegetation, especially for C_3 plants, as increased levels leads to higher assimilation rates and to an increase in stomatal resistance resulting in a decline in transpiration and improved WUE of crops. Simulation studies have been conducted to study the effect of increased concentration of CO_2 on yield of crops. This was also found that under elevated CO_2 levels, yield of rice and wheat increased by 15 and 28 per cent, respectively for a doubling in CO_2 concentration in NW India. There was also compared to base level of 330 ppm CO_2, grain yield of rice would increase by 1.5, 6.6 and 8.7 per cent with enhanced CO_2 concentrations of 400, 500 and 600 ppm, respectively.

Effect of elevated CO_2 on C_3 and C_4 plants

C_3 plants respond more favorably to increasing CO_2 than C_4 plants because they tend to suppress rates of photorespiration. Productivity of a crop community depends mainly on its photosynthetic capacity, which is highly dependent on climatic conditions. There are numerous reports, which suggest that environmental carbon dioxide concentration is increasing and as result temperature is also increasing. Both of these environmental variables play an important role in determining photosynthesis rates of the crop canopy. Carbon dioxide is a substrate for photosynthesis reactions. Atmospheric carbon dioxide reaches the chloroplasts, the site of photosynthesis reactions, by diffusion process through the pores on leaf surface called stomata. The temperature, at which a crop is exposed, influences activity of various enzymes including those responsible for photosynthesis. The difference in photosynthesis efficiency between C_3 and C_4 plants is due to photorespiration.

Causes for difference in photorespiration

Crop plants are broadly classified into two groups on the basis of the photosynthetic mechanism followed by them. Majority of cultivated species show first stable product of photosynthesis as a 3 carbon molecule called glyceric acid-3-phosphate and are called as C_3 plants. Whereas some crops like sugarcane, maize and sorghum have an additional mechanism of photosynthetic reaction in which the first stable product is a 4 carbon molecule. These plants are called as C_4 plants which are considered to be more efficient than C_3 plants. The advantage of C_4 mechanism over C_3 is most pronounced in the conditions of low carbon dioxide availability, high light intensity and high temperature. These conditions favour photorespiratory carbon loss leading to reduced net photosynthesis of C_3 plants. This photorespiration is negligible in C_4 plants due to leaf anatomy and enzymes involved in carboxylation. In C_3 plants, all the photosynthetic reactions take place in mesophyll cells which get atmospheric air having both carbon dioxide and oxygen in their natural proportion of approximately 20 per cent O_2 and 0.03 per cent CO_2. The enzyme Rubisco (ribulose-1, 5-bisphosphate carboxylase-oxygenase) has both properties of carboxylation (carbon fixation) as well as oxygenation responsible for photorespiration. Due to high availability of oxygen in mesophyll cells (where Rubisco is present), C_3 plants show considerable rate of photorespiration. Whereas in case of C_4 plants, carboxylation process of photosynthesis takes place in presence of enzyme phosphoenol pyruvate (PEP Case) in mesophyll cells in which both carbon dioxide and oxygen are available. This enzyme has very high affinity to CO_2 with no oxygenation property and hence, it can not utilize oxygen. The carbon dioxide fixed in mesophyll cells in form of 4 carbon acid is decarboxylated and pumped into bundle sheath cells where it is again carboxylated by the enzyme Rubisco. These bundle sheath cells are surrounded by mesophyll cells thereby limiting oxygen availability. Moreover, pumping of carbon dioxide from mesophyll cells increase carbon dioxide concentration in bundle sheath cells so much that all the reaction sites of Rubisco get sufficient carbon dioxide molecules to carry out carboxylation leaving negligible/no site for reaction with oxygen. The differences in leaf anatomy and enzyme involved in primary CO_2 fixation are responsible for differences in photorespiration rates between C_3 and C_4 plants.

Response to elevated CO_2 concentration

Increase in carbon dioxide concentration causes an increase in photosynthesis rate in both C_3 and C_4 plants up to a certain CO_2 concentration, which is variable with crop species. But very high CO_2 concentration causes reduction in leaf photosynthesis due to partial closure of stomata. The response of C_3 plants to elevated CO_2 concentration for carbon fixation is more than that of C_4 plants. It is because of the reason that at elevated CO_2 concentration, the ratio of chloroplastic $CO_2 : O_2$ increases causing more carbon dioxide molecules competing with oxygen for reaction site of Rubisco resulting in increased photosynthesis and reduced respiration leading to high net photosynthesis rate in C_3 plants. Low response of C_4 plants to elevated CO_2 concentration is because of two reasons:

» The enzyme PEP carboxylase has about 100 times more affinity to CO_2 than Rubisco and therefore, PEP Case can react well even with low amount of CO_2 present in the chloroplast. This enzyme is incapable of reacting with oxygen. Therefore, increased ratio of $CO_2 : O_2$ in mesophyll cells has low utility.

» The carbon dioxide liberated from C_4 acids is pumped into bundle sheath cells (where Rubisco is present) causing CO_2 concentration much higher than atmospheric concentration. Therefore, increased atmospheric carbon dioxide concentrations do not make proportional increase in CO_2 concentration near the site of action of Rubisco enzyme in C_4 plants.

Differential response of C_3 and C_4 plants to elevated carbon dioxide concentration is reported on the basis of a large number of FACE (Free Air CO_2 enrichment) experiments. There was higher increase in photosynthesis of C_3 species than C_4 species in FACE experiments. They also found that peak leaf area index and crop yields increase in C_3 plants under elevated CO_2 concentration but in case of sorghum, a C_4 plant, both these parameters reduced. The decrease in stomatal conductance and evapo-transpiration under elevated CO_2 concentration may be due to partial stomatal closure at elevated levels of carbon dioxide. The benefit of CO_2 enrichments are more pronounced in a short period study but in long term, the effects on photosynthesis is less, may be due to accumulation of photosynthates is source leaf because of less demand by sink.

Effect of elevated CO_2 on fertilization

The total amount of N-fertilizer required by C_3 crops to support optimal productivity at elevated CO_2 concentration is likely to remain same, despite the increase in biomass. This was found in wheat and sorghum some studies conducted in 2004. While an increase in losses of nitrogen can leads to an increase in demand of its fertilizers, secondly in areas where favorable environmental conditions will be created due to altered rainfall and temperature regimes the need for fertilizers may increase. The increase in CO_2 in the atmosphere could enhance plant growth by increasing the rate of photosynthesis leading to more leaf expansion and a large canopy. Photosynthesis is the net accumulation of carbohydrates formed by uptake of CO_2.

Effect of elevated CO_2 on water requirement of plants

Warmer air temperatures will influence leaf evaporation by affecting vapour pressure diffusion because warmer air has a greater water vapor holding capacity so it will increase the evaporative gradient at the leaf surface. It is possible that vapour pressure diffusion will rise with global warming, particularly in areas where precipitation is expected to decrease. In another study found when soybean grown in outdoor chambers that elevated CO_2 reduced stomatal conductance, by 33% at an average growth temperature of 27°C, and by only 17 at 40°C. Increases in leaf area index can balance any changes that reduced canopy evaporation has on canopy water use, while having the potential to reduce soil evaporation. In field grown, wheat and sorghum crops exposed to FACE, ET decreased by 9 and 7 per cent, respectively, relative to control plots. In a nutrient-poor, water-limited grassland system, there were small reductions in ET at elevated CO_2. These small but insignificant differences in ET accumulated over time, leading to marginal but significant water savings at elevated CO_2. Modeled ET for soybean and maize and found that, at the simplest level with no feedback operating, ET was reduced by 15.1 and 24.7 per cent in soybean and maize, respectively, at double ambient (CO_2). With maximum complexity accounting for atmospheric feedbacks, soil fluxes, and physiological responses, ET decreased by 5.4 and 8.6 per cent, in soybean and maize respectively, at high CO_2. Introduction of water stress to the model increased the reductions in ET for both crops. Grant *et al* (2001) reported from a wheat experiment, in which FACE decreased ET by 7 and 19 per cent under high and low N supplies, respectively. In contrast, soil water stress can enhance the reductions in ET under CO_2 enrichment as has been observed in sorghum. Warmer air temperatures can nullify the effects of elevated CO_2 on ET. There was found in soybean when grown in outdoor chambers that double ambient CO_2 reduced ET by an average of 9 at 23°C, while high CO_2 had no effect on ET above 35°C. In a nut shell, the available data suggests that elevated CO_2 reduces ET of well-watered canopies.

Effect of elevated CO_2 on weeds

Apart from the direct CO_2 fertilization effect, climatic change, particularly precipitation and temperature, will have effects on weed biology. Temperature and precipitation are primary abiotic factors, which control distribution of vegetation on the globe, and as such will impact the geographical distribution of weeds with subsequent effects on their growth, reproduction, and competitive abilities. Increasing temperature may mean expansion of weeds into higher latitudes or higher altitudes. Many of the weeds associated with warm-season crops originated in tropical or warm temperature areas; consequently, northward expansion of these weeds may accelerate with warming. This was found in itchgrass (*Rottboelliia cochinchinensis*) a warming of 3°C (day night temperature increase from 26/20 to 29/23°C) increased biomass and leaf area by 88 and 68 per cent, respectively. This was also reported a Northward expansion of weeds, such as *Imperata cylindrica* and witchweed (*Striga asiatica*) due to warming. However warming may restrict the southern expansion of some plants such as wild proso millet (*Panicum miliaceum*) due to increased competition. Most of the crop species are C_3 plants, while many weed species are C_4 plants. C_3 plants are expected to benefit more from elevated CO_2 than C_4 plants which suggests that crops will gain a competitive advantage over most weeds. This could result in changes in herbicide efficacy because at elevated temperatures, metabolic activity tend to increase uptake, translocation, and efficacy of many herbicides, while moisture deficit, especially when severe, tends to decrease efficacy of post emergence herbicides, which generally perform good when plants are actively growing. Also found that elevated CO_2 levels reduced the efficacy of the widely used herbicide glyphosate. In controlled-environment studies, herbicide efficacy was mostly reduced by elevated CO_2, and effects were dependent on the mode of action of herbicides, on weed species, and on competition. Double-ambient CO_2 caused a decrease of 57 per cent in efficacy of the herbicide fluazifopbutyl + fenoxyprop (blocks the activity of (AC Case) applied to *Avena fatua* (C_3), no effects of elevated CO_2 were found when the herbicide was applied to *S. viridis* (C_4). Differences in growth response and effects of CO_2 on herbicide efficacy between *A. fatua* (C_3) and *S. viridis* (C_4) serve to illustrate the complexity of the issue.

Effect of elevated CO_2 on nutrient cycling

Higher biomass productivity by plants under elevated CO_2 will ultimately increase organic matter inputs to soils. Increased soil carbon under elevated CO_2 could lead to higher soil microbial biomass and immobilization of nutrients. This will have negative effect on plant growth and has been demonstrated in a pot CO_2 experiment where microbial biomass N increased and plant responses to elevated CO_2 were negative mainly due to increased input of high C: N compounds to soils. On the other hand, extra C inputs to soils can increase microbial activity and thus enhance the mineralization of organic matter. Its positive effect on plant growth has been proposed in other study. Two different mechanisms may explain these observations: First, microbial N may have been primed by extra C inputs under elevated CO_2 (This mechanism was proposed in 1993), second, increased soil moisture at elevated CO_2 may have led to increased N mineralization, at least in water limited ecosystems. Microbial immobilization of extra N under elevated CO_2 may be restricted to systems where N supply is abundant and nutrient cycles are not in equilibrium with plant demand. Phosphorus and sulphur may respond to elevated CO_2 in a similar way as that of N because increases in soil microbial biomass will also be accompanied by the immobilization of these nutrients and the decomposition of soil organic matter will release the mineral nutrients that it contained. Higher soil moisture has been reported in many ecosystems exposed to elevated CO_2 and this will enhance the leaching of nutrients because more water will drain through the soil profile, when saturation is exceeded. This was also observed increased NO_3^- leaching from tropical communities. While many studies reported decrease in NO_3^- leaching and also found decreased NO_3^- concentrations below the rooting zone. There was observed reductions in soil NO_3^- concentrations in calcareous grassland communities exposed to elevated CO_2 for several years. Soil NO_3^- concentrations are regulated by many interacting processes, including nitrification and denitrification, immobilization of NH_4^+ and NO_3^- by soil microbes, and rooting patterns and root uptake of NH_4^+ and NO_3^-. Therefore, predictions are difficult, but a principal control is certainly plant uptake of mineral N, which will reduce NH_4^+ available for nitrification or remove the NO_3^- produced.

Effects of Climate Change on Ecosystem

Climate change has evident effects on ecosystems and people. Here's a list of the climate change phenomena we're already experiencing first-hand.

Ice melting

One of the most evident consequences is melting ice – the melting of the cryosphere, those portions of Earth's surface where water is in solid form, including ice caps, glaciers, and permafrost (those areas where soil is permanently frozen). According to predictions, Arctic ice could completely melt during the hottest periods of the year by the end of the century. The cryosphere naturally plays a crucial role in the global climate system and a change in its extension could cause a change in the system itself. Fragile ecosystems like oceans, mountains and wetlands could be damaged permanently.

Sea level rise

Melting ice caps in Antarctica and Greenland have most likely led sea level to rise by 3.1 millimetres per year between 1993 and 2003, according to IPCC. The rise is expected to reach 15 to 95 centimetres by 2100.

Ocean acidification

Increased levels of CO_2 in the atmosphere will also lead to ocean acidification, causing irreparable damage to marine ecosystems like the Great Barrier Reef that is a UNESCO heritage site as it is home to more than 400 types of coral, 1,500 species of fish and 4,000 types of mollusc. It also holds great scientific interest as the habitat of species such as the dugong and the large green turtle, which are threatened with extinction. Professional services firm Deloitte estimated the economic value of this treasure: 56 billion Australian dollars, and 64,000 jobs.

Desertification

Desertification (and, thus, heat waves) will expand to areas that currently boast a temperate climate such as the areas north and south of the Sahara desert, including the Mediterranean countries, causing severe damage to agriculture. Crops will significantly drop while more and more people will face undernourishment. In particular, yields from maize and wheat crops could drop by 50 per cent over the next 35 years due to global warming. It's a risk that has to be prevented considering that people suffering from hunger are currently slightly decreasing. The study State of Food Insecurity in the World 2015 conducted by the International Fund for Agricultural development (IFAD) and the World Food Programme (WFP) estimates that nearly 795 million people don't eat enough food. They amounted to 1 billion in 1990-1992.

Events like El Niño – a variation in the southern oscillation that causes significant changes in climate including hurricanes, storms, flooding in Central America and severe drought linked to wild fires in western Pacific areas will be more frequent and intense causing casualties and economic loss. This could lead to the outburst of diseases, like malaria, in areas previously unaffected.

Biodiversity loss

It's not only due to climate change, but also because of humanity, that the Earth is facing a relentless mass extinction, the sixth, resulting in a significant drop in our Planet's biodiversity. Species extinction rate is extremely high and half the living species could become extinct by the end of the century. This biodiversity loss has grave and far-reaching implications for human well-being, said John Knox, a human rights expert and professor of international law at Wake Forest University. Knox is also a UN Special Rapporteur on environment and human rights and the author of the first report of the United Nations that recognises that healthy biodiverse ecosystems are essential for human rights. As it happens with desertification, biodiversity loss in particular of plant species could slow down disease control and increase the spreading of infectious and autoimmune diseases.

Effect of climate change on crop production

"Climate change is acting as a brake. We need yields to grow to meet growing demand, but already climate change is slowing those yields," Michael Oppenheimer, professor at Princeton University and co-author of the fifth report by the IPCC (Intergovernmental Panel on Climate Change, which brings together scientists from all around the world). It is in this report that the scientific community came together to point out that decrease in crop yields is already taking place due to global warming.

Interactive effect of changing climatic factors on crop production

The ultimate productivity of crops is determined by the interaction of genotypes, soil constituents, water, temperature, day length etc. According to temperature, solar radiation and water directly affects the physiological processes involved in grain development and indirectly affects the grain yield by influencing the incidence of disease of insect and diseases. They found that the rice grain yield was correlated positively with average solar radiation and negatively with average daily mean temperature during reproductive stage. Relatively low temperature and high solar radiation during reproductive stage had positive effect on number of spikelets and hence increased the grain yield. Solar radiation had positive influence on grain filling during the ripening period. The simulation results indicate that warm climate with decreasing radiation levels will affect the growth and yield of cereal crops. However, the harmful effects of increasing temperature on growth and yield are likely to be counter-balanced by the increasing levels of CO_2 concentrations to some extent in the near future. There was also observed that in India the adverse effects of 1-2°C rise in temperature could be absorbed with 5-10 per cent increase in precipitation. The grain yield increase of 20-30 per cent may be possible on about 70 per cent area under rice and wheat. In northern India, warming could offset some losses in yield by early pod set in winter grain legumes like chickpea and lentil. All India estimates of production based on current relative contribution of different states in total production, showed decline in production from the current levels by 3.16 and 13.72 per cent in the year 2020 and 2050, respectively. Currently the winters are severe in Punjab, Haryana and western UP witnessing frosting in December and January. In future climate scenarios warming may ease the chilling conditions in these regions to favour potato productivity, while in other regions with cooler winter season the warming from current levels may prove detrimental.

The maximum LAI, biomass and grain yield of wheat and rice declined when the radiation decreased by 10 per cent from the normal but increased when the radiation was enhanced by 10 per cent from the normal. The simulation results suggest that the growth and yield of wheat and rice would be influenced by increasing temperature. The adverse affects generated by high temperature scenario may be lessened to some extent by decrease in radiation amounts. But this aspect is still uncertain because the radiation is expected to be on the lower side under the influence of greenhouse effect. In the Punjab state, there are indications that the amount of radiation is likely to decrease. As a result, the production of wheat and rice may be adversely affected depending upon the degree of change in the coming years. The past increase in CO_2 experienced to date and the projections of its increase in the future will no doubt counter balance the negative effects of rise in temperature on the crop productivity.

Effect of elevated temperature and rainfall

Temperature is a very important factor affecting crop productivity right from seed germination to harvest all phonological stages of crop are affected by temperature. Any variation in temperature will affect crop productivity adversely. It concluded that a decline in rainfall with anticipated thermal stress (maximum temperature is increased by 0.18 °C and minimum temperature is increased by 1.58 °C leads, on an average, to a reduction in crop yield by nearly 5 per cent for every 10% decline. The positive effects of elevated CO_2 almost cancel out with enhanced thermal stress (maximum temperature is increased by 0.28 °C and minimum temperature is increased by 2.58 °C and reduction in rainfall by 40 and 50 per cent such that soybean yield is up by only 0.1 per cent and down by 6 per cent, respectively. This suggests that significantly deficient monsoon rainfall conditions combined with thermal stress should adversely affect the positive effect of elevated CO_2 on the soybean crop in Madhya Pradesh, India. The significant increase in rate of photosynthesis and reduction in stomatal conductance was observed in both the chickpea cultivars grown at high temperature. Pusa 1053 showed significant increase (64 %) in rate of photosynthesis during flowering stage. Similarly, Pusa 1108 plants showed 15.5 and 24.3 per cent increase in rate of photosynthesis during vegetative and podding stages. Stomatal conductance decreased in both the cultivars under elevated temperature significantly at all the growth stages. Among two cultivars, Pusa 1053 showed higher reductions (43%) in stomatal conductance during vegetative stage. The response of crop species to temperature depends upon the temperature optima of photosynthesis, growth and yield. When the level of temperature is below the optimum for photosynthesis, a small increase in temperature can greatly increase the rate of photosynthesis and crop growth and the reverse is true when the level of temperature is near the maximum for growth and photosynthesis. The enhancement in photosynthesis rate under elevated temperature indicated that ambient temperature during the crop growth period was below optimum for photosynthesis. As a result, exposure of the plants to high temperature increased rate of photosynthesis.

Field studies indicated that increased temperature hastened the rate of senescence resulting in reduced LAI and total biomass in wheat, the decreased crop duration with increased temperature resulted in reduction in gain yield. The maximum LAI in wheat decreased by 4.5 to 33.8 per cent, in rice by 1.5 to15.9 per cent and in groundnut by 1.2 to 5.3 per cent when the temperature increased from 0.5 to3.0°C above normal, the study further revealed that with similar increase in temperature the grain yield of wheat, rice and maize declined by 5.5 to 25.7 per cent, 2.4 to 25.1 per cent and 7.4 to 21.4 per cent, respectively from the normal yields. In another study conducted using CERES wheat model to assess the effect of intra seasonal increase in temperature from normal on yield of wheat sown on different dates revealed that in general an increase in temperature from mid February to mid March severely affected the yield of early, normal and late sown wheat.

Amongst the three cropping systems, cotton-wheat will be affected more adversely than maize-wheat and rice-wheat systems. The adverse effect of increased temperature was more for maximum temperature than minimum temperature. Though increase CO_2 would increase the crop productivity but the magnitude of increase in crop yields was less than that of decrease by the increased temperature. Validate the CERES-Rice model under Punjab conditions and give the % deviation over normal scenario. The results showed a decline in crop duration, grains m^{-2}, grain yield, maximum LAI, grains per ear, biomass and straw yield with each 0.5 °C increase in temperature over the normal during the crop season. Under the warm climatic scenarios, the reduced source size (leaf area) coupled with poor sink strength (as depicted by the number of grains per ear) reduced number of effective tillers (as indicated through lesser number of grains m^{-2}) and shorter period of harvesting solar radiation (crop duration) resulted in considerable decline in biomass and grain yield of rice crop over the normal.

Effect of elevated temperature and CO_2 on C_3 and C_4 plants

Impact assessment of climate change on crop productivity is very difficult due to complicity of the response of different plants processes to changes in temperature and carbon dioxide concentration. The rise in carbon dioxide concentration may favor C_3 photosynthesis because rise in atmospheric CO_2 would not reach the level in foreseeable future at which photosynthesis is reduced due to stomatal closure. The C_4 plants being less responsive to elevated CO_2 concentration will have less or no improvement in their photosynthetic rate. The rise in temperature may have both beneficial and harmful effects on crop photosynthesis depending upon location, crop species, crop growth stage and level of temperature. The temperate reasons, the species and crop growth stages, which require warm climate may benefit from temperature rise. On the other hand, crops, which are adapted for low temperature and crops grown in tropical climate may be adversely affected due to rise in temperature. The productivity of crops will depend upon relatives gain in photosynthesis compared to respiratory loss. The latter is to play more important role because photosynthetic gain is limited by day light but increase in temperature associated respiration continues for whole of the crop life cycle. The plants like any other organism, have inbuilt mechanism to make adjustments in their different processes to enable to survive in a given set of conditions (acclimation). How and to what extent plants will adapt to environmental change, is not properly under stood. Therefore, the effects of short term (hours and crop season) experiments may not hold true in long term of decades on century changes in climate which is a slow but continuous change. Rice crop is sensitive to changes in temperature and carbon dioxide concentration. Being a C_3 plant, rice holds an edge over C_4 plants due to increase in photosynthetic rates under expected enhanced CO_2 concentrations. The results of the simulation study for interactive effects of increasing temperature and CO_2 concentration revealed that the adverse effect of increase in temperature on growth and yield of crop was counter-balanced by favorable effect of increasing CO_2 levels up to some particular combination. With temperature increase of 1.0°C from normal, CO_2 concentration of only more than 500 ppm was able to nullify the negative deviations in growth and yield, but when temperature increased by 2.0°C from normal, even 600 ppm CO_2 was unable to nullify the adverse effect of temperature.

Effect of elevated temperature and CO_2 on rice-wheat system productivity

Climate change may have serious direct and indirect effects on the rice-wheat system and food security of India. This may be aggravated by water scarcity, drought, flood, and decline in soil organic C content. Simulation models for rice production indicate a reduction in yield of about 5% per degree rise in mean temperature above 32°C. This would counter balance any increase in yield due to increased CO_2 concentration. Rice is sensitive to hot temperature at anthesis; sterility in some varieties occurs if temperatures exceed 35°C at anthesis for only about one hour. At anthesis, spikelet fertility declines from 90 to 20 per cent after only 2 hour exposure to 38°C, and to 0 per cent by less than one hour exposure to 41°C. The critical temperature for spikelet fertility (defined as when fertility exceeds 80%) varies between genotypes, but it is about 32-36°C. Below 20°C and above about 32°C, spikelet sterility becomes a major factor, even if growth is satisfactory. The higher temperatures and reduced radiation associated with increased cloudiness caused spikelet sterility and reduced yields to such an extent that any increase in dry-matter production as a result of CO_2 fertilization proved to be of no advantage in grain productivity of rice. There was also reported that if all other climate variables remain constant, a temperature increase of 1, 2, and 3°C would reduce the grain yield of rice by 5.4, 7.4 and 25.1 per cent, respectively. This was also suggested that rice production in the Asian region may decline by 3.8 per cent under changed climate. Under elevated CO_2 yields of rice increased significantly (28% for a doubling of CO_2), however, 2°C increase in temperature cancelled out the positive effect of elevated CO_2 on rice. The rice yield declined by 10 per cent for each 1°C increase in growing season minimum temperature in the dry season, whereas, the effect of maximum temperature was insignificant. The warmer temperature hastens crop development, shortens the growth period and thus finally lowers the grain yield. Impact of high temperature on crop growth and yield is largely determined by the duration and coincidence of it with sensitive crop growth phase. Period from panicle initiation to flowering stage is found to be more sensitive to high temperature stress in wheat and rice. Exposure to high temperature from seedling stage to panicle initiation stage affected yield predominantly causing tiller mortality and reduced number of spikes. Coincidence of high temperature stress with panicle initiation to flowering phase of crop affect grain yield by reducing dry matter accumulation, productive tillers, number of spikes, grain weight and increased floret sterility. In case crop is exposed to heat stress from flowering to maturity, then the reduction in yield is predominantly caused by floret sterility leading to reduced number of grains per spike and also due to reduced grain weight.

Effects of climate change on wheat production include reduced grain yield over most of India, with the greatest impacts in lower potential areas such as the eastern IGP. Physiological traits that are associated with wheat yield in heat-prone environments are canopy temperature depression, membrane thermostability and leaf chlorophyll content during grain filling, leaf conductance and photosynthesis and senescence. Grain growth is shorter with heat stress, thereby influencing grain filling and resulting in lower yield. Wheat cultivars capable of maintaining high test weight under heat stress are more tolerant to high temperature. In Punjab (India), a temperature increase of 1, 2 and 3°C from present-day conditions, would reduce the grain yield of wheat by 8.1, 18.7 and 25.7 per cent, respectively. Under elevated CO_2, yields of wheat increased significantly (28% for a doubling of CO_2), however, 3°C rice in temperature nullified the positive effect of elevated CO_2 on wheat. An increase in yield of wheat to the extent of 29-37 and 16-28 per cent for different genotype were recorded under rainfed and irrigated conditions, respectively, for a temperature rise coupled with elevated CO_2 (T_{max} + 1.0°C, T_{min} + 1.5°C and 460 ppm CO_2) compared with the current climate. An increase in temperature on the order of 3°C or more, however, cancelled out the beneficial effects of elevated CO_2. The impact of modified climate was observed to be higher under rainfed conditions than under irrigated conditions for all genotypes. Conducted simulation studies to find the impact of climate change on wheat yields for several locations in India using a modeling approach the results indicated that, in northern India, a 1°C rise in the mean temperature had no significant effect on potential yields, though an increase of 2°C reduced potential grain yields at most place. In another study using the CERES-Wheat model, showed that wheat yields were lower than those in the current climate, even with the beneficial effects of CO_2 on crop yield. Yield reductions were due to a shortening of the wheat growing season, resulting from an increase in temperature. A 2°C increase resulted in a 15-17 per cent decrease in grain yield of rice and wheat but, beyond that, the decrease was very high in wheat. The grain filling of wheat is seriously impaired by heat stress due to reductions in leaf and ear photosynthesis at high temperatures. Studied the impact of temperature increased in March 2004 on the productivity of wheat. A temperature increase above normal ranged from 1 to 12°C in different parts of northern India, resulting in a wheat production loss of 4.6 million tonnes due to increased incidences of pests and diseases, and advanced maturity of wheat by 10-20 days further reduced grain weight. Studied the impacts of rainfall variability on wheat yield in Northwest India and showed that the years with scarce rainfall resulted in only 34 per cent (Ludhiana) and 35 per cent (Delhi) of the baseline yield. In Ludhiana, high rainfall years resulted in 200 per cent yield as compared with the baseline yield, whereas these years resulted in only 105 per cent yield in Delhi.

Effect of elevated temperature and rainfall on nutrient cycling

Elevated CO_2 concentration along with other green house active atmospheric gases will lead to higher mean atmospheric temperatures and altered precipitation patterns (IPCC, 2001). Studied N mineralization impacts in vitro experiments at different temperatures and soil moisture contents. The reduction in mineralization rate when reducing soil moisture from 60 to 10 per cent of field capacity was much larger relatively at 25°C (≈70%) than at 5°C (≈50%). Warming both experimentally and naturally, is accompanied by increased evapo-transpiration and, therefore, decreased soil moisture and plant water availability, which results in a further reduction of microbial activity. Contrary to this, several whole-ecosystem studies have shown that increased air temperatures can actually result in decreased soil temperatures. It could be due to increased biomass of plants at elevated ambient temperatures can effectively insulate soils from solar radiation, further, taller species will absorb solar radiation farther off the ground and convective heating of soils will effectively be reduced. Heated forest floor by 3 to 5°C using electric cables found increased NO_3^- and NH_4^+ in runoff. There were also found higher soil NO^{3-}, Mg and Al concentrations at the warmer spots. High temperatures during summer can increase evapo-transpiration particularly in areas with relatively dry climate, drier soils may have higher rates of erosion by wind and rain, also, the frequency of droughts may increase, further enhancing erosive losses. Plant canopies reduce erosive power of rain by interception, subsurface roots hold the soil in place, and crop residues and surface mulch reduce rill erosion rates. Increased rainfall may also increase erosion because of increased amount of precipitation and due to extreme rainfall events as is predicted for many areas. Soil erosion by rain is determined by intensity of rain, a 1.5 to 2.0 per cent increase in erosion rates can be expected by per percent increase in precipitation.

Effect of elevated temperature, light and humidity on herbicide efficacy

Studied the effects of environmental conditions on herbicide efficacy on *Agropyron repens* and found that increased light, temperature and humidity immediately following application increased the efficacy of fluazifop-butyl, but that prolonged exposure of plants to increased temperature and light decreased efficacy. In this study, increased daytime temperature also caused decreased herbicide efficacy. Herbicide efficacy was found to increase, decrease, or remain the same when plants were subjected to changed environmental conditions (climate change) and efficacy changes were species specific.

Strategies to Deal with Changed Climatic Scenario

According to the recent IPCC assessment, agricultural production in South Asia could fall by 30 per cent by 2050 if no action is taken to combat the effects of increasing temperatures and hydrologic changes. Adaptive options to deal with the impact of climate change are:

- » Minimum tillage or zero tillage with residue cover in surface for improving soil quality
- » Cover cropping, in-situ residue management and restoration of degraded lands for soil moisture conservation and improved C-sequestration
- » Agroforestry with multipurpose trees, crops and animal components for improving

- » Integrated farming systems and watershed development with animal, fishery and hedge row cropping for soil and moisture conservation and nutrient recycling.

- » Screening short duration varieties for their drought resistance.

- » Popularization of technologies like system of rice cultivation (SRI) and aerobic rice cultivation for water saving and mitigation of green house gas (GHG) emission.

- » In-*situ* biomass management instead of biomass burning to reduce CO_2 emission and improve hydrology.

- » Promotion of technologies that enhance biological N fixation and improve nutrient and water use efficiency to reduce N_2O emission and dependence on non-renewable energy.

- » Change in planting dates and crop varieties are another adaptive measure to reduce impacts of climate change to some extent. For example, the Indian Agricultural Research Institute study indicates that losses in wheat production in future can be reduced from 4 - 5 million tons to 1-2 million tons if a large percentage of farmers could change to timely planting and changed to better adapted varieties.

Nutrient management

- » Precise N application (dose, time and place)
- » Use of slow release N fertilizers or nitrification inhibitors
- » Applying N when least susceptible to loss or prior to plant uptake
- » Integrated nutrient management (INM) and Site-Specific Nutrient Management (SSNM) have the potential to mitigate effects of climate change.

Water management

Efficient water use leads to more grain and residue production which ultimately results in more carbon sequestration and reduces GHGs emission to a considerable level. Proper drainage improves the aeration in the soil and reduces the CH_4 and N_2O emission from rice fields.

Researchable issues identified for future

- » Breeding for improved crop varieties with specific reference to growth and flowering phenology, photo sensitivity/insensitivity, stability in response to inputs viz., lodging resistant, optimum tillering, harvest index etc.

- » Evolving efficient water and soil management practices in addition to identification of crops and varieties with high water use efficiency, dry matter conversion ratio, positive response to temperature extremes and elevated CO_2.

- » Identifying new intercropping and novel farming system combinations including livestock and fisheries, which can withstand predicted climate change situations and can be economically viable

- » Identifying cost effective methods for reducing greenhouse gas emission from rice paddies and also from cropping systems with livestock components

- » Promoting conservation agriculture practices especially in water harvesting, nutrient, pest and disease management.

It can be concluded that:

- » Climate change is a reality.
- » C_3 plants will be benefited more than C_4 plants at elevated CO_2.
- » Weeds will become more competitive from carbon fertilization.
- » Mitigation strategies need to be studied to meet the challenge posed by climate change on agriculture productivity
- » Water management practices such as alternate wetting and drying, mid-season drainage helps in reducing CH_4 emission from rice fields
- » Increase humidity and higher temperature will result in more infestation of diseases.

CONTRIBUTION OF AGRICULTURE TO CLIMATE CHANGE

At the same time, agriculture especially intensive agriculture, characterised by monocultures and aimed at feeding farm animals is one of the sectors that generates the highest amount of emissions of CO_2 (the main greenhouse gas). This quantity can be compared only to the sum total of the CO_2 emitted by all forms of transportation.

By looking deeper, we can observe that agriculture and the deforestation it causes were responsible for one fifth (21%) of all CO_2 emissions in the decade from 2000 to 2010 (approximately 44 billion tonnes). This occurs because agriculture needs an increasing amount of space alongside massive amounts of chemical fertilisers now that the demand for meat and its products has increased dramatically in developing countries. This is damaging forests, which in turn would be able to absorb CO_2 and mitigate anthropic (man-made) emissions. A vicious cycle that makes agriculture both a victim (have the negative effects of global warming on food supply) and a perpetrator (one of the main causes of climate change).

Most of the time, when agriculture perpetrates its crimes, it isn't even contributing to feeding the ever-increasing world population. In fact, 95 per cent of the soy produced in the world is consumed by farm animals mostly bovines, which demonstrates this conflict. Also, according to a study conducted by the Chalmers University of Technology in Goteborg, Sweden this means that producing one kilogramme of bovine meat require 200 kilos of CO_2 emissions. There are 700 million pigs in China alone, one for every two citizens, half of the global population of farm pigs. In order to feed these animals, forced to live in cages inside industrial warehouses, Beijing imports 80 million tonnes of soy, especially from Latin America and more specifically from the Brazilian Amazon where endless fields of soy are destroying one of the most biodiverse places in the world, one of the world's green lungs.

Agriculture and climate change is called as agroecology

The Food and Agriculture Organization (FAO) seems to have a clear idea of what should be done and is promoting sustainable practices in various countries through agroecology. This is a series of social and environmental measures aimed at creating a sustainable agricultural system that optimises and stabilises crop yields. These practices also tackle the effects of climate change, such as desertification and the rise in sea levels, and among them organic agriculture plays an essential role as it respects natural cycles, drastically reducing human impact.

According to the latest Eurostat data, from 2010 to today organic agriculture in Europe has grown by 2 million hectares, reaching a total of 11 million hectares of land (more that 6 per cent of the European total). If we want to continue the comparison with China which was until recently one of the least evolved countries with regards to organic practices this type of agriculture occupies 1.6 million hectares and generates 4.7 billion euros, according to data presented by Federbio, the Italian Federation of organic and biodynamic agriculture.

Effect of Agriculture on Environment, Eating Habits Matter, Especially in Europe

Agriculture and climate change. Concluding our world tour in the Old Continent, the aforementioned Chalmers University of Technology in Goteborg points us in a specific direction so that we can meet the CO_2 emission reduction targets set by the European Union: we must eat less bovine meat and dairy products. We can't protect the environment without changing our eating habits. Agricultural industries and intensive farming are in fact responsible for about one quarter of CO_2 emissions in Europe.

The Paris Agreement has set a clear objective: limiting the global temperature rise to "well below 2 ^0C", and to do everything in our power to "limit the temperature increase to 1.5 ^0C". In addition to the impact of energy (we of course can't ignore the terrible damage caused by fossil fuels combustion), making agriculture and all the activities connected to it sustainable is the answer to win the battle against global warming, as well as accelerate the transition to a healthier and more just society.

Efforts to fight climate change

The COP_{24} climate change conference is taking place in Katowice, Poland with the aim of establishing regulations that will bring the Paris Agreement into effect and averting a planetary catastrophe.

The concentration of CO_2 in the atmosphere continues increasing: the highest levels of the past three million years have just been recorded. Greenhouse gas emissions aren't simply failing to decrease, they keep growing: over the past 22 years we've experienced 20 of the hottest years since records began, according to the World Meteorological Organisation. There's no more time left, the abyss is widening underneath our feet and one misstep will spell tragedy for us all. The Special Report 15 (SR15), a document produced by the Intergovernmental Panel on Climate Change (IPCC) published in October once again underlines the absolutely urgent necessity for effective provisions to limit the growth of global average temperatures and climate change. Otherwise we will have to face a "climate catastrophe". According to the report, we have twelve years to halve worldwide emissions so that we remain below the plus 1.5 ^0C limit established by the IPCC. With these objectives in mind, United Nations representatives are gathered in Katowice, Poland for the COP24 from the 3rd to the 14th of December, the 2018 UN climate change conference.

The goal of COP$_{24}$

The primary goal of the negotiations, which will end on the 14th of December, is to approve the Paris Rulebook, a set of regulations which will bring the 2015 Paris Agreement into effect. The treaty, ratified by 184 countries, sets forth objectives for global action on climate change and proposes that the increase in global temperatures be kept "well below 2 ^{0}C". During the talks, initiatives to counter climate change set out by the 198 member states will be analysed and evaluated.

The world's future is at stake

Collective action by all states is necessary to reach the main goal but many aspects of the Paris Agreement aren't binding for the nations who signed it, which means the UN can't force them to comply. The gap between the commitments that have been made and what is actually needed is still too big, as the IPCC report highlights. "Key outcomes from this meeting of UN climate negotiators will lay the foundations for continued multilateral progress in tackling climate change at the scale and speed necessary to match what science tells us is needed". These are the hopeful words of Manuel Pulgar Vidal, responsible for climate and energy at the WWF. "The scale of the challenge and the opportunity – that this meeting presents should help focus minds".

No more excuses for not cutting CO_2

We have green, clean energy at our disposal as well as the technology to make use of this energy effectively and efficiently. We could, without a doubt, completely stop extracting fossil fuels and truly try to stop climate change, but the opposite is happening. The global consumption of petrol has come close, for the first time in history, to a hundred million barrels a day, and investment in the extraction of highly polluting fuels such as lignite and tar sands continues apace.

Brazil and the United States take a step in the wrong direction

During the campaign that led him to become president of the United States and then once in office, Donald Trump attacked the Paris Agreement, announcing that he doesn't wish to respect the climate commitments made by the Obama administration. However, Article 28 of the treaty establishes that no country can abandon its commitment before three years have passed since its coming into effect. Thus, the US continues to send delegations to summits like the COP$_{24}$ while it waits for 2020, when the accord kicks in. Meanwhile, Brazil announced a few days ago that it intends to retract its proposal to organise the COP$_{25}$, planned for the 11th to 22nd of November 2019, thus dispelling any doubts about President-elect Jair Bolsonaro's environmental policy orientation.

For some, half a degree will mean extinction

Changes to our climate are already altering entire ecosystems, forcing the organisms that live in them to move to more favourable climates. For thousands of animal and plant species the increase in temperatures could turn out to be fatal, and the difference between 1.5 to 2 ^0C increases is, literally, a matter of life and death. Coral reefs, for example, are invaluably precious troves of biodiversity and are in grave danger because of rising sea temperatures. This phenomenon is causing coral bleaching, which, according to researchers, will mean coral reefs are the first ecosystems to become extinct in the modern age.

Directly or indirectly global warming threatens many animal species, such as koalas, who need to drink more because of the heat, many reptiles whose sex is determined by the temperature before eggs hatch and thus risk extinction, sterns, forced to migrate to colder climates, steinbocks, who are seeing the alpine pastures they depend on greatly diminish. The disappearance of these species, along with tens of thousands of others, would be an incalculable loss for our planet and our species. According to a recent study, 3 to 5 million years will need to pass before current levels of biodiversity are re-established. We, on the other hand, might not even still exist.

World wildlife fund (WWF)

Other than bringing the Paris Agreement into effect, the COP_{24} should also mark the start of a series of initiatives, including the strengthening of climate action both before and after 2020, accompanied by financing and other types of support for developing countries, according to the WWF. It should also make nations take more responsibility, increasing the transparency of their actions to fight climate change and addressing the inadequacy of the measures that have already been undertaken, providing precise indications of what needs to be done, including better integration of nature-based measures.

Only 12 years to avoid a climate catastrophe

Need to act now or the consequences of global warming will be catastrophic. The scientists of the latest IPCC climate change report make the last call for humanity to take action. The increase in the average global temperature could hit 1.5 degrees as early as 2030, compared to pre-industrial levels. The Special Report 15 (SR_{15}) of the Intergovernmental Panel on Climate Change (IPCC) is more than a warning call. It's the definitive confirmation and climate change deniers should now give in of the catastrophic consequences we will face if we don't act to limit global warming. This means cutting greenhouse gas emissions in record time.

IPCC's latest report

The report (published in South Korea, where negotiations among governments took place, at 10 a.m. local time) is a detailed analysis. 250 pages curated by a task force of 91 experts from 40 countries who analysed more than 6,000 studies and whose work has been in turn examined by tens of scientists. The document explains that the average temperature of the surface of the world's lands and oceans increased by 0.17 °C/decade, since 1950.

If not curbed, this trend will lead the Earth to exceed the threshold of +1.5 °C between 2030 and 2052 (according to the different scenarios the SR_{15} took into consideration). This means that in just 12 years we could reach the temperature rise that the Paris Agreement hypothesised for 2100. And if the trend remains unvaried for the rest of the century, a catastrophic scenario is certain, as we'd reach a +3 °C rise at this pace.

Need to reach the threshold of 1.5 °C before 2030

The IPCC underlined how we already reached +1 degree between 2017 and 2018. An almost deperate situation, but according to the IPCC our Planet isn't doomed yet. "Our role isn't determining if the goal of the Paris Accord is feasible or not. But nothing in the scientific literature tells we can't reach it. What we explained are the conditions needed to meet that goal. Now it is responsibility of the policy makers," said Henri Waisman, researcher at the Institute for Sustainable Development and International Relations (IDDRI) and co-author of the report. The special report, in fact, was commissioned to the IPCC during the COP_{21} in Paris in 2015, with the aim to outline the current trend in reference to the 1.5-degree goal on the basis of the commitments made by each government to reducing greenhouse gas emissions.

New technologies to reduce CO_2 in the atmosphere

This is why the IPCC underlined the fact that it is now crucial to introduce new technologies to remove CO_2 in the atmosphere, and doing it on a large scale. The aim is to produce the so called "negative emissions". "Using biomass alone (forests, woods and green areas, editor's note) to intercept CO_2 would mean conflict with industries like agriculture, whose areas would be limited. This would exacerbate land grabbing," Waisman added. The problem is that artificial techniques to remove carbon dioxide are still in their early stages. The SR_{15} not only outlines the climate evolution trend, it also explains what we'll face even if the Paris

Agreement targets are met. Even with a rise of 1.5 °C, the most vulnerable countries are likely to not have enough time to adapt to the consequences of climate change. This is especially true for atolls as sea level will increase for centuries due to the melting of perennial ice.

Sea level will rise for centuries. The Arctic will be free of ice during summer. Yet, with a rise of 1.5°C sea level would be 10 centimetres lower than in a - 2 °C scenario, in 2100. Moreover, with a rise of 1.5 °C the hypothesis of an Arctic ocean free of sea ice will be highly probable (once a century), while it would be way more frequent (once in a decade) with a -2°C rise. Similarly, with +2 °C 99 per cent of coral reefs would disappear, while the percentage would decrease to 70-90 per cent in a more favourable scenario. "Every extra bit of warming matters, especially since warming of 1.5 °C or higher increases the risk associated with long-lasting or irreversible changes, such as the loss of some ecosystems," said Hans-Otto Pörtner, chairman of one of IPCC working groups.

The IPCC also states that numerous changes in the oceans will cause the death of the species that have more difficulty to move. And overcoming the chemical changes due to the acidification process will take thousands of years. Moreover, heat waves are likely to multiply in the northern hemisphere. "The risk is that southern Europe will face desertification by the end of the century," underlines WWF. Flooding and drought will not only hit the old continent, but also northern America and Asia. And hurricanes will become stronger.

The solution to limit damages is to zero net emissions by 2050

The report dedicates an entire chapter to possible solutions. Different countries collided precisely on this point due to their opposite interests. The IPCC repeatedly underlined the need of drastically reducing the energy demand of industries, transport and buildings. To save the Planet we need to reduce global CO_2 emissions by 45 per cent by 2030 (compared to 2010 levels) and zero "net emissions" by 2050. Renewable energies will also hit 70-85 per cent by 2050.

Predicted Climate Change Impacts on Agriculture

The predicted changes to agriculture vary greatly by region and crop. Findings for wheat and rice are reported here:

Wheat production

The study found that increases in temperature (by about 2°C) reduced potential grain yields in most places. Regions with higher potential productivity (such as northern India) were relatively less impacted by climate change than areas with lower potential productivity (the reduction in yields was much smaller);

- » Climate change is also predicted to lead to boundary changes in areas suitable for growing certain crops.

- » Reductions in yields as a result of climate change are predicted to be more pronounced for rain fed crops (as opposed to irrigated crops) and under limited water supply situations because there are no coping mechanisms for rainfall variability.

- » The difference in yield is influenced by baseline climate. In sub tropical environments the decrease in potential wheat yields ranged from 1.5 to 5.8 per cent, while in tropical areas the decrease was relatively higher, suggesting that warmer regions can expect greater crop losses.

Rice production

- » Overall, temperature increases are predicted to reduce rice yields. An increase of 2-4 °C is predicted to result in a reduction in yields.

- » Eastern regions are predicted to be most impacted by increased temperatures and decreased radiation, resulting in relatively fewer grains and shorter grain filling durations.

- » By contrast, potential reductions in yields due to increased temperatures in Northern India are predicted to be offset by higher radiation, lessening the impacts of climate change.

- » Although additional CO_2 can benefit crops, this effect was nullified by an increase of temperature.

The policy implications of these predictions

The policy implications for climate change impacts in agriculture are multi-disciplinary, and include possible adaptations to:

- » **Food security policy**: to account for changing crop yields (increasing in some areas and decreasing in others) as well as shifting boundaries for crops, and the impact that this can have on food supply.

- » **Trade policy**: changes in certain crops can affect imports/exports; depending on the crop (this is particularly relevant for cash crops such as chillies).

- » **Livelihoods**: With agriculture contributing significantly to GNP, it is critical that policy addresses issues of loss of livelihood with changes in crops, as well as the need to shift some regions to new crops, and the associated skills training required.

- » **Water policy**: Because impacts vary significantly according to whether crops are rain fed or irrigated, water policy will need to consider the implications for water demand of agricultural change due to climate change.

- » **Adaptive measures**: Policy-makers will also need to consider adaptive measures to cope with changing agricultural patterns. Measures may include the introduction of the use of alternative crops, changes to cropping patterns, and promotion of water conservation and irrigation techniques.

Needs for further research

Due to the complex interaction of climate impacts, combined with varying irrigation techniques, regional factors, and differences in crops, the detailed impacts of these factors need to be investigated further. Specific recommendations for further research include:

- » Precision in climate change prediction with higher resolution on spatial and temporal scales;

- » Linking of predictions with agricultural production systems to suggest suitable options for sustaining agricultural production;

- » Preparation of a database on climate change impacts on agriculture;

- » Evaluation of the impacts of climate change in selected locations; and

- » Development of models for pest population dynamics.

Greenhouse gas emissions and its management

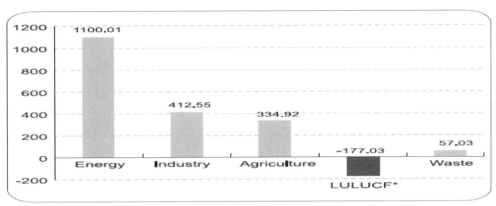

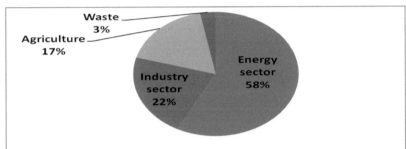

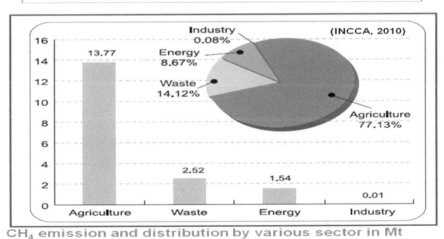

Emission of GHG (CO_2 equivalent emissions) from different Agriculture sector of India INCCA, 2010

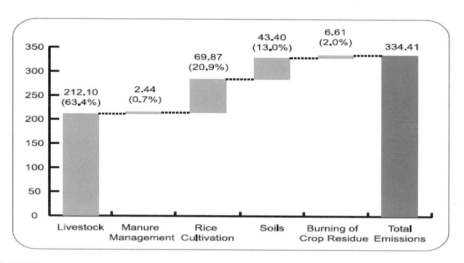

GHGS emissions sources in agricultural practices

1. Primary sources of C emissions (direct emission)

A) Mobile Operations

Tillage   Sowing   Transplanting   Pesticide spray   Harvesting and Transport

B) Stationary Operations

Pumping water   Grain processing and Drying

C) Emission from soil

CH_4 from paddy field and cattle, N_2O from fertilized field and manure and CO_2 (Lal, 2004)

2. Secondary sources of C emission (indirect emission)
Manufacturing, Packaging, Storing and transport of Fertilizers, Pesticides, other Chemicals, diesel and electricity

Manufacturing   Packaging and storing   Transport

Sources of methane in agriculture

Rice field

Ruminant

Manure management

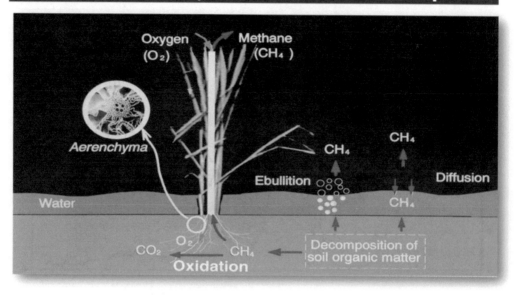

Potential of Agriculture Sector to Mitigate Climate Change

Agriculture can mitigate climate change by either reducing GHG emissions or by sequestering CO_2 from the atmosphere in the soil.

1. **Carbon sequestration**: Long-term storage of carbon in the terrestrial biosphere, underground or the oceans so that the buildup of carbon dioxide concentration in the atmosphere will reduce or slow down.

2. **Resource conserving techniques:** Resource conservation techniques are the practices, when followed results in saving of energy, cost and also reduce the environmental pollution over the conventional practices.

Advantages

- » Increased soil organic carbon
- » Saving in irrigation water
- » Increase in energy efficiency
- » Promotes timely sowing
- » Saving of resources
- » Yield advantage
- » Lower production cost
- » Improved soil health
- » Increase in profitability

Various RCTs are given below:

1. No-tillage
2. Laser land leveling
3. DSR/SRI
4. Leaf colour chart for N

5. Tensiometer
6. Raised Bed planting
7. FIRBS
8. Irrigation management like Micro irrigation
9. Crop diversification and Intercropping

Mitigation option

1. Carbon Sequestration
   - Zero tillage
   - Reduce tillage
   - Residue retention

2. Reducing GHGs emission
   - SRI, DSR
   - N management
   - Nitrification inhibitor

3. Reducing Fossil fuel Consumption
   - improved machineries & equipments

4. Bio-diesel or Bio-fuel

Adaptation options

1. Timely Sowing
   - zero tillage

2. Water Management/ reducing water loss
   - Laser land levelling

- » FIRBS
- » Micro irrigation

3. Suitable crop/cultivar for moisture stress, high temperature

4. Crop diversification & rotation

- » Intercropping
- » Introduction of legume

Potential benefits of key RCTs in terms of climate change adaptation relative to conventional practices

RCTs	Potential benefits
Zero tillage	C sequestration, reduced fuel consumption and GHG emission, timely sowing
Laser levelling	Reduced water use and fuel consumption, increased area for cultivation
DSR	Less requirement of water, time saving, better soil condition, deeper root growth, reduced methane emission
Diversification	Efficient use of water, increased income, increased nutritional security, conserve soil fertility, reduced risk
Raised bed planting	Less water use, improved drainage, better residue management, less lodging of crop, more tolerant to water stress
Leaf colour chart	Reduces fertilizer N requirement, reduced N loss and environmental pollution

8. Causes And Impact Of Hurricanes And Tropical Storms

Tropical Storms Prone Places

Hurricanes	North Atlantic
Typhoons	Pacific
Tropical Cyclones	South East Asia
Willy-Willies	Australia

Conditions for formation tropical storms

- » Over Oceans
- » Ocean temperature over 27°C
- » Water heated to a depth of several metres
- » Close to the East Coast of continents
- » Late summer or early autumn, when sea temperatures are at their highest (noticed how hurricanes always hit America around September/October!)

Causes of tropical storms

1. Air on surface of ocean is heated (it also contains lots of moisture)
2. Hot, humid air raises, cools and condenses. Clouds form.
3. Rising air creates low pressure. Air rushes in to fill gap left by rising air.

4. Rotation of the earth means winds do not blow straight. Winds circle towards the centre.

5. The storm continues to feed itself.

6. Whole system moves westwards towards land.

7. When the system crosses the land it loses its source of heat and moisture. The tropical storm losses its energy and dies out.

Difference between tropical disturbance, tropical depression, tropical Storm and hurricane

- **A tropical disturbance** is a discrete weather system of apparently organized convection, originating in the tropics or subtropics and existing for a period of over 24 hours. Disturbances are characteristically approximately 200-600 km in diameter.

- **A tropical depression** is a tropical cyclone displaying a closed circulation pattern, in which the maximum sustained wind speed reaches up to but does not exceed 17m/s.

- **A tropical Sstorm** is also a tropical cyclone, but with faster winds speeds. Tropical cyclones are classified as tropical storms when the maximum sustained surface wind speed ranges between 17.5 and 32.5 m/s.

- **A hurricane** occurs when a tropical cyclone reaches or exceeds maximum sustained wind speeds of 33 m/s. Hurricanes can be further classified using the Saffir- Simpson Hurricane Wind Scale, which is a 1 -5 classification of the hurricanes intensity at the indicated time.

Occurrence of hurricanes and tropical cyclones: These hazards affect certain areas at different times during the season. The official Atlantic Hurricane Season extends from June 1st to November 30th, but Trinidad is most likely to be affected during the period of August to September.

Difference between hurricane watch and hurricane warning

- **A hurricane watch** is issued when hurricane conditions (maximum sustained winds of 33m/s or higher) are possible within a specified area. A watch is typically issued 48 hours prior to the anticipated hazard onset.

- **A hurricane warning** is issued when hurricane conditions are expected within a specified area. A warning is typically issued 36 hours prior to the anticipated hazard onset.

Vulnerable to Hurricanes and Tropical Cyclones

Trinidad and Tobago is located in the extreme south of the Caribbean, a position, which implies that the country is safe from the impacts of hurricanes and tropical cyclones. This however is not true. While in Trinidad and Tobago the risk to hurricanes is minimal when comparison to other islands in the Caribbean, the potential to be hit by hurricanes still remain, as they have in the past.

Vulnerable

Both the population and the natural and built environment are vulnerable to the impacts of hurricanes and Tropical Storms. Vulnerability depends on:

- Low lying areas: Intense rainfall associated with hurricanes and tropical cyclones can rigger flooding as a secondary hazard.
- Poorly constructed buildings: such as Squatter settlements. Strong winds are another characteristic of hurricanes and tropical cyclones and at times can become powerful enough to damage roofs and infrastructure such as electricity and telecommunication lines.
- Unstable Land: Intense rainfall can also trigger landslides as secondary hazards, this can affected settlements on slopes or on unstable soils.

Prevention of impacts of hurricanes and tropical cyclones

Adequate preparation can reduce and even prevent the impacts of hurricanes and tropical cyclones. Here are a few things you can do:

Prior to a hurricane

- Trim trees growing near electrical wires and telephone lines.
- Secure windows and doors; if possible acquire hurricane straps for the roof.
- Cleans drains and gutters regularly
- Avoid dumping garbage in rivers
- Prepare an emergency kit containing first aid items , battery operated radio, food and water for at least seven days, small tools such as plastic sheeting, gloves, flashlights etc

- » Secure important documents such as passports, birth certificates, marriages certificates, exam cards etc. These should be stored ideally in a water proof bag at a safe location.

- » Identify an alternative location for temporary stay in the event that evacuation is necessary, such as a family member or close friend. If no such place is found, then the nearest possible shelter location should be identified.

- » Secure household items which may fall and break causing damage such as glass items and other sharp or heavy objects.

During a hurricane

- » Remain Calm
- » Stay indoors away from windows
- » Keep updated on the situation via radio/television/ internet
- » After the Hurricane
- » Ensure all members of your party (family) are accounted for; if someone is missing immediately contact the relevant authorities.
- » Boil water unless you are told it's safe.
- » Stay clear of downed electricity lines, report to the relevant authorities
- » Avoid /limit contact with flood waters
- » Lend assistance to injured or special population such as children, elderly or differently able.

Managing storm damaged woods

Strong winds, ice, snow and tornadoes are natural occurrences in Wisconsin forests. When severe, storms can cause extensive damage to forests by uprooting, wounding, bending and breaking trees. Storm damage management should involve a quick assessment to determine the extent of the damage, the need and potential for salvage, and woodland management efforts to return the woodland to a productive status. When evaluating woodland damage, be safe. Watch for hanging branches or broken limbs which may fall when in the woods. Wear a hard-hat when working and stay out of the woods when windy. Broken branches are easier to see when the leaves have fallen. Unless you are experienced with chainsaws, do not attempt to fell storm damaged trees yourself.

Assessment of damage area

Consider mapping the damaged area. Walk your property and note the extent of the damage on your maps or photos. Draw boundaries to help determine the size of the area impacted. Note species, size, type of damage, quality of trees, etc. Types of damage are:

Breakage: This is the most common type of storm damage. Its impact depends on the degree and pattern of damage as well as the species involved. Trees with less than 50% crown (branches and leaves) loss will most likely recover; trees with more than 75% crown top loss are likely to die and be a greater risk for both insects and diseases; trees with 50% to 75% crown loss should be maintained but may develop stain and decay loss to the wood and should reevaluated in 4 to 6 years. Trees with structural damage to the main trunk, including splits and fractures, should be removed.

Uprooted: Trees that are completely uprooted will be degraded quickly by insects, stain and fungi. Trees which are partially uprooted and their crowns are still green with leaves will last longer.

Major wounds: Storms often cause major wounding. If these wounds are more than two inches deep and affect more than 25% of the circumference of the trees trunk, they are major sites for stain and decay and should be salvaged. Smaller wounds do not represent major damage to trees.

Bent: Trees are often bent over after major storms. These trees often have cracks or fractures in the trunk and major limbs. If the cracks or fractures extend down more than 25% of the trees trunk, harvesting is recommended. Trees less than 15 feet tall with small cracks will usually straighten and recover.

Salvage potential

The potential for salvaging the damaged woodland parallels the marketability of non damaged forests in Wisconsin. Tree value is determined by species, size quality. Generally, trees less than 10 inches in diameter have no saw timber value, and instead are utilized as pulpwood or other products. Large trees are more valuable than small trees and trees with fewer defects are more valuable than trees with more defects. If salvageable trees are still standing and have branches with green leaves, they will not degrade significantly in the next 6 to 12 months. Trees which have blown over or are not standing should be salvaged before next spring. Wood on the ground begins to degrade immediately; there are some differences in species as to how fast stain and decay enter the wood. Loggers are not interested in removing small number of trees because of the costs of bringing in equipment and labor. There needs to be sufficient quantity as well as quality of timber to attract buyers. If less than 50 trees are damaged, consider salvaging for your own use by transporting to a sawmill for custom sawing or using a small portable sawmill. For larger number of trees, consider working with a consulting forester to mark the salvage as a timber sale and seek competitive bids for optimum prices.

There may be some affects on log markets in Wisconsin due to these storms. Normally, prices go down as supplies increase, although this should be mostly a temporary price trend. It is wise to not rush into salvage, but talk with neighbors, foresters and loggers about timber prices.

Woodland Management

Don't abandoned good forestry practices when working with damaged woodlands. Don't remove too many trees; keep the stocking up in stands even if this means leaving some damaged trees to occupy the sites. Storms often cause damage in small areas or patches. If damage is severe in small patches, consider small group clear cutting to both remove the damage vegetation and provide sunlight for seedlings to grow and reoccupy the site. This may reduce future storm damage by removing exposed trees susceptible to blow down. Work with your forester to evaluate reproduction needs before harvesting. Initially, work first in saw timber stands for salvaging; often young pole sized or smaller stands will recover better from storm damage. When salvaging tree avoid causing additional damage during logging. The storms may provide some opportunity to improve wildlife habitat in woodlands. Small clearings may benefit some species. Trees with broken tops and little economic value will probably develop into good den or snag trees during the next few years. Each stand is unique and each landowner has special goals and objectives for their forest property. How the woodland responds and recovers is dependent on both its natural ecological characteristics and how the damage is handled. Remember that Wisconsin's woodlands will respond with woody vegetation to fill vacant positions caused by the storm and will remain a woodland; woodlands are resilient and recover from damage through additional growth and reproduction. Through proper harvesting and removal of damaged trees, the speed of recovery will be increased and how the woodland responds can be partially directed.

Managing the effects of tropical storms

Reducing the effects of tropical storms includes:

- » Studying tropical storms once they form
- » Providing an early warning system

» Long-term planning in areas prone to tropical storms

Storm Water Management

Storm water is rainwater and melted snow that runs off streets, lawns, and other sites. When storm water is absorbed into the ground, it is filtered and ultimately replenishes aquifers or flows into streams and rivers. In developed areas, however, impervious surfaces such as pavement and roofs prevent precipitation from naturally soaking into the ground. Instead, the water runs rapidly into storm drains, sewer systems, and drainage ditches and can cause:

- » Downstream flooding
- » Stream bank erosion
- » Increased turbidity (muddiness created by stirred up sediment) from erosion
- » Habitat destruction
- » Changes in the stream flow hydrograph (a graph that displays the flow rate of a stream over a period of time)
- » Combined sewer overflows
- » Infrastructure damage
- » Contaminated streams, rivers, and coastal water

Manage of storm water

Traditional storm water management design has been focused on collecting storm water in piped networks and transporting it off site as quickly as possible, either directly to a stream or river, to a large storm water management facility (basin), or to a combined sewer system flowing to a wastewater treatment plant.

Low impact development (LID) and wet weather green infrastructure address these concerns through a variety of techniques, including strategic site design, measures to control the sources of runoff, and thoughtful landscape planning.

LID aims to restore natural watershed functions through small-scale treatment at the source of runoff. The goal is to design a hydrologically functional site that mimics predevelopment conditions.

Wet weather green infrastructure encompasses approaches and technologies to infiltrate, evapotranspire, capture, and reuse stormwater to maintain or restore natural hydrologies.

Storm water management goals and strategies

Environmental Protection Agency's (EPA's) goal is to transform its conventional building storm water management approach by using low impact development (LID), wet weather green infrastructure, and best practices that will decrease the storm water impact of its facilities on the environment. To accomplish this and meet federal storm water management requirements, EPA has developed two broad strategies, one for new construction and major renovations, and the other for existing facilities.

Strategy for new construction and major renovations

In new construction and major renovations, EPA will meet the requirements contained in the Technical Guidance on Implementing the Storm water Runoff Requirements for Federal Projects.

In 2010, EPA adopted this guidance for all its new construction and renovation projects and incorporated the requirements into its green building standards, including the Architecture and Engineering (A&E) Guidelines, Best Practice (Environmental) Lease Provisions, and Green Check. In new construction and major renovations, low impact development (LID)/wet weather green infrastructure can often be implemented at comparable or lower costs to conventional storm water management (curb and gutter and pipe and pond systems).

Strategy for existing facilities

EPA strives to reduce storm water impacts and install retrofits that use LID/wet weather green infrastructure approaches wherever opportunities exist. The Agency gives the highest priority for retrofits to existing sites without storm water management. Its next priority is to retrofit existing sites that use conventional or non-LID/wet weather green infrastructure storm water management approaches.

The Agency is currently compiling technical data on its inventory of storm water management practices at its facilities. This inventory will help quantify the benefits of existing storm water practices and identify cost-effective opportunities for future green infrastructure retrofits.

Federal storm water management requirements

The following federal requirements drive EPA and other federal agencies to practice storm water management and implement sustainable storm water practices:

- » Energy Independence and Security Act of 2007 (EISA)
- » Technical Guidance on Implementing the Storm water Runoff Requirements for Federal Projects under Section 438 of the Energy Independence and Security Act of 2007 and Section 438
- » Executive Order 13514
- » Guiding Principles for Federal Leadership in High Performance and Sustainable Buildings

Energy Independence and Security Act of 2007 (EISA)

Section 438 of EISA instructs federal agencies to "use site planning, design, construction, and maintenance strategies for the property to maintain or restore, to the maximum extent technically feasible, the predevelopment hydrology of the property with regard to the temperature, rate," for any project with a footprint that exceeds 5,000 square feet.

Technical guidance on implementing the storm water runoff requirements for Federal Projects under Section 438 of EISA

EPA's Office of Water (OW) coordinated the development of these federal guidelines and issued Technical Guidance on Implementing the Storm water Runoff Requirements for Federal Projects under Section 438 of the Energy Independence and Security Act on December 4, 2009.

Executive Order 13514: Federal Leadership in Environmental, Energy, and Economic Performance

Executive Order (EO) 13514 reiterates the requirement of EO 13423 that federal agencies implement the Guiding Principles for Federal Leadership in High Performance and Sustainable Buildings (Guiding Principles) in at least 15 percent of their existing buildings by the end of FY 2015. Section 2(g) of EO 13514 calls for agencies to "implement high performance sustainable federal building design, construction, operation and management, maintenance, and deconstruction." It instructs federal agencies to develop and implement strategies:

» Ensuring that all new construction, major renovation, or repair and alteration of Federal buildings comply with the Guiding Principles.

» Ensuring that at least 15 percent of the agency's existing buildings (above 5,000 gross square feet) and building leases (above 5,000 gross square feet) meet the Guiding Principles by fiscal year 2015 and that the agency makes annual progress toward 100-percent conformance with the Guiding Principles for its building inventory.

Guiding principles for federal leadership in high performance and sustainable buildings

The Guiding Principles for Federal Leadership in High Performance and Sustainable Buildings (Guiding Principles), which incorporate requirements from EISA, require agencies to employ design and construction strategies that reduce storm water runoff, polluted site water runoff, and the use of potable water for irrigation. They promote the use of decentralized stormwater management design strategies to maintain or restore site hydrology to pre-development conditions and promote water-efficient landscaping and irrigation strategies.

The *Guiding Principles*, last revised on December 1, 2008, contain two sets of principles: one for new construction and major renovation of buildings, the other for existing buildings. The new building principles focus primarily on design and construction, while the existing building principles emphasize sustainable operations, maintenance, and management.

Guiding Principle III, Protect and Conserve Water, is present in both the new construction and major renovations set of *Guiding Principles* and the existing buildings set of *Guiding Principles*, and in both sets it contains the following language regarding storm water management:

"Employ design and construction strategies that reduce storm water runoff and discharges of polluted water offsite. Per EISA Section 438, to the maximum extent technically feasible, maintain or restore the predevelopment hydrology of the site with regard to temperature, rate, volume, and duration of flow using site planning, design, construction, and maintenance strategies."

Practices for storm water management

EPA facilities draw on the following best practices, also called Integrated Management Practices (IMPs), to design, implement, and evaluate their storm water management efforts.

Best Practices

- Bioretention cells
- Curb and gutter elimination
- Grassed swales
- Green parking design
- Infiltration trenches
- Inlet protection devices
- Permeable pavement
- Permeable pavers
- Rain barrels and cisterns
- Riparian buffers
- Sand and organic filters
- Soil amendments
- Storm water planters
- Tree box filters
- Vegetated filter strips
- Vegetated roofs

Bioretention cells: A bioretention cell or rain garden is a depressed area with porous backfill (material used to refill an excavation) under a vegetated surface. These areas often have an under drain to encourage filtration and infiltration, especially in clayey soils. Bioretention cells provide groundwater recharge, pollutant removal, and runoff detention. Bioretention cells are an effective solution in parking lots or urban areas where green space is limited.

Curb and gutter elimination: Curbs and gutters transport flow as quickly as possible to a stormwater drain without allowing for infiltration or pollutant removal. Eliminating curbs and gutters can increase sheet flow and reduce runoff volumes. Sheet flow, the form runoff takes when it is uniformly dispersed across a surface, can be established and maintained in an area that does not naturally concentrate flow, such as parking lots. Maintaining sheet flow by eliminating curbs and gutters and directing runoff into vegetated swales or bioretention basins helps to prevent erosion and more closely replicate predevelopment hydraulic conditions. A level spreader, which is an outlet designed to convert concentrated runoff to sheet flow and disperse it uniformly across a slope, may also be incorporated to prevent erosion.

Grassed swales: Grassed swales are shallow grass-covered hydraulic conveyance channels that help to slow runoff and facilitate infiltration. The suitability of grassed swales depends on land use, soil type, slope, imperviousness of the contributing watershed, and dimensions and slope of the grassed swale system. In general, grassed swales can be used to manage runoff from drainage areas that are less than 4 hectares (10 acres) in size, with slopes no greater than 5 percent. Use of natural, low-lying areas is encouraged and natural drainage courses should be preserved and utilized

Green parking refers to several techniques that, applied together, reduce the contribution of parking lots to total impervious cover. Green parking lot techniques include: setting maximums for the number of parking lots created; minimizing the dimensions of parking lot spaces; utilizing alternative pavers in overflow parking areas; using bioretention areas to treat stormwater; encouraging shared parking; and providing economic incentives for structured parking.

Infiltration trenches: Infiltration trenches are rock-filled ditches with no outlets. These trenches collect runoff during a storm event and release it into the soil by infiltration (the process through which storm water runoff penetrates into soil from the ground surface). Infiltration trenches may be used in conjunction with another storm water management device, such as a grassed swale, to provide both water quality control and peak flow attenuation. Runoff that contains high levels of sediments or hydrocarbons (for example, oil and grease) that may clog the trench are often pretreated with other techniques such as water quality inlets (series of chambers that promote sedimentation of coarse materials and separation of free oil from storm water), inlet protection devices, grassed swales, and vegetated filter strips.

Inlet protection devices: Inlet protection devices, also known as hydrodynamic separators, are flow-through structures with a settling or separation unit to remove sediments, oil and grease, trash, and other storm water pollutants. This technology may be used as pre-treatment for other storm water management devices. Inlet protection devices are commonly used in potential storm water "hot spots"- areas where higher concentrations of pollutants are more likely to occur, such as gas stations.

Permeable pavement is an alternative to asphalt or concrete surfaces that allows storm water to drain through the porous surface to a stone reservoir underneath. The reservoir temporarily stores surface runoff before infiltrating it into the subsoil. The appearance of the alternative surface is often similar to asphalt or concrete, but it is manufactured without fine materials and instead incorporates void spaces that allow for storage and infiltration. Under drains may also be used below the stone reservoir if soil conditions are not conducive to complete infiltration of runoff.

Permeable pavers promote groundwater recharge. Permeable interlocking concrete pavements (PICP) are concrete block pavers that create voids on the corners of the pavers (pictured to the right). Concrete grid paver (CGP) systems are composed of concrete blocks made porous by eliminating finer particles in the concrete which creates voids inside the blocks; additionally, the blocks are arranged to create voids between blocks. Plastic turf reinforcing grids (PTRG) are plastic grids that add structural support to the topsoil and reduce compaction to maintain permeability. Grass is encouraged to grow in PTRG, so the roots will help improve permeability due to their root channels.

Rain barrels and cisterns harvest rainwater for reuse. Rain barrels are placed outside a building at roof downspouts to store rooftop runoff for later reuse in lawn and garden watering. Cisterns store rainwater in significantly larger volumes in manufactured tanks or underground storage areas. Rainwater collected in cisterns may also be used in non-potable water applications such as toilet flushing. Both cisterns and rain barrels can be implemented without the use of pumping devices by relying on gravity flow instead. Rain barrels and cisterns are low-cost water conservation devices that reduce runoff volume and, for very small storm events, delay and reduce the peak runoff flow rates. Both rain barrels and cisterns can provide a source of chemically untreated "soft water" for gardens and compost, free of most sediment and dissolved salts.

Riparian buffers: A riparian, or forested, buffer is an area along a shoreline, wetland, or stream where development is restricted or prohibited. The primary function of aquatic buffers is to physically protect and separate a stream, lake, or wetland from future disturbance or encroachment. If properly designed, a buffer can provide storm water management and can act as a right-of-way during floods, sustaining the integrity of stream ecosystems and habitats.

Sand and organic filters: Sand and organic filters direct storm water runoff through a sand bed to remove floatables, particulate metals, and pollutants. Sand and organic filters provide water quality treatment, reducing sediment, biochemical oxygen demand (BOD), and fecal coli form bacteria, although dissolved metal and nutrient removal through sand filters is often low. Sand and organic filters are typically used as a component of a treatment train to remove pollution from storm water before discharge to receiving waters, to groundwater, or for collection and reuse. Variations on the traditional surface sand filter (such as the underground sand filter, perimeter sand filter, organic media filter, and multi-chamber treatment train) can be made to fit sand filters into more challenging design sites or to improve pollutant removal.

Soil amendments increase the soil's infiltration capacity and help reduce runoff from the site. They have the added benefit of changing physical, chemical, and biological characteristics so that the soils become more effective at maintaining water quality. Soil amendments, which include both soil conditioners and fertilizers, make the soil more suitable for the growth of plants and increase water retention capabilities. The use of soil amendments is conditional on their compatibility with existing vegetation, particularly native plants.

Storm water planters: Storm water planters are small landscaped storm water treatment devices that can be placed above or below ground and can be designed as infiltration or filtering practices. Storm water planters use soil infiltration and biogeochemical processes to decrease storm water quantity and improve water quality, similar to rain gardens and green roofs but smaller in size—storm water planters are typically a few square feet of surface area compared to hundreds or thousands of square feet for rain gardens and green roofs. Types of storm water planters include contained planters, infiltration planters, and flow-through planters.

Tree box filters: Tree box filters are in-ground containers used to control runoff water quality and provide some detention capacity. Often pre manufactured, tree box filters contain street trees, vegetation, and soil that help filter runoff before it enters a catch basin or is released from the site. Tree box filters can help meet a variety of storm water management goals, satisfy regulatory requirements for new development, protect and restore streams, control combined sewer overflows (CSOs), retrofit existing urban areas, and protect reservoir watersheds. The compact size of tree box filters allows volume and water quality control to be tailored to specific site characteristics. Tree box filters provide the added value of aesthetics while making efficient use of available land for storm water management. Typical landscape plants (for example, shrubs, ornamental grasses, trees and flowers) are an integral part of the bioretention system. Ideally, plants should be selected that can withstand alternating inundation and drought conditions and that do not have invasive root systems, which may reduce the soil's filtering capacity.

Vegetated filter strips: Filter strips are bands of dense vegetation planted downstream of a runoff source. The use of natural or engineered filter strips is limited to gently sloping areas where vegetative cover can be established and channelized flow is not likely to develop. Filter strips are well suited for treating runoff from roads and highways, roof downspouts, very small parking lots, and impervious surfaces. They are also ideal components for the fringe of a stream buffer, or as pretreatment for a structural practice.

Green roofs/Vegetated roofs consist of an impermeable roof membrane overlaid with a lightweight planting mix with a high infiltration rate and vegetated with plants tolerant of heat, drought, and periodic inundations. In addition to reducing runoff volume and frequency and improving runoff water quality, a green roof can reduce the effects of atmospheric pollution, reduce energy costs, and create an attractive environment. They have reduced replacement and maintenance costs and longer life cycles compared to traditional roofs.

EPA's storm water management results

EPA has begun to make significant progress in increasing Agency storm water management efforts and implementing sustainable storm water management, low impact development (LID) practices, and wet weather green infrastructure throughout its facilities. These efforts have resulted in both environmental and economic benefits:

Environmental benefits

- » Controls runoff pollutants (quality) and volume (quantity).
- » Promotes infiltration, groundwater recharge, runoff storage, filtration, and storm water reuse.
- » Minimizes grading and impervious surfaces.
- » Maintains or restores the site's pre-development conditions.
- » Minimizes erosion and protects downstream waterways.

Economic and land value benefits

- » Reduces the need for storm pipes, curbs, gutters, and large ponds which can maximize open space and save on construction costs.
- » Reduces the potential for downstream flooding and combined sewer overflows.
- » Increases real estate value and site aesthetics in some cases.

» Reduces building cooling costs in some cases.

» Reduces storm water fees if the local jurisdiction charges fees based on a site's impervious surface area.

As EPA continues to pursue sustainable storm water management, LID, and wet weather green infrastructure opportunities, the Agency is investigating methods of quantifying the benefits of these opportunities. For example, EPA is working to inventory existing storm water management practices at its facilities, which will allow it to quantify the potential environmental benefits of completed projects.

As a result of its storm water management and LID projects at Headquarters, EPA has received the following awards:

» GSA Real Property Management award for the William Jefferson Clinton Building South Courtyard.

» 2007 Landscape Contractors Association Grand Award for Commercial Landscape Installation for the William Jefferson Clinton Building South Courtyard.

Classification, Depletion, Management Of Natural Resources

NATURAL RESOURCES AND THEIR CLASSIFICATION

Natural resources occur naturally within environments that exist relatively undisturbed by humanity, in a natural form. A natural resource is often characterized by amounts of biodiversity and geo-diversity existent in various ecosystems.

Natural resources are derived from the environment. Some of them are essential for our survival while most are used for satisfying our wants. Natural resources may be further classified in different ways.

Natural resources are materials and components (something that can be used) that can be found within the environment. Every man-made product is composed of natural resources (at its fundamental level). A natural resource may exist as a separate entity such as fresh water, and air, as well as a living organism such as a fish, or it may exist in an alternate form which must be processed to obtain the resource such as metal ores, oil, and most forms of energy.

There is much debate worldwide over natural resource allocations; this is partly due to increasing scarcity (depletion of resources) but also because the exportation of natural resources is the basis for many economies (particularly for developed nations such as Australia).

Some natural resources such as sunlight and air can be found everywhere, and are known as ubiquitous resources. However, most resources only occur in small sporadic areas, and are referred to as localized resources. There are very few resources that are considered inexhaustible (will not run out in foreseeable future) – these are solar radiation, geothermal energy, and air (though access to clean air may not be). The vast majority of resources are exhaustible, which means they have a finite quantity, and can be depleted if managed improperly.

Classification

There are various methods of categorizing natural resources, these include source of origin, stage of development, and by their renewability. These classifications are described below. On the basis of origin, resources may be divided into:

- » **Biotic:** Biotic resources are obtained from the biosphere (living and organic material), such as forests and animals, and the materials that can be obtained from them. Fossil fuels such as coal and petroleum are also included in this category because they are formed from decayed organic matter.

- » **Abiotic:** Abiotic resources are those that come from non-living, non-organic material. Examples of abiotic resources include land, freshwater, air and heavy metals including ores such as gold, iron, copper, silver, etc.

Considering their stage of development, natural resources may be referred to in the following ways:

- » **Potential resources:** Potential resources are those that exist in a region and may be used in the future. For example petroleum occurs with sedimentary rocks in various regions, but until the time it is actually drilled out and put into use, it remains a potential resource.

- » **Actual resources:** Actual resources are those that have been surveyed, their quantity and quality determined and are being used in present times. The development of an actual resource, such as wood processing depends upon the technology available and the cost involved.

- » **Reserve resources:** The part of an actual resource which can be developed profitably in the future is called a reserve resource.

- » **Stock resources:** Stock resources are those that have been surveyed but cannot be used by organisms due to lack of technology. For example: hydrogen.

Renewability is a very popular topic and many natural resources can be categorized as either renewable or non-renewable:

- » **Renewable resources**: Renewable resources can be replenished naturally. Some of these resources, like sunlight, air, wind, etc., are continuously available and their quantity is not noticeably affected by human consumption. Though many renewable resources do not have such a rapid recovery rate, these resources are susceptible to depletion by over-use. Resources from a human use perspective are classified as renewable only so long as the rate of replenishment/recovery exceeds that of the rate of consumption.

» **Non-renewable resources**: Non-renewable resources either form slowly or do not naturally form in the environment. Minerals are the most common resource included in this category. By the human perspective, resources are non-renewable when their rate of consumption exceeds the rate of replenishment/recovery; a good example of this are fossil fuels, which are in this category because their rate of formation is extremely slow (potentially millions of years), meaning they are considered non-renewable. Some resources actually naturally deplete in amount without human interference, the most notable of these being radio-active elements such as uranium, which naturally decay into heavy metals. Of these, the metallic minerals can be re-used by recycling them, but coal and petroleum cannot be recycled.

Extraction of Natural Resources

Resource extraction involves any activity that withdraws resources from nature. This can range in scale from the traditional use of preindustrial societies, to global industry. Extractive industries are, along with agriculture, the basis of the primary sector of the economy. Extraction produces raw material which is then processed to add value. Examples of extractive industries are hunting, trapping, mining, oil and gas drilling, and forestry. Natural resources can add substantial amounts to a country's wealth, however a sudden inflow of money caused by a resource boom can create social problems including inflation harming other industries ("Dutch disease") and corruption, leading to inequality and underdevelopment, this is known as the "resource curse".

Extractive industries represent a large growing activity in many less-developed countries but the wealth generated does not always lead to sustainable and inclusive growth. Extractive industry businesses often are assumed to be interested only in maximizing their short-term value, implying that less-developed countries are vulnerable to powerful corporations. Alternatively, host governments are often assumed to be only maximizing immediate revenue. Researchers argue there are areas of common interest where development goals and business cross. These present opportunities for international governmental agencies to engage with the private sector and host governments through revenue management and expenditure accountability, infrastructure development, employment creation, skills and enterprise development and impacts on children, especially girls and women.

Depletion of Natural Resources

In recent years, the depletion of natural resources has become a major focus of governments and organizations. This is evident in the UN's Agenda 21 Section Two, which outlines the necessary steps to be taken by countries to sustain their natural resources. The depletion of natural resources is considered to be a sustainable development issue.

The term sustainable development has many interpretations, most notably the Brundtland Commission's 'to ensure that it meets the needs of the present without compromising the ability of future generations to meet their own needs', however in broad terms it is balancing the needs of the planet's people and species now and in the future. In regards to natural resources, depletion is of concern for sustainable development as it has the ability to degrade current environments and potential to impact the needs of future generations.

The conservation of natural resources is the fundamental problem. Unless it solves that problem, it will avail little to solve all others.

Depletion of natural resources is associated with social inequity. Considering most biodiversity are located in developing countries, depletion of this resource could result in losses of ecosystem services for these countries. Some view this depletion as a major source of social unrest and conflicts in developing nations.

At present, with it being the year of the forest, there is particular concern for rainforest regions which hold most of the Earth's biodiversity. According to Nelson deforestation and degradation affect 8.5 per cent of the world's forests with 30 per cent of the Earth's surface already cropped. If we consider that 80 per cent of people rely on medicines obtained from plants and ¾ of the world's prescription medicines have ingredients taken from plants, loss of the world's rainforests could result in a loss of finding more potential life saving medicines.

The depletion of natural resources is caused by 'direct drivers of change, such as Mining, petroleum extraction, fishing and forestry as well as 'indirect drivers of change' such as demography, economy, society, politics and technology. The current practice of Agriculture is another factor causing depletion of natural resources. For example, the depletion of nutrients in the soil might be due to excessive use of nitrogen and desertification. The depletion of natural resources is a continuing concern for society. This is seen in the cited quote given by

Theodore Roosevelt, a well-known conservationist and former United States president, was opposed to unregulated natural resource extraction.

Natural Resources Management

Natural resources management is a discipline in the management of natural resources such as land, water, soil, plants and animals, with a particular focus on how management affects the quality of life for both present and future generations.

Management of natural resources involves identifying who has the right to use the resources and who does not for defining the boundaries of the resource. The resources are managed by the users according to the rules governing of when and how the resource is used depending on local condition.

A successful management of natural resources should engage the community because of the nature of the shared resources the individuals who are affected by the rules can participate in setting or changing them. The users have the rights to devise their own management institutions and plans under the recognition by the government. The right to resources includes land, water, fisheries and pastoral rights.[The users or parties accountable to the users have to actively monitor and ensure the utilisation of the resource compliance with the rules and to impose penalty on those peoples who violates the rules. These conflicts are resolved in a quick and low cost manner by the local institution according to the seriousness and context of the offence. The global science-based platform to discuss natural resources management is the World Resources Forum, based in Switzerland.

As discussed above, our natural resources viz., water, soil, climate and biodiversity are under serious threat. Rational and scientific use of these limited resources will largely determine the prospectus of agricultural production in the 21st century. This is the right time to introspect the issues concerning their degradation, depletion and pollution. Future strategies must focus on up gradation of soil and water quality and their use efficiency.

Water management

Water will be a major constraint in food production - both in quantity and quality particularly in rainfed regions. The problem of ground water depletion is becoming serious in the states of Punjab, Haryana, Gujarat, Rajasthan and Tamil Nadu. For example, during 1989 to 1995, the number of blocks having dark and over-exploited ground water increased from 3 to 14 per cent in Gujarat, from 33 to 47 percent in Haryana. The options available to check ground water decline include: recharging of the ground water through conservation of precious rainwater during monsoon season, diversification from high water demanding crops to less water requiring crops, increasing water use efficiency through agronomic manipulations and efficient tillage practices.

Future gains in agricultural productivity of the country shall be critically determined by integrated development and utilization of surface and groundwater resources. Of 400 million ha-m of annual precipitation, only 29 per cent is harnessed in the country. About 92 million ha-m of surplus monsoon runoff is lost to the sea and could be stored in sub-surface aquifers for augmenting the water resources. The indiscriminate use of canal water is leading to water is logging and secondary salinization in the major irrigation commands of the country. The increased groundwater extraction has resulted in a decline in water table at an alarming rate, putting an additional burden on farmers in terms of equipment and energy etc. In north-west India alone, about one million centrifugal pumps require replacement with submersible ones at a staggering cost of about Rs. 7,000 crores.

It is estimated that even after achieving the full irrigation potential, nearly 50 per cent of the total cultivated area will remain rainfed and important source of livelihood. Therefore, integrated and holistic development of rainfed areas within the participatory perspective of watershed management programmes will constitute one of the key elements of future agricultural production in the country. We need to develop model watersheds in different agro-ecological regions of the country ensuring full participation of local people to demonstrate their usefulness inconserving natural sources and ameliorating the socio-economic conditions of local populace. Conjunctive use of canal and groundwater is essential to save canal water for recharging of ground aquifers. The creation of a combined canal-tank system of irrigation could have multiple productions of fish, poultry, piggery and other small enterprises. The priority needs to be given to the following issues in the future.

- » In-situ and ex-situ conservation of rainwater and its efficient recycling in rainfed areas

- » Multiple use of water for increasing water productivity

- » Conjunctive use of rain-, surface- and ground-water for maintaining sustainable hydrologic regime

- » Increasing water use efficiency through efficient utilization of available irritation water in dry areas through large-scale promotion of microirrigation techniques such as drip and sprinklers

- » Ground water recharge and management
- » Use of poor quality waters including industrial effluents and sewage waters

Water Conservation

Water conservation encompasses the policies, strategies and activities to manage fresh water as a sustainable resource to protect the water environment and to meet current and future human demand. Population, household size and growth and affluence all affect how much water is used. Factors such as climate change will increase pressure on natural water resources especially in manufacturing and agricultural irrigation.

Aims

The aims of water conservation efforts include as follows:

To ensure availability for future generations, the withdrawal of fresh water from an ecosystem should not exceed its natural replacement rate.

- » Energy conservation. Water pumping, delivery and waste water treatment facilities consume a significant amount of energy. In some regions of the world over 15% of total electricity consumption is devoted to water management.
- » Habitat conservation. Minimizing human water use helps to preserve fresh water habitats for local wildlife and migrating waterfowl, as well as reducing the need to build new dams and other water diversion infrastructures.

Strategies

In implementing water conservation principles there are a number of key activities that may be beneficial.

1. Any beneficial reduction in water loss, use and waste
2. Avoiding any damage to water quality.
3. Improving water management practices that reduce or enhance the beneficial use of water.

Social solutions

Water conservation programs are typically initiated at the local level, by either municipal water utilities or regional governments. Common strategies include public outreach campaigns, tiered water rates (charging progressively higher prices as water use increases), or restrictions on outdoor water use such as lawn watering and car washing. Cities in dry climates often require or encourage the installation of xeriscaping or natural landscaping in new homes to reduce outdoor water usage.

One fundamental conservation goal is universal metering. The prevalence of residential water metering varies significantly worldwide. Recent studies have estimated that water supplies are metered in less than 30 per cent of UK households, and about 61 per cent of urban Canadian homes (as of 2001). Although individual water meters have often been considered impractical in homes with private wells or in multifamily buildings, the U.S. Environmental Protection Agency estimates that metering alone can reduce consumption by 20 to 40 percent. In addition to raising consumer awareness of their water use, metering is also an important way to identify and localize water leakage. Water metering would benefit society in the long run it is proven that water metering increases the efficiency of the entire water system, as well as help unnecessary expenses for individuals for years to come. One would be unable to waste water unless they are willing to pay the extra charges, this way the water department would be able to monitor water usage by public, domestic and manufacturing services.

Some researchers have suggested that water conservation efforts should be primarily directed at farmers, in light of the fact that crop irrigation accounts for 70 per cent of the world's fresh water use. The agricultural sector of most countries is important both economically and politically, and water subsidies are common. Conservation advocates have urged removal of all subsidies to force farmers to grow more water-efficient crops and adopt less wasteful irrigation techniques.

New technology poses a few new options for consumers, features such and full flush and half flush when using a toilet are trying to make a difference in water consumption and waste. Also available in our modern world is shower heads that help reduce wasting water, old shower heads are said to use 5-10 gallons per minute. All new fixtures available are said to use 2.5 gallons per minute and offer equal water coverage.

Household applications

Contract to popular view, experts suggest the most efficient way is replacing toilets and retrofitting washers.

Water-saving technology for the home includes

- Low-flow shower heads sometimes called energy-efficient shower heads as they also use less energy,

- Low-flush toilets and composting toilets. These have a dramatic impact in the developed world, as conventional Western toilets use large volumes of water.

- Dual flush toilets created by Caroma include two buttons or handles to flush different levels of water. Dual flush toilets use up to 67% less water than conventional toilets.

Saline water (sea water) or rain water can be used for flushing toilets.

- Faucet aerators, which break water flow into fine droplets to maintain "wetting effectiveness" while using less water. An additional benefit is that they reduce splashing while washing hands and dishes.

- Raw water flushing where toilets use sea water or non-purified water

- Wastewater reuse or recycling systems, allowing:
  - Reuse of graywater for flushing toilets or watering gardens
  - Recycling of wastewater through purification at a water treatment plant.

- Rainwater harvesting

- High-efficiency clothes washers

- Weather-based irrigation controllers

- Garden hose nozzles that shut off water when it is not being used, instead of letting a hose run.

- Using low flow taps in wash basins

- Swimming pool covers that reduce evaporation and can warm pool water to reduce water, energy and chemical costs.

» Automatic faucet is a water conservation faucet that eliminates water waste at the faucet. It automates the use of faucets without the use of hands.

Commercial applications

Many water-saving devices (such as low-flush toilets) that are useful in homes can also be useful for business water saving. Other water-saving technology for businesses includes:

- » Waterless urinals
- » Waterless car washes
- » Infrared or foot-operated taps, which can save water by using short bursts of water for rinsing in a kitchen or bathroom
- » Pressurized water brooms, which can be used instead of a hose to clean sidewalks
- » X-ray film processor re-circulation systems
- » Cooling tower conductivity controllers
- » Water-saving steam sterilizers, for use in hospitals and health care facilities
- » Rain water harvesting
- » Water to Water heat exchangers.

Agricultural applications

For crop irrigation, optimal water efficiency means minimizing losses due to evaporation, runoff or subsurface drainage while maximizing production. An evaporation pan in combination with specific crop correction factors can be used to determine how much water is needed to satisfy plant requirements. Flood irrigation, the oldest and most common type, is often very uneven in distribution, as parts of a field may receive excess water in order to deliver sufficient quantities to other parts. Overhead irrigation, using center-pivot or lateral-moving sprinklers, has the potential for a much more equal and controlled distribution pattern. Drip irrigation is the most expensive and least-used type, but offers the ability to deliver water to plant roots with minimal losses. However, drip irrigation is increasingly affordable, especially for the home gardener and in light of rising water rates. There are also cheap effective methods similar to drip irrigation such as the use of soaking hoses that can even be submerged in the growing medium to eliminate evaporation.

As changing irrigation systems can be a costly undertaking, conservation efforts often concentrate on maximizing the efficiency of the existing system. This may include chiseling compacted soils, creating furrow dikes to prevent runoff, and using soil moisture and rainfall sensors to optimize irrigation schedules. Usually large gains in efficiency are possible through measurement and more effective management of the existing irrigation system.

Minimum water network target and design

The cost effective minimum water network is a holistic framework/guide for water conservation that helps in determining the minimum amount of freshwater and wastewater target for an industrial or urban system based on the water management hierarchy i.e. it considers all conceivable methods to save water. The technique ensures that the designer desired payback period is satisfied using Systematic Hierarchical Approach for Resilient Process Screening (SHARPS) technique.

Water conservation methods

Our ancient religious texts and epics give a good insight into the water storage and conservation systems that prevailed in those days. Over the years rising populations, growing industrialization, and expanding agriculture have pushed up the demand for water. Efforts have been made to collect water by building dams and reservoirs and digging wells; some countries have also tried to recycle and desalinate (remove salts) water. Water conservation has become the need of the day. The idea of ground water recharging by harvesting rainwater is gaining importance in many cities. In the forests, water seeps gently into the ground as vegetation breaks the fall. This groundwater in turn feeds wells, lakes, and rivers. Protecting forests means protecting water 'catchments'. In ancient India, people believed that forests were the 'mothers' of rivers and worshipped the sources of these water bodies.

Some ancient Indian methods of water conservation

The Indus Valley Civilization, that flourished along the banks of the river Indus and other parts of western and northern India about 5,000 years ago, had one of the most sophisticated urban water supply and sewage systems in the world. The fact that the people were well acquainted with hygiene can be seen from the covered drains running beneath the streets of the ruins at both Mohenjodaro and Harappa. Another very good example is the well-planned city of Dholavira, on Khadir Bet, a low plateau in the Rann in Gujarat. One of the oldest water harvesting systems is found about 130 km from Pune along Naneghat in the Western Ghats. A large number of tanks were cut in the rocks to provide drinking water to tradesmen who used to travel along this ancient trade route. Each fort in the area had its own water harvesting and storage system in the form of rock-cut cisterns, ponds, tanks and wells that are still in use today. A large number of forts like Raigad had tanks that supplied water.

» In ancient times, houses in parts of western Rajasthan were built so that each had a rooftop water harvesting system. Rainwater from these rooftops was directed into underground tanks. This system can be seen even today in all the forts, palaces and houses of the region.

» Underground baked earthen pipes and tunnels to maintain the flow of water and to transport it to distant places, are still functional at Burhanpur in Madhya Pradesh, Golkunda and Bijapur in Karnataka, and Aurangabad in Maharashtra.

Rainwater harvesting

In urban areas, the construction of houses, footpaths and roads has left little exposed earth for water to soak in. In parts of the rural areas of India, floodwater quickly flows to the rivers, which then dry up soon after the rains stop. If this water can be held back, it can seep into the ground and recharge the groundwater supply.

Town planners and civic authority in many cities in India are introducing bylaws making rainwater harvesting compulsory in all new structures. No water or sewage connection would be given if a new building did not have provisions for rainwater harvesting. Such rules should also be implemented in all the other cities to ensure a rise in the groundwater level. Realizing the importance of recharging groundwater, the CGWB (Central Ground Water Board) is taking steps to encourage it through rainwater harvesting in the capital and elsewhere. A number of government buildings have been asked to go in for water harvesting in Delhi and other cities of India.

All you need for a water harvesting system is rain, and a place to collect it! Typically, rain is collected on rooftops and other surfaces, and the water is carried down to where it can be used immediately or stored. You can direct water run-off from this surface to plants, trees or lawns or even to the aquifer.

Some of the benefits of rainwater harvesting are as follows:

» Increases water availability

» Checks the declining water table

- » Environmentally friendly
- » Improves the quality of groundwater through the dilution of fluoride, nitrate, and salinity
- » Prevents soil erosion and flooding especially in urban areas

Rainwater Harvesting: A Success Story

Once Cherrapunji was famous because it received the largest volume of rainfall in the world it still does but ironically, experiences acute water shortages. This is mainly the result of extensive deforestation and because proper methods of conserving rainwater are not used. There has been extensive soil erosion and often, despite the heavy rainfall and its location in the green hills of Meghalaya, one can see stretches of hillside devoid of trees and greenery. People have to walk long distances to collect water.

In the area surrounding the River Ruparel in Rajasthan, the story is different - this is an example of proper water conservation. The site does not receive even half the rainfall received by Cherrapunji, but proper management and conservation have meant that more water is available than in Cherrapunji.

The water level in the river began declining due to extensive deforestation and agricultural activities along the banks and, by the 1980s, a drought-like situation began to spread. Under the guidance of some NGOs (non-government organizations), the women living in the area were encouraged to take the initiative in building johads (round ponds) and dams to hold back rainwater. Gradually, water began coming back as proper methods of conserving and harvesting rainwater were followed. The revival of the river has transformed the ecology of the place and the lives of the people living along its banks. Their relationship with their natural environment has been strengthened. It has proved that humankind is not the master of the environment, but a part of it. If human beings put in an effort, the damage caused by us can be undone.

Agriculture

Conservation of water in the agricultural sector is essential since water is necessary for the growth of plants and crops. A depleting waters table and a rise in salinity due to overuse of chemical fertilizers and pesticides has made matters serious. Various methods of water harvesting and recharging have been and are being applied all over the world to tackle the problem. In areas where rainfall is low and water is scarce, the local people have used simple techniques that are suited to their region and reduce the demand for water.

In India's arid and semi-arid areas, the 'tank' system is traditionally the backbone of agricultural production. Tanks are constructed either by bunding or by excavating the ground and collecting rainwater.

Rajasthan, located in the Great Indian Desert, receives hardly any rainfall, but people have adapted to the harsh conditions by collecting whatever rain falls. Large bunds to create reservoirs known as khadin, dams called johads, tanks, and other methods were applied to check water flow and accumulate run-off. At the end of the monsoon season, water from these structures was used to cultivate crops. Similar systems were developed in other parts of the country. These are known by various local names ¾ jal talais in Uttar Pradesh, the haveli system in Madhya Pradesh, ahar in Bihar, and so on.

Reducing water demand

Simple techniques can be used to reduce the demand for water. The underlying principle is that only part of the rainfall or irrigation water is taken up by plants, the rest percolates into the deep groundwater, or is lost by evaporation from the surface. Therefore, by improving the efficiency of water use, and by reducing its loss due to evaporation, we can reduce water demand. There are numerous methods to reduce such losses and to improve soil moisture. Some of them are listed below:

- » Mulching, i.e., the application of organic or inorganic material such as plant debris, compost, etc., slows down the surface run-off, improves the soil moisture, reduces evaporation losses and improves soil fertility.

- » Soil covered by crops, slows down run-off and minimizes evaporation losses. Hence, fields should not be left bare for long periods of time.

- » Ploughing helps to move the soil around. As a consequence it retains more water thereby reducing evaporation.

- » Shelter belts of trees and bushes along the edge of agricultural fields slow down the wind speed and reduce evaporation and erosion.

- » Planting of trees, grass, and bushes breaks the force of rain and helps rainwater penetrate the soil.

- » Fog and dew contain substantial amounts of water that can be used directly by adapted plant species. Artificial surfaces such as netting-surfaced traps or polyethylene sheets can be exposed to fog and dew. The resulting water can be used for crops.

- Contour farming is adopted in hilly areas and in lowland areas for paddy fields. Farmers recognize the efficiency of contour-based systems for conserving soil and water.

- Salt-resistant varieties of crops have also been developed recently. Because these grow in saline areas, overall agricultural productivity is increased without making additional demands on freshwater sources. Thus, this is a good water conservation strategy.

- Transfer of water from surplus areas to deficit areas by inter-linking water systems through canals, etc.

- Desalination technologies such as distillation, electro-dialysis and reverse osmosis are available.

- Use of efficient watering systems such as drip irrigation and sprinklers will reduce the water consumption by plants.

Solution of water and environment conservation

The most important step in the direction of finding solutions to issues of water and environmental conservation is to change people's attitudes and habits¾this includes each one of us. Conserve water because it is the right thing to do. There can follow some of the simple things that have been listed below and contribute to water conservation:

- Try to do one thing each day that will result in saving water. Don't worry if the savings are minimal¾every drop counts! You can make a difference.

- Remember to use only the amount you actually need.

- Form a group of water-conscious people and encourage your friends and neighbours to be part of this group. Promote water conservation in community newsletters and on bulletin boards. Encourage your friends, neighbours and co-workers to also contribute.

- Encourage your family to keep looking for new ways to conserve water in and around your home.

- Make sure that your home is leak-free. Many homes have leaking pipes that go unnoticed.

- Do not leave the tap running while you are brushing your teeth or soaping your face.

- See that there are no leaks in the toilet tank. You can check this by adding colour to the tank. If there is a leak, colour will appear in the toilet bowl within 30 minutes. (Flush as soon as the test is done, since food colouring may stain the tank.).

» Avoid flushing the toilet unnecessarily. Put a brick or any other device that occupies space to cut down on the amount of water needed for each flush.

» When washing the car, use water from a bucket and not a hosepipe.

» Do not throw away water that has been used for washing vegetables, rice or dals¾use it to water plants or to clean the floors, etc.

» There can store water in a variety of ways. A simple method is to place a drum on a raised platform directly under the rainwater collection source. You can also collect water in a bucket during the rainy season.

The Challenges for agricultural water conservation

New pressures on regional water budgets, particularly in the Western States, have raised important questions concerning the sustainability of water resources for irrigated agriculture:

» Can irrigated agriculture adapt to climate-adjusted water supplies and emerging water demands through conventional means alone (i.e., the adoption of more efficient irrigation technologies, improved water management practices, and/or cropland allocation shifts)

» What changes in water institutions may be needed to complement and drive water conservation policy to more effectively manage increasingly scarce water supplies for agriculture?

» How will these changes impact irrigated agriculture, land and water resource use, the environment, and rural economies?

Efficiency of irrigated agriculture

Prior to the 1970s, gravity-fed furrow and flood irrigation systems were the dominant production systems for irrigated crop agriculture. By 1978, sprinkler irrigation, including center-pivot systems, accounted for about 35 percent of crop irrigation in the Western States. Virtually all of this transition involved adoption of high-pressure sprinkler irrigation. While the center-pivot system improved field irrigation efficiency, water conservation was not the primary motivation for its widespread adoption. Other factors, such as yield enhancement from uniform water application and irrigation's expansion into productive lands that were not suitable for a gravity system due to topography, soils, or distance from traditional riparian boundaries, were the primary drivers behind the early transition from gravity-flow irrigation to center-pivot sprinkler irrigation. The expansion of irrigated crop agriculture, along with increasing water demands from nonagricultural users, significantly intensified the competition for available water resources. At the same time, large-scale water supply enhancement was becoming more restricted for fiscal and environmental reasons. Water conservation in irrigated agriculture became an increasingly important focus of water policy to address water allocation concerns. Various water policy analyses as early as the late 1960s recognized the merits of new regulatory, conservation, and water market policies designed to mitigate water resource allocation conflicts as stated in several studies. At the same time, producers receiving assistance from Federal and State resource conservation programs adopted more efficient irrigation systems to improve irrigation returns, enhance the health and productivity of their resource base, and ensure a more sustainable future for their livelihoods. Adoption of more efficient irrigation systems and water management practices has been examined extensively, particularly within the 17 Western States as reported by some studies.

Between 1984 and 2008 a substantial shift has occurred across the Western States away from gravity irrigation to pressure irrigation systems. In 1984, for example, 71 percent of all crop agricultural water in the West was applied using gravity irrigation systems. By 2008, operators used gravity systems to apply just 48 percent of water for crop production, while pressure irrigation systems accounted for 51.5 per cent or an increase of 23 percentage points from 1984. By 2008, much of the acreage in more efficient pressure irrigation systems included drip, low-pressure sprinkler, or low-energy precision application systems. Improved pressure systems contributed to reduce agricultural water use, as fewer acre-feet were required to irrigate a greater number of acres using these systems. From 1984 to 2008, total irrigated acres across the West increased by 2.1 million acres, while total agricultural water applied declined by nearly 100,000 acre-feet. On farm crop irrigation efficiency is measured as the fraction of applied water beneficially used by the crop, including the quantity of water required for crop ET (consumptive use) and water to leach salts from the crop root zone as found in few studies. Water applied to crops but not used for beneficial purposes is generally regarded as field loss, including water lost through excess evaporation and transpiration by non cropped biomass as well as surface runoff and percolation below the crop-root zone. Some portion of water loss to surface runoff and deep percolation may eventually return to the hydrologic system through surface return flow and/or aquifer recharge and may be available for other economic and environmental uses. Improving on farm irrigation efficiency, while generally recognized as conserving water on the farm, may or may not conserve water within the watershed or river basin. What happens to irrigation water that leaves the farm (i.e., water not beneficially consumed through crop production) and its ultimate impact on local or regional water supplies depend on the many factors that influence the hydrologic water balance for the watershed or river basin. Water balance accounts for where all the water within a watershed (or river basin) comes from and where it goes and is significantly influenced by soils, plants, climate, water source, topography, and hydrologic characteristics both on and off the farm. The literature indicates that if conserved water at the watershed or river basin level is the important policy issue, then water conservation programs that emphasize on farm irrigation efficiency must consider the fate of applied irrigation water in a regional water-balance context. Several studies were discussed situations where improved on farm irrigation efficiency may or may not contribute to watershed or river basin water conservation. The conservation potential of improved irrigation efficiency reflects the share of field losses that are "irrecoverable" for additional uses in the basin (e.g., agricultural and non-agricultural diversions and environmental flows) or otherwise unusable (or detrimental) due to impaired water quality. These studies suggest that improved on farm irrigation can conserve water beyond the farm by:

» Reducing unnecessary evaporation and unwanted transpiration, particularly by weeds and other non cropped biomass within waterlogged parts of irrigated fields, along water supply ditches and canals, and within and along irrigation drainage pathways.

» Improving rainfall use with precipitation capture and moisture retention techniques (e.g., land grading, snow fences, plant-row mulches, and furrow diking techniques).

» Reducing deep percolation water that becomes severely degraded in quality or is uneconomic to recover.

» Reducing field runoff that is lost to the hydrologic system (i.e., runoff water that is not reusable because of salinization or entry to a saline body).

» Reducing crop ET requirements associated with downstream irrigated agriculture (i.e., by reducing saline return flows, which allows downstream irrigators to reduce their salt leaching requirements).

» Reducing normal crop ET associated with crop stress under deficit irrigation (i.e., the irrigator intentionally provides the crop with less than its full ET requirement, resulting in reduced yield but higher net economic returns).

The research also indicates that, in many cases, conserved water to augment water supply in the watershed or river basin may not be the primary policy concern. Water conservation programs may also focus on enhancing the viability and sustainability of the regional agricultural economy, improving the quality and availability of water supplies locally, improving the quality of return flows, and reducing environmental degradation of existing regional supplies. Numerous USGS National Water-Quality Assessment studies have identified irrigated agriculture as a key contributor to many of the Nation's degraded surface water bodies and groundwater aquifers because irrigation often makes heavier use of agricultural chemicals and because excess irrigation increases the hydrologic transport of agricultural chemicals, salts, and other soil-based chemicals potentially detrimental to water-based ecosystems. Thus, without adding to regional water supplies, water conservation programs encouraging improved on farm irrigation efficiency may purposefully serve local and regional economic, water-quality, and environmental goals that contribute to farmer and societal welfare, improve and wildlife habitat, and reduce ecosystem and human health risks associated with environmental pollution.

According to the National Research Council report, Toward Sustainable Agricultural Systems in the 21st Century and the recent USDA Research, Education, and Economics Action Plan achieving a more sustainable future for irrigated agriculture through agricultural water conservation involves three elements:

» Continue to encourage adoption of high-efficiency irrigation application systems.

» Place greater emphasis on adoption of more efficient irrigation production systems (a farming systems approach) that better manage when and how much water is applied at the field level, enhancing producer ability to respond to water shortages as well as promote agricultural water conservation through deficit irrigation while improving farm profits.

» Better integrate on farm water conservation with watershed-level water management mechanisms that help facilitate optimal allocation of limited water supplies among competing demands (e.g., use of conserved water rights, drought-year water banks, water-option markets, contingent water markets, reservoir management, irrigated The box "Irrigation Production Systems and Agricultural Water Conservation" expands upon the concept of accounting for basin-level water balance by considering the fate of farm-level water savings/losses and strategies underway to improve onfarm irrigation water-use efficiency and integrate water conservation and watershed water-management tools.

Watershed-level water management tools can create more efficient water allocations by encouraging water-resource stakeholders to recognize the opportunity value of water across competing uses and by facilitating water transfers through varying degrees of market-based trading and reallocation schemes. USDA presently participates in watershed-scale agricultural water conservation and water-management activities through Federal, State and local partnership agreements established under its Agricultural Water Enhancement Program (AWEP). With the 2008 Food, Conservation, and Energy Act's establishment of AWEP, USDA's water conservation program was effectively extended to embody a focus on farms as well as on a watershed/ regional/institutional conservation. Since 2009, USDA's Natura Resources Conservation Service, (NRCS) has entered into 101 AWEP partnership agreements designed to enhance agricultural water conservation.

Irrigation Production Systems and Agricultural

Water conservation

Agricultural water conservation has long been recognized as key to providing water resources to meet the increasing demands of competing uses. Historically, promoting producer adoption of irrigation technologies that increase farm level irrigation efficiency has been a principal policy focus. Through the use of conservation-incentive programs, Federal and State agencies have funded improvements in irrigation system efficiency to help meet the needs of competing water demands. In recent years, however, the appropriateness of this policy approach has been challenged. Concerns have been raised about the effect of irrigation technology adoption on irrigation consumptive water use and the amount of water actually "conserved."

Researchers believe that public promotion of more efficient irrigation technologies can unintentionally increase irrigated crop consumptive water use at the basin level by encouraging wider adoption of crop irrigation and/or the production of more water-intensive crops. This policy concern suggests that improving irrigation efficiency may not always "conserve" water for off-farm uses. The potential of irrigation efficiency improvements to achieve water savings within a basin depends partly on the nature of irrigation system losses and rates of irrigation return flow to surface streams and aquifers. Potential water savings also depend on whether water-use efficiency gains are offset by increases in crop consumptive water use. Where return flows are high from irrigation systems, real water conservation may require reduced crop consumptive water use within the basin.

Improved technology alone may not be enough. Producer adoption of more efficient irrigation technology may increase agricultural water consumption in several ways

» More efficient irrigation systems allow the producer to reduce the quantity of water applied to a field, often through improved uniformity of field-water distribution and timing of water applications to meet crop growth-stage requirements. These improvements may also result in higher crop yields, which generally increase crop consumptive water use.

» In the absence of defined "conserved" water rights, water savings from irrigation efficiency improvements on one field may be applied to additional crop acreage under irrigation. Unless restricted, "water spreading" over an expanded acreage base generally increases aggregate agricultural water consumption.

» Improved irrigation technologies can alter the economics of irrigation enough to entice producers to adjust traditional cropping patterns, potentially shifting to more water-intensive irrigated crops. In the High Plains, for example, higher yields and reduced irrigation pumping costs with improved irrigation efficiency have prompted a shift from irrigated wheat and sorghum production to increased acreage in irrigated corn.

These types of cropping pattern adjustments may increase aggregate crop consumptive water use. Improved irrigation efficiency can also have off-farm implications. Specifically, upstream water savings may be claimed by downstream (junior water-right) irrigators, increasing basin-wide agricultural consumptive water use. Where seasonal in stream water flows are limiting, increases in crop consumptive water use may adversely impact water allocation objectives for environmental and other purposes at the basin level. Similarly, groundwater savings from reduced irrigation pumping requirements under high-efficiency systems may be offset by expanded aquifer withdrawals for additional irrigated acreage and other purposes off the farm.

Benefits of technological improvements

While the effects of irrigation efficiency on crop consumptive use and net water savings are an important policy concern, investment in more efficient on farm irrigation technologies has additional benefits:

» Improved irrigation technologies are generally productivity enhancing, requiring less land and water inputs for a given level of yield.

» Enhanced irrigation efficiency produces on farm water savings through reduced applications that also reduce farm water costs. Higher productivity and reduced water costs are important to the economic viability of a sustainable irrigated agriculture sector over the long term.

» Improved on farm irrigation efficiency generally results in significant water quality and environmental benefits. Efficient irrigation "production systems" allow producers to improve their nutrient management practices through chemical application efficiencies, reduced soil erosion runoff, improved salinity control, and improved drainage water quality. Improving on farm irrigation efficiency also reduces nutrient loads, pesticides, and trace elements in irrigation runoff to surface waters, as well as leaching of agrichemicals into groundwater supplies, producing off-farm benefits for ecosystem habitats, endangered species recovery, biodiversity, and human health.

Policy to achieve conservation goal

Public water conservation programs that encourage producer adoption of efficient "irrigation production systems," integrated within basin-level institutional water management initiatives (e.g., conserved water-right provisions, groundwater withdrawal restrictions, water banks, and option and contingent water markets), could enhance the potential for real agricultural water conservation. Integrated farm and basin-level institutional conservation initiatives could encourage both irrigators and conservation program managers to consider the alternative opportunity values of water across competing demands, improving allocation of scarce water resources. In doing so, irrigation efficiency improvements, as part of a broader basin-level conservation plan, may be combined with other practices, such as deficit irrigation, acreage idling, and off-farm water transfers, that reduce crop water consumption in water-deficit years while allowing producers to maximize farm income. Through the adoption of highly efficient irrigation production systems, in combination with water allocation frameworks that encourage producers and program managers to jointly consider the opportunity values of water within the basin, the overall efficiency in water allocations can be improved while enhancing real agricultural water conservation.

Facts of water conservation

» Less than 2 per cent of the Earth's water supply is fresh water.

» Of all the earth's water, 97 per cent is salt water found in oceans and seas.

» Only 1 per cent of the earth's water is available for drinking water.

» Two percent is frozen.

- The human body is about 75 per cent water.
- A person can survive about a month without food, but only 5 to 7 days without water.
- Showering and bathing are the largest indoor uses (27%) of water domestically.
- A leaky faucet can waste 100 gallons a day.
- One flush of the toilet uses 6 ½ gallons of water.
- An average bath requires 37 gallons of water.
- An average family of four uses 881 gallons of water per week just by flushing the toilet.
- The average 5-minute shower takes 15-25 gallons of water--around
- 40 gallons are used in 10 minutes.
- Take short showers instead of baths.
- A full bathtub requires about 36 gallons of water.
- You use about 5 gallons of water if you leave the water running while brushing your teeth.
- If you water your grass and trees more heavily, but less often, this saves water and builds stronger roots.
- Each person needs to drink about 2 ½ quarts (80 ounces) of water every day.
- Water your lawn only when it needs it. If you step on the grass and it springs back up when you move, it doesn't need water. If it stays flat, it does need water.
- Run your dishwasher and washing machine only when they are full.
- When washing a car, use soap and water from a bucket. Use a hose with a shut-off nozzle for rinsing.
- Never put water down the drain when there may be another use for it such as watering a plant or garden, or cleaning.
- Avoid flushing the toilet unnecessarily.
- Dispose of tissues, insects and other such waste in the trash rather than the toilet.

- » When washing dishes by hand, fill one sink or basin with soapy water.
- » Quickly rinse under a slow-moving stream from the faucet.
- » An automatic dishwasher uses 9 to 12 gallons of water while hand washing dishes can use up to 20 gallons.
- » Store drinking water in the refrigerator rather than letting the tap run every time you want a cool glass of water.
- » Water lawns during the early morning hours, or evening when temperatures and wind speed are the lowest.
- » This reduces losses from evaporation.
- » Do not hose down your driveway or sidewalk.
- » Use a broom to clean leaves and other debris from these areas.
- » Using a hose to clean driveway wastes hundred of gallons of water.
- » Don't leave the water running when brushing your teeth or shaving. Get in the habit of turning off the water when it's not being used.
- » Use of bowl of water to clean fruits & vegetables rather than running water over them. You can reuse this for your house plants.
- » Public water suppliers process 38 billion gallons of water per day or domestic and public use.
- » Approximately 1 million miles of pipelines and aqueducts carry waterin the U.S. & Canada. That's enough pipes to circle the earth 40 times.
- » About 800,000 water wells are drilled each year in the United States for domestic, farming, commercial, and water testing purposes.
- » More than 13 million households get their water from their own private wells and are responsible for treating and pumping the water themselves.
- » Industries released 197 million pounds of toxic chemicals into waterways in 1990.
- » You can refill an 8-oz glass of water approximately 15,000 times for the same cost as a six-pack of soda pop.

- » A dairy cow must drink four gallons of water to produce one gallon of milk.

- » 300 million gallons of water are needed to produce a single day's supply of U.S. newsprint.

- » One inch of rainfall drops 7,000 gallons or nearly 30 tons of water on a 60' by 180' piece of land.

Soil Management

Issues related to soil

Soil is the second important natural resource, which has been over-exploited. The soil related issues of current concern are: increased pace of degradation due to physical, chemical and biological stresses; decreasing organic matter and deteriorating quality; heavy metal accumulation and nutrient imbalance; low efficiency of applied nutrients and site-specific micronutrient deficiencies. Recent figures on nutrient mining in India clearly indicate that nutrient additions are unable to match the nutrient removal by crops. Increased ratio of N:P:K (at present about 7:3:1) from its balanced composition of 4:2:1 has resulted in nutrient imbalance in general and appearance of deficiency of micronutrients in particular. These issues need to be addressed through focusing our research on: -

- » Inventorizatian, characterization and monitoring of various natural resources using modern tools and techniques, especially remote sensing and Geographic Information System (GIS) in different rainfed regions bf the country.

- » Development of sustainable land use plans for each agro-ecological sub-region in-the country taking into consideration the biophysical and socio-economic conditions of the target groups at panchayat/watershed/district level.

- » Development of Integrated Plant Nutrient System (IPNS) models for dominant cropping patterns under different rainfed regions for improving fertilizer use efficiency and soil health.

- » Determination of soil organic carbon pools, capacity for carbon sequestering and quantification of carbon stocks in diverse soils/agro-eco regions. Mapping of existing carbon stocks needs to be carried out to be eligible for futuristic carbon trading. Refinement of technology for conservation tillage and residue management is required for achieving synergy between nutrient, water and energy, particularly tinder rainfed conditions.

- » Inventorization, augmentation and quality assessment of available bio/organic resources like nutrient carriers, bio-pesticides, weedicides, soil amendments having potential for organic/biodynamic farming in rainfed regions.

- » Promotion of organic farming through demonstration and training on system of organic farming, biofertilizers, vermin-composting, enriched farm yard manure/compost etc.

Soil conservation

Soil conservation is a set of management strategies for prevention of soil being eroded from the Earth's surface or becoming chemically altered by overuse, acidification, salinization or other chemical soil contamination. It is a component of environmental soil science.

- » Decisions regarding appropriate crop rotation, cover crops, and planted wind breaks are central to the ability of surface soils to retain their integrity, both with respect to erosive forces and chemical change from nutrient depletion. Crop rotation is simply the conventional alternation of crops on a given field, so that nutrient depletion is avoided from repetitive chemical uptake/deposition of single crop growth.

Erosion prevention practices

- » There are conventional practices that farmers have invoked for centuries. These fall into two main categories: contour farming and terracing, standard methods recommended by the US Natural Resources Conservation Service, whose Code 330 is the common standard. Contour farming was practiced by the ancient Phoenicians, and is known to be effective for slopes between two and ten percent. Contour plowing can increase crop yields from 10 to 50 percent, partially as a result from greater soil retention. There are many erosion control examples such as conservation tillage, crop rotation, and growing cover crops.

- » Keyline design is an enhancement of contour farming, where the total watershed properties are taken into account in forming the contour lines. Terracing is the practice of creating benches or nearly level layers on a hillside setting. Terraced farming is more common on small farms and in underdeveloped countries, since mechanized equipment is difficult to deploy in this setting.

- » Human overpopulation is leading to destruction of tropical forests due to widening practices of slash-and-burn and other methods of subsistence farming necessitated by famines in lesser developed countries. A sequel to the deforestation is typically large scale erosion, loss of soil nutrients and sometimes total desertification.

Perimeter runoff control

» Trees, shrubs and ground-covers are effective perimeter treatment for soil erosion prevention, by insuring any surface flows are impeded. A special form of this perimeter or inter-row treatment is the use of a "grass way" that both channels and dissipates runoff through surface friction, impeding surface runoff, and encouraging infiltration of the slowed surface water.

Windbreaks

» Windbreaks are created by planting sufficiently dense rows of trees at the windward exposure of an agricultural field subject to wind erosion. Evergreen species are preferred to achieve year-round protection; however, as long as foliage is present in the seasons of bare soil surfaces, the effect of deciduous trees may also be adequate.

Salinity management

» Salinity in soil is caused by irrigating the crops with salty water. During the evaporation process the water from the soil evaporates leaving the salt behind causing salinization. Salinization causes the soil structure to break down causing infertility and the plants cannot grow.

» The ions responsible for salination are: Na^+, K^+, Ca^{2+}, Mg^{2+} and Cl^-. Salinity is estimated to affect about one third of all the earth's arable land.[4] Soil salinity adversely affects the metabolism of most crops, and erosion effects usually follow vegetation failure. Salinity occurs on drylands from over irrigation and in areas with shallow saline water tables. In the case of over-irrigation, salts are deposited in upper soil layers as a byproduct of most soil infiltration; excessive irrigation merely increases the rate of salt deposition. The best-known case of shallow saline water table capillary action occurred in Egypt after the 1970 construction of the Aswan Dam. The change in the groundwater level due to dam construction led to high concentration of salts in the water table. After the construction, the continuous high level of the water table led to soil salination of previously arable land.

» Use of humic acids may prevent excess salination, especially in locales where excessive irrigation was practiced. The mechanism involved is that humic acids can fix both anions and cations and eliminate them from root zones. In some cases it may be valuable to find plants that can tolerate saline conditions to use as surface cover until salinity can be reduced; there are a number of such saline-tolerant plants, such as saltbush, a plant found in much of North America and in the Mediterranean regions of Europe.

Soil organisms

» When worms excrete egesta in the form of casts, a balanced selection of minerals and plant nutrients is made into a form accessible for root uptake. US research shows that earthworm casts are five times richer in available nitrogen, seven times richer in available phosphates and eleven times richer in available potash than the surrounding upper150 mm of soil. The weight of casts produced may be greater than 4.5 kg per worm per year. By burrowing, the earthworm is of value in creating soil porosity, creating channels enhancing the processes of aeration and drainage.

Mineralization

» To allow plants full realization of their phytonutrient potential, active mineralization of the soil is sometimes undertaken. This can be in the natural form of adding crushed rock or can take the form of chemical soil supplement. In either case the purpose is to combat mineral depletion of the soil. There are a broad range of minerals that can be added including common substances such as phosphorus and more exotic substances such as zinc and selenium. There is extensive research on the phase transitions of minerals in soil with aqueous contact.

» The process of flooding can bring significant bedload sediment to an alluvial plain. While this effect may not be desirable if floods endanger life or if the eroded sediment originates from productive land, this process of addition to a floodplain is a natural process that can rejuvenate soil chemistry through mineralization and macronutrient addition.

Various methods of soil conservation

» Soil conservation measures should aim at preventing or at least minimising the soils loss. In order to do this proper land utilisation coupled with agricultural practices should be adopted.

» Broadly categorizing there are two methods of soil conservation. These are biological and mechanical. The biological measures are again divided into Agronomic, Agrostilogical and Dry farming we shall study these measures in some detail.

Measures

1. **Agronomic practices:** Normally, the land will possess a vegetational cover so as to prevent erosion. The measures to be followed must be patterned along the nature's own methods of conservation. The following are some of the methods.

2. **Contour farming:** Crops are cultivated along the contour of the land. The plough marks will be on level and can hold the rain. Even in heavy rain, the runoff is checked by the plants growing along the contour. Tillage: contour tilling will prevent the excess run of water.

3. **Mulching:** Interculturing operations will kill weeds and soil mulches help the plants to be rooted firmly in the soil.

4. **Crop rotations:** Alternatively growing a cereal and a legume in the same field will not only increase the yield, but also increase the fertility of the soil. They also help in checking soil erosion.

5. **Strip cropping:** This is an agricultural practice of growing plants in suitable strips in the field. This is of the following types.

6. **Contour strip cropping:** This is cultivation of soil protecting crops in strips alternating with erosion permitting crops. The strips should be across the slope.

7. **Field strip cropping:** Plants are cultivated in parallel strips across the slopes. Wind strip cropping: Crops are planted across the slopes to prevent soil loss. These may be legumes or grasses.

8. **Agrostological measures:** Cultivation of grass in a land which is heavily eroded is called an agrostological measure. This is of two types. In ley farming grass is cultivated in rotation with regular crops. This helps in soil protection as well as produce fodder to cattle. If a land is heavily eroded it is best to allow it to the growth of grasses for few years. This will help in the checking of erosion.

9. **Dry farming method:** This may be practised where rainfall is low, indefinite and variable. In dry farming methods only crops are grown that can sustain even a very low rainfall. The most important aspects of dry farming are conservation of soil moisture and fertility.

10. **Mechanical measures:** The main aims of mechanical measures are to allow for the absorption of run off, dividing the slope into short ones and protection against run off. A few of the mechanical measures are discussed below:

11. **Basin listing:** Small basins are formed along the contour with an implement called basin lister. These will hold water for some time.

12. **Sub soiling:** Soil is broken with a sub soiler into fine grains to increase their absorptive capacity.

13. **Contour terracing:** Along the contour, series of ridges or bunds of mud are formed to check the run off. This is of four types. In channel terrace a shallow channel is dug and the mud is deposited along the lower edge of the canal. In broad base ridge terrace a canal is formed on the contour by exavating the mud. The canal is wide. If it is narrow it is called narrow based ridge terrace. In bench terracing a series of platforms are formed along the contour across the general slope of the plant.

14. **Contour trenching:** Several 2 feet by one foot trenches are formed across the slopes at suitable intervals. Tree seedlings are to be planted above the trench.

15. **Terrace outlet:** Outlets are to be constructed for the safe disposal of runoff water.

16. **Gully control:** Suitable water conservation measures are to be taken so as to prevent the formation of gullies.

17. **Ponds:** Construction of small ponds at suitable places to store water is a good practice.

18. **Stream bank protection:** Banks of channels or rivers usually cave in during floods. To prevent this, construction of stone or concrete protective walls should be undertaken. In addition to this, planting some useful tree species will also prevent stream bank erosion.

Climate

Climate change and global warming caused by the emission of greenhouse gasses have emerged as important issue in the last two decades. Carbon dioxide (CO_2), methane (CR_4), nitrous oxide (N_2O), hydrofluorocarbons (HFCs), perfluorocarbons (PFCs) and sulphur hexafluouride (SFs) are the six important gases, which are responsible for global warming. The recent projections are that South Asia may have an increase in temperature from 0.1 to 0.3°C by 2012 and 0.4 to 2.0°C by 2070. Sea level rise between 0.15 to 0.94 meter, submergence of islands/coastal areas and change in rainfall pattern are predicted over the next century. This will alter biodiversity and demand a new set of land use pattern. Similarly, stratospheric ozone depletion in water months has been observed during 1969 to 1988. This kind of ozone depletion may lead to increased ultra-violet (UV) radiation with far reaching adverse impact on earth's environment and human as well as livestock population. Decreasing trend in the density of glaciers may have serious repercussions.

These weather related issues call for greater understanding of crop-weather relationships and developing crop-weather models to devise efficient agricultural production strategic. The concerns of rising temperature and decreasing ozone protection demand minimizing/moderating the emission of GHGs. Adaptations to predicted climate change impacts on productivity include: linking weather forecasting with climate forecasting, land use change, breeding varieties that can adjust to such changes, agronomic manipulations, CO, sequestration, substitution of fossil fuels with bio-fuels, moderating methane emission from agricultural fields and weather- based forewarning of pests and disease incidences.

Biodiversity

With wide variety of ecological habitants, India supports an enormous diversity of flora and fauna. A significant part of this biodiversity exists in rainfed regions. Biodiversity is generally distinguished at generic, species and ecosystem level. For the sake of sustainable survival of mankind, conservation of diversity is of paramount importance. Out of 18 hot spots of biodiversity in the world, two are in the Western Ghats and North Eastern Region of India. The main areas of genetic diversity in wild types fall under Malabar region, Deccan Peninsula and eastern Himalayas with 113, 96, 51 species, respectively. India has about 8 per cent of the total existing plants and animals of the world. Of 45000 plant species, which exist in India, about 15000 are flowering plant species. There are nearly 81000 species of animals in India which form about 6.5 per cent of the total world fauna. Fungi followed by flowering plants constitute the largest group of the total plant components of the Indian biota. A survey conducted in the Himalayan region in the altitudinal range between 1800 and 3599 meter, showed that the area under traditional species has declined form 85 per cent in 1970 to 55 percent in 1990. The factors responsible for erosion of biodiversity include: monocropped agriculture, introduction of traditional crops to non-traditional areas, introduction of irrigation in desert areas, increased pace of land degradation such as erosion, salinity, acidity etc; indiscriminate use of inorganic inputs including pesticides and weedicides and deforestation etc. Biodiversity conservation should be one of the major concerns of efficient resource management for sustainable production in general and rainfed regions in particular.

Agroforestry

A sizeable area in the country, particularly in rainfed regions, faces severe degradation due to deforestation. The present area under forest ever is unable to support demand for tree based products. Fuel wood is the dominant source of energy in rural areas and small and marginal farmers in rainfed regions mainly depend upon trees to meet this requirement. These issues call for developing area specific agroforestry models not only to increase the forest cover but also as diversification and export generating venture. Accordingly, the future strategic planning for agroforestry should focus on the following:

- Development of site-specific agroforestry models for rehabilitation and reclamation of all kinds of waste and abandoned lands.

- Identification and upgradation of agroforestry species, which can with stand drought, floods and other biotic and abiotic stresses.

- In addition to already existing dominant tree species like *Khejri* (*Prosopis cineraria*) in arid areas of Rajasthan, the top feed species such as *Ailanthus excelsa, Salvadora bleoldes, Prosopis juliflora* and edible *Opuntiaficus indica* have been identified to meet fodder shortage especially during drought situation. Research on such plants should get the highest priority.

- Major thrust in future will be on exploiting agroforestry as an option for carbon sequestration; bio-drainage of waterlogged saline areas; option for environmentally safe disposal of industrial effluents and detoxification of soils loaded with heavy metals; and upgradation of productivity of rangelands/grasslands.

- Identification, collection, conservation and improvement of forage crops, grasses and top feed fodder resources and development of agro-techniques for their successful cultivation in rainfed regions of the country.

- Upgradation of productivity, carrying capacity and sustainability of grasslands/rangelands with major emphasis on alpine and temperate regions.

Agro-forestry or agro-sylviculture is an integrated approach of using the interactive benefits from combining trees and shrubs with crops. It combines agricultural and forestry technologies to create more diverse, productive, profitable, healthy, and sustainable land-use systems. A narrow definition of agro-forestry is "trees on farms."

Agro-forestry is any sustainable land-use system that maintains or increases total yields by combining food crops (annuals) with tree crops (perennials) and/or livestock on the same unit of land, either alternately or at the same time, using management practices that suit the social and cultural characteristics of the local people and the economic and ecological conditions of the area.

Agro-forestry is a collective name for a land-use system and technology whereby woody perennials are deliberately used on the same land management unit as agricultural crops and/or animals in some form of spatial arrangement or temporal sequence. In an agro-forestry system there are both ecological and economical interactions between the various components.

Agro-forestry systems include both traditional and modern land-use systems where trees are managed together with crops and/or animal production systems in agricultural settings.

FAO promotes agro-forestry using multi-sectoral approaches. Agro-forestry is in the heart of many initiatives such as watershed management, non wood forest products and enterprises, climate change mitigation and adaptation, waste water reuse, landscape restoration, food systems through integrated territorial development, urban agriculture, and trees outside forests assessments.

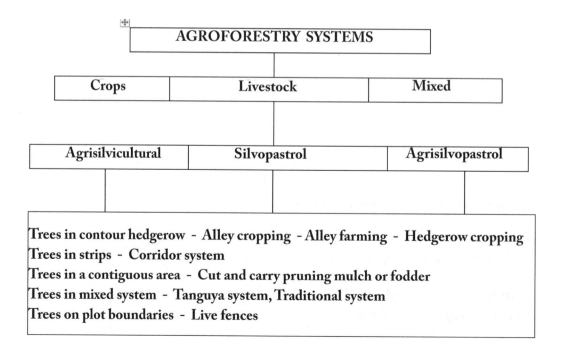

The theoretical base for agroforestry comes from ecology, via agroecology. From this perspective, agroforestry is one of the three principal land-use sciences. The other two are agriculture and forestry.

The efficiency of photosynthesis drops off with increasing light intensity, and the rate of photosynthesis hardly increases once the light intensity is over about one tenth that of direct overhead sun. This means that plants under trees can still grow well even though they get less light. By having more than one level of vegetation, it is possible to get more photosynthesis than with a single layer.

Agroforestry has a lot in common with intercropping. Both have two or more plant species (such as nitrogen-fixing plants) in close interaction both provide multiple outputs, as a consequence, higher overall yields and, because a single application or input is shared, costs are reduced. Beyond these, there are gains specific to agroforestry. Agroforestry is also defined as the chicken land management unit whereby, woody perennials are deliberately grown on same piece of land along with agricultural crop and or livestock in some form of spatial arrangement or temporal sequence.

Difference between social forestry and agroforestry

Social forestry is defined as "Forestry outside the conventional forests, which primarily aim at providing continuous flow of goods and services for the benefit of people. This definition implies that the production of forest goods for the needs of the local people is Social forestry. Thus, social forestry aims at growing forests of the choice of the local population. Other stated that Conceptually Social forestry deals with poor people to produce goods such as fuel, fodder etc. to meet the needs of the local community particularly underprivileged section.

Tree cultivation in non-forest areas

1. **Farm forestry:** Farm forestry is the name given to programmes, which promote commercial tree growing by farmers on their own land. Farm forestry was defined as the practice of forestry in all its aspects in and the around the farms or village lands integrated with other farm operations.

2. **Extension forestry:** It is the practice of forestry in areas devoid of tree growth and other vegetation situated in places away from the conventional forest areas with the object of increasing the area under tree growth.

It includes the following:

a. **Mixed forestry:** It is the practice of forestry for raising fodder grass with scattered fodder trees, fruit trees and fuel wood trees on suitable wastelands, panchayat lands and village commons

b. **Shelterbelts:** Shelterbelt is defined as a belt of trees and or shrubs maintained for the purpose of shelter from wind, sun, snow drift, etc.

c. **Linear strip plantations:** These are the plantations of fast growing species on linear strips of land.

   1. **Rehabilitation of degraded forests:** The degraded area under forests needs immediate attention for ecological restoration and for meeting the socio economic needs of the communities living in and around such areas.

   2. **Recreation forestry:** It is the practice of forestry with the object of raising flowering trees and shrubs mainly to serve as recreation forests for the urban and rural population. This type of forestry is also known as Aesthetic forestry which is defined as the practice of forestry with the object of developing or maintaining a forest of high scenic value.

Types of Agroforestry Systems

1. Structural basis

A. Nature of components

I) Agri-silvicultural systems

In this system, agricultural crops are intercropped with tree crops in the inter-space between the trees. Under this system agricultural crops can be grown up to two years under protective irrigated condition and under rainfed farming up to four years. The crops can be grown profitably upto the above said period beyond which it is uneconomical to grow grain crops. However fodder crops, shade loving crops and shallow rooted crops can be grown economically. Wider spacing is adopted without sacrificing tree population for easy cultural operation and to get more sunlight to the intercrop. Performance of the tree crops is better in this system when compared to monoculture.

II) Silvo-pastoral systems

The production of woody plants combined with pasture is referred to Silvipasture system. The trees and shrubs may be used primarily to produce fodder for livestock or they may be grown for timber, fuel wood and fruit or to improve the soil.

This system is classified in to three categories

a. **Protein bank**: In this Silvi-pastoral system, various multipurpose trees (protein rich trees) are planted in or around farmlands and range lands for cut and carry fodder production to meet the feed requirement of livestock during the fodder deficit period in winter. Ex. *Acacia nilotica, Albizia lebbeck, Azadirachta indica, Leucaena leucocephala, Gliricidia sepium, Sesbania grandiflora*

b. **Live fence of fodder trees and hedges**: In this system, various fodder trees and hedges are planted as live fence to protect the property from stray animals or other biotic influences. Ex. *Gliricidia sepium, Sesbania grandiflora, Erythrina sp, Acacia sp.*

c. **Trees and shrubs on pasture**: In this system, various tree and shrub species are scattered irregularly or arranged according to some systemic pattern to supplement forage production. Ex. *Acacia nilotica, Acacia leucophloea, Tamarindus indica, Azadirachta indica.*

III) Agro-silvopastoral systems

The production of woody perennials combined with annuals and pastures is referred Agri-silvopastural system.

This system is grouped into two categories

A) Home gardens: This system is found extensively in high rainfall areas in tropical South and South East Asia. This practice finds expression in the states of Kerala and Tamil Nadu with humid tropical climates where coconut is the main crop. Many species of trees, bushes , vegetables and other herbaceous plants are grown in dense and in random or spatial and temporal arrangements. Most home gardens also support a variety of animals. Fodder grass and legumes are also grown to meet the fodder requirement of cattle. In India, every homestead has around 0.20 to 0.50 ha land for personal production.

Home gardens represent land use systems involving deliberate management of multipurpose trees and shrubs in intimate association with annual and perennial agricultural crops and livestock within the compounds of individual houses. The whole tree- crop- animal units are being intensively managed by family labour. Home gardens can also be called as Multitier system or Multitier cropping.

Home gardens are highly productive, sustainable and very practicable. Food production is primary function of most home gardens.

Structure of Home Gardens: Home gardens are characterized by high species diversity and usually 3-4 vertical canopy strata. The layered configuration and compatible species admixture are the most conspicuous characteristics of all home gardens. Generally all home gardens consist of an herbaceous layer near the ground, a tree layer at the upper levels and an intermediate layer. The lower layer can be partitioned into two, the lowermost being at less than 1.0m in height, dominated by different vegetables and the second layer of 1.0 -3.0/m height comprising food crops such as banana, papaya and so on. The upper tree layer can also be divided into two, consisting of emergent, full grown timber and fruit trees occupying the upper most layer of 25m height and medium size trees of 10-20m occupying the next lower layer. The intermediate layer of 5-10m height is dominated by various fruit trees.

Choice of species

a. **Woody species:** *Anacardium occidentale, Artocarpus heterophyllus, Citrus spp, Psiduim guajava, Mangifera indica, Azadirachta indica, Cocus nucifera*

b. **Herbaceous species:** Bhendi, Onion, cabbage, Pumpkin, Sweet potato, Banana, Beans etc.

B) Woody Hedgerows:

In this system various woody hedges, especially fast growing and coppicing fodder shrubs and trees are planted for the purpose of browse, mulch, green manure, soil conservation etc. The following species viz., *Erythrina sp, Leucaena luecocephala, Sesbania grandiflora* are generally used.

IV) Other systems

a) Apiculture with trees: In this system various honey (nector) producing trees frequently visited by honeybees are planted on the boundary of the agricultural fields.

b) Aquaforestry: In this system various trees and shrubs preferred by fish are planted on the boundary and around fish ponds. Tree leaves are used as feed for fish. The main role of this system is fish production and bund stabilization around fish ponds

c) Mixed wood lots: In this system, special location specific Multipurpose Trees (MPTs) are grown mixed or separately planted for various purposes such as wood, fodder, soil conservation , soil reclamation etc.

B. Arrangement of components

Two types of arrangements:

a. **Spatial Arrangement:** Spatial arrangement of plants in an agroforestry mixture may result in dense mixed stands (as in home gardens) or in sparse mixed stands (as in most systems of trees in pastures).

b. **Temporal Arrangement:** Temporal arrangements of plants in Agroforestry may also take various forms. An extreme example is the conventional shifting cultivation cycles involving 2-4 years of cropping and more than 15 years of fallow cycle, when a selected woody species or mixtures of species may be planted. Similarly, some silvi-pastoral systems may involve grass leys in rotation with some species of grass remaining on the land for several years. These temporal arrangement of components in agroforestry are termed coincident, concomitant, overlapping, separate and interpolated.

2. Functional basis

All agroforestry systems have two functions:

A) Productive functions

    The Productive functions are:
        I) Food
        II) Fodder
        III) Fuel wood
        IV) Cloths
        V) Shelter
        VI) NTFPs

B) Protective functions

    The Protective functions are:
    i) Wind breaks
    II) Shelterbelts
    III) Soil conservation
    IV) Soil improvement

3. Socio-economic classification: Based on socioeconomic criteria as scale of production and level of technology input and management, agroforestry systems have been grouped in to three categories.

A) Commercial Agroforestry systems: The term commercial is used whenever the scale of the production of the output is the major aim of the system.

Examples:

a. Commercial production of plantation crops such as rubber, oilpalm, and coconut with permanent underplanting of food crops, pastures.

b. Commercial production shade tolerating plantation crops such as coffee, tea and cocoa under overstorey of shade trees.

B) Intermediate Agroforestry systems: Intermediate systems are those between commercial and subsistence scale of production and management.

Examples:

Production of perennial cash crops and subsistence food crops undertaken on farms wherein the cash crops fulfill the cash needs and the food crops meet the family's food needs.

C) Subsistence Agroforestry systems: Subsistence AF systems are those wherein the use of land is directed towards satisfying basic needs and is managed mostly by the owner and his family.

4. Ecological classification

A) Humid / sub humid
B) Semiarid / arid
C) Highlands

A) Agroforestry systems in Humid / Subhumid lowlands

Examples: Homegardens, Trees on rangelands and pastures, improved fallow in shifting cultivation and Multipurpose woodlots.

B) Agroforestry systems in Semiarid and arid lands

Examples: Various forms of silvopastoral systems, wind breaks and shelterbelts.

C) Agroforestry systems in Tropical High lands

Examples:

Production systems involving plantation crops such as coffee, tea, use of woody perennials in soil conservation and improved fallow.

Benefits of Agroforestry System

A) Environmental benefits

i. Reduction of pressure on natural forests.

ii. More efficient recycling of nutrients by deep rooted trees on the site

iii. Better protection of ecological systems

iv. Reduction of surface run-off, nutrient leaching and soil erosion through impeding effect of tree roots and stems on these processes

v. Improvement of microclimate, such as lowering of soil surface temperature and reduction of evaporation of soil moisture through a combination of mulching and shading

vi. Increment in soil nutrients through addition and decomposition of litterfall.

vii. Improvement of soil structure through the constant addition of organic matter from decomposed litter.

B) Economic benefits

i. Increment in an outputs of food, fuel wood, fodder, fertiliser and timber;

ii. Reduction in incidence of total crop failure, which is common to single cropping or monoculture systems

iii. Increase in levels of farm income due to improved and sustained productivity

C) Social benefits

i. Improvement in rural living standards from sustained employment and higher income

ii. Improvement in nutrition and health due to increased quality and diversity of food outputs

iii. Stabilization and improvement of communities through elimination of the need to shift sites of farm activities.

Agroforestry systems can be advantageous over conventional agricultural, and forest production methods. They can offer increased productivity, economic benefits, and more diversity in the ecological goods and services provided.

Biodiversity in agroforestry systems is typically higher than in conventional agricultural systems. With two or more interacting plant species in a given land area, it creates a more complex habitat that can support a wider variety of birds, insects, and other animals. Depending upon the application, impacts of agroforestry can include:

» Reducing poverty through increased production of wood and other tree products for home consumption and sale

» Contributing to food security by restoring the soil fertility for food crops

» Cleaner water through reduced nutrient and soil runoff

» Countering global warming and the risk of hunger by increasing the number of drought-resistant trees and the subsequent production of fruits, nuts and edible oils

» Reducing deforestation and pressure on woodlands by providing farm-grown fuel wood

» Reducing or eliminating the need for toxic chemicals (insecticides, herbicides, etc.)

» Through more diverse farm outputs, improved human nutrition

» In situations where people have limited access to mainstream medicines, providing growing space for medicinal plants.

Agroforestry practices may also realize a number of other associated environmental goals, such as:

» Carbon sequestration

- » Odour, dust, and noise reduction
- » Green space and visual aesthetics
- » Enhancement or maintenance of wildlife habitat

Adaptation to climate change

There is some evidence that, especially in recent years, poor smallholder farmers are turning to agroforestry as a mean to adapt to the impacts of climate change. A study from the CGIAR research program on Climate Change, Agriculture and Food Security (CCAFS) found from a survey of over 700 households in East Africa that at least 50% of those households had begun planting trees on their farms in a change from their practices 10 years ago. The trees ameliorate the effects of climate change by helping to stabilize erosion, improving water and soil quality and providing yields of fruit, tea, coffee, oil, fodder and medicinal products in addition to their usual harvest. Agroforestry was one of the most widely adopted adaptation strategies in the study, along with the use of improved crop varieties and intercropping.

Applications of agroforestry

Agroforestry represents a wide diversity in application and in practice. One listing includes over 50 distinct uses. The 50 or so applications can be roughly classified under a few broad headings. There are visual similarities between practices in different categories. This is expected as categorization is based around the problems addressed (countering winds, high rainfall, harmful insects, etc.) and the overall economic constraints and objectives (labor and other inputs costs, yield requirements, etc.). The categories include:

- » Parklands
- » Shade systems
- » Crop-over-tree systems
- » Alley cropping
- » Strip cropping
- » Fauna-based systems
- » Boundary systems
- » Taungyas

- » Physical support systems
- » Agroforests
- » Wind break and shelterbelt

Parkland: Parklands are visually defined by the presence of trees widely scattered over a large agricultural plot or pasture. The trees are usually of a single species with clear regional favorites. Among the benefits, the trees offer shade to grazing animals, protect crops against strong wind bursts, provide tree prunings for firewood, and are a roost for insect or rodent-eating birds.

There are other gains. Research with *Faidherbia albida* in Zambia showed that mature trees can sustain maize yields of 4.1 tonnes per hectare compared to 1.3 tonnes per hectare without these trees. Unlike other trees, Faidherbia sheds its nitrogen-rich leaves during the rainy crop growing season so it does not compete with the crop for light, nutrients and water. The leaves then regrow during the dry season and provide land cover and shade for crops.

Shade systems: With shade applications, crops are purposely raised under tree canopies and within the resulting shady environment. For most uses, the understory crops are shade tolerant or the over story trees have fairly open canopies. A conspicuous example is shade-grown coffee. This practice reduces weeding costs and increases the quality and taste of the coffee.

Crop-over-tree systems: Not commonly encountered, crop-over-tree systems employ woody perennials in the role of a cover crop. For this, small shrubs or trees pruned to near ground level are utilized. The purpose, as with any cover crop, is to increase in-soil nutrients and/or to reduce soil erosion.

Alley cropping: With alley cropping, crop strips alternate with rows of closely spaced tree or hedge species. Normally, the trees are pruned before planting the crop. The cut leafy material is spread over the crop area to provide nutrients for the crop. In addition to nutrients, the hedges serve as windbreaks and eliminate soil erosion.

Alley cropping has been shown to be advantageous in Africa, particularly in relation to improving maize yields in the sub-Saharan region. Use here relies upon the nitrogen fixing tree species *Sesbania sesban, Tephrosia vogelii, Gliricidia sepium* and *Faidherbia albida*. In one example, a ten-year experiment in Malawi showed that, by using fertilizer trees such as *Tephrosia vogelii* and *Gliricidia sepium*, maize yields averaged 3.7 tonnes per hectare as compared to one tonne per hectare in plots without fertilizer trees or mineral fertilizer.

Strip cropping: Strip cropping is similar to alley cropping in that trees alternate with crops. The difference is that, with alley cropping, the trees are in single row. With strip cropping, the trees or shrubs are planted in wide strip. The purpose can be, as with alley cropping, to provide nutrients, in leaf form, to the crop. With strip cropping, the trees can have a purely productive role, providing fruits, nuts, etc. while, at the same time, protecting nearby crops from soil erosion and harmful winds.

Fauna-based systems

There are situations where trees benefit fauna. The most common examples are the silvopasture where cattle, goats, or sheep browse on grasses grown under trees. In hot climates, the animals are less stressed and put on weight faster when grazing in a cooler, shaded environment. Other variations have these animals directly eating the leaves of trees or shrubs.

There are similar systems for other types of fauna. Deer and hogs gain when living and feeding in a forest ecosystem, especially when the tree forage suits their dietary needs. Another variation, aquaforestry, is where trees shade fish ponds. In many cases, the fish eat the leaves or fruit from the trees.

Boundary systems

There are a number of applications that fall under the heading of a boundary system. These include the living fences, the riparian buffer, and windbreaks.

» A living fence can be a thick hedge or fencing wire strung on living trees. In addition to restricting the movement of people and animals, living fences offer habitat to insect-eating birds and, in the case of a boundary hedge, slow soil erosion.

» Riparian buffers are strips of permanent vegetation located along or near active watercourses or in ditches where water runoff concentrates. The purpose is to keep nutrients and soil from contaminating surface water.

» Windbreaks reduce the velocity of the winds over and around crops. This increases yields through reduced drying of the crop and/or by preventing the crop from toppling in strong wind gusts.

Taungya

Taungya is a system originating in Burma. In the initial stages of an orchard or tree plantation, the trees are small and widely spaced. The free space between the newly planted trees can accommodate a seasonal crop. Instead of costly weeding, the underutilized area provides an additional output and income. More complex taungyas use the between-tree space for a series of crops. The crops become more shade resistant as the tree canopies grow and the amount of sunlight reaching the ground declines. If a plantation is thinned in the latter stages, this opens further the between-tree cropping opportunities.

Physical support systems

In the long history of agriculture, trellises are comparatively recent. Before this, grapes and other vine crops were raised atop pruned trees. Variations of the physical support theme depend upon the type of vine. The advantages come through greater in-field biodiversity. In many cases, the control of weeds, diseases, and insect pests are primary motives.

Agroforests

These are widely found in the humid tropics and are referenced by different names (forest gardening, forest farming, tropical home gardens and, where short-statured trees or shrubs dominate, shrub gardens). Through a complex, disarrayed mix of trees, shrubs, vines, and seasonal crops, these systems, through their high levels of biodiversity, achieve the ecological dynamics of a forest ecosystem. Because of the internal ecology, they tend to be less susceptible to harmful insects, plant diseases, drought, and wind damage. Although they can be high yielding, complex systems tend to produce a large number of outputs. These are not utilized when a large volume of a single crop or output is required.

Various challenges

Agroforestry is relevant to almost all environments and is a potential response to common problems around the globe, and agroforestry systems can be advantageous compared to conventional agriculture or forestry. Yet agroforestry is not very widespread, at least according to current but incomplete USDA surveys as of November, 2013.

As suggested by a survey of extension programs in the United States, some obstacles (ordered most critical to least critical) to agroforestry adoption includes:

» Lack of developed markets for products

- » Unfamiliarity with technologies
- » Lack of awareness of successful agroforestry examples
- » Competition between trees, crops, and animals
- » Lack of financial assistance
- » Lack of apparent profit potential
- » Lack of demonstration sites
- » Expense of additional management
- » Lack of training or expertise
- » Lack of knowledge about where to market products
- » Lack of technical assistance
- » Cannot afford adoption or start up costs, including costs of time
- » Unfamiliarity with alternative marketing approaches (e.g. web)
- » Unavailability of information about agroforestry
- » Apparent inconvenience
- » Lack of infrastructure (e.g. buildings, equipment)
- » Lack of equipment
- » Insufficient land
- » Lack of seed/seedling sources

Some solutions to these obstacles have already been suggested although many depend on particular circumstances which vary from one location to the next.

Crop diversification

Continuous cultivation of same crop/cropping system in a given area results in Fatigue of resources particularly the soil and water. Recent example is of rice and wheat cultivation for more than 3 decades in the Indo-gangetic plain. Higher input use to maintain yield levels many a time does not help, rather it results in problems of groundwater contamination and environmental degradation. This calls for diversification to more remunerative cropping systems or resorting to other farm enterprises, such as animal husbandry, horticulture, poultry, fish, high value industrial crops etc.

Medicinal and aromatic crops such as *Isabol, Matricaria, Dil*, etc. have been identified, which can be grown using saline water up to 8 deci siemens/m, Their exploitation as commercial crops is being explored through linking it with post- harvest processing, value addition and agri-based business ventures. More of such industrial high value crops need to he identified and promoted to generate export.

Identification of biologically more productive, sustainable, profitable and environment friendly crops/cropping systems is required for different agroecological regions of the country in general and as alternative to rice-wheat in particular.

In the changing agricultural scenario during globalization, agriculture in India has to face new challenges to compete at the global level in many agricultural commodities. Indian agriculture is now facing second generation problems like raising or lowering of water-table, nutrient imbalance, soil degradation, salinity, resurgence of pests and diseases, environmental pollution and decline in farm profit. Crop diversification shows lot of promise in alleviating these problems through fulfilling the basic needs and regulating farm income, withstanding weather aberrations, controlling price fluctuation, ensuring balanced food supply, conserving natural resources, reducing the chemical fertilizer and pesticide loads, environmental safety and creating employment opportunity. Diversification is gradually taking place as a consequence of either launching macro-economic reforms in agriculture sector or rising domestic demand due to urbanization and increasing income levels. Crop diversification has been recognized as an effective strategy for achieving the objectives of food security, nutrition security, income growth, poverty alleviation and employment generation, judicious use of land and water resources, sustainable agricultural development and environmental improvement. The ability of the country to diversify the cropping pattern for attaining various goals depends on the opportunities available for diversification, the need for diversification and responsiveness of the farmers to these needs and opportunities.

10. Crop Diversification For Maximizing Productivity

CROP DIVERSIFICATION

An effective strategy for achieving crop diversification

- » Food & nutrition security
- » Income growth
- » Poverty alleviation
- » Employment generation
- » Judicious use of land and water resources
- » Sustainable agricultural development
- » Environmental improvement

The opportunities for crop diversification emerge from technological breakthroughs, changes in demand pattern, development of irrigation, availability of marketing infrastructure and new trade arrangements. The necessity for crop diversification for keeping in view:

- » Reducing risks associated with yield, market and prices
- » Arresting the degradation of natural resources and the environment
- » Attaining national goals like employment generation, self-reliance in critical crop products and for earning foreign exchange

Diversification is the process to take advantage of emerging opportunities created by technology, new markets, changes in policy etc. to meet certain goals, challenges and threats and to reduce risk. Crop diversification is one of the major components of diversification in agriculture. Crop diversification may be adopted as a strategy for profit maximization through reaping the gains of complementary and supplementary relationships or in equating substitution and price ratios for competitive products. It also acts as a powerful tool in minimization of risk in farming.

These considerations make a strong case for farm/crop diversification in India. Crop diversification in India is generally viewed as a shift from traditionally grown less remunerative crops to more remunerative crops. Market infrastructure development and certain other price related supports also induce crop shift. Higher profitability and also the resilience/stability in production also induce crop diversification. Crop diversification and large number of crops are practiced in rainfed areas to reduce the risk factor of crop failures due to drought. Crop substitution and shift are also taking place in the areas with distinct soil problems for example, the growing of rice in high water-table areas replacing pulses and cotton, promotion of soybean in place of sorghum in vertisols (medium and deep black soils). The crop diversification also takes place due to government policies and thrust in some crops over a given time, eg. Establishment of Technology Mission on Oilseeds (TMO) to give thrust on oilseed production as a national need for country's requirement to reduce imports.

Benefits of Crop Diversification

The crop diversification may result in enhanced profitability, reduce pests, spread out of labour more uniformly, different planting and harvesting times can reduce risks from weather and new crops can be renewable resources of high value products. Crop diversification in rainfed regions aims to make rainfed agriculture achieving nutritional security, more employment and income generating, eco-friendly, poverty alleviation and comparative advantage in new trade regime (Vittal et al., 2007). Some of the issues and functions provided by diversification in rainfed regions are mentioned in Table 1. Perennial species play an important role in areas where cropping of annual plants has reduced total water use and allowed water-table to rise, with a resultant salinization. In such areas, an appropriate density of trees in agroforestry systems can help reestablish a hydrological balance that keeps the water-table and its salt content below the root zone of the crops. The emphasis in agro-ecological analysis is on the processes and balance of resource supply and capture and on the competitive and complimentary relationships between the planned and unplanned biodiversity (Table 2).

Crop diversification in rainfed regions of India

Land degradation and climate change are the twin problems challenging rainfed agriculture in India. Kinds, degree and extent of land degradation are of immediate concern in sustaining production system, reducing cost of production and natural resource management and conservation. The crops are grouped into rice, oilseeds, pulses and coarse cereals. In each production system based upon diversification index and severity of soil degradation, horizontal and vertical diversification is suggested. Horizontal diversification is advantageous in effective utilization for natural resources, viz. soil, light water and conservation, employment generation and risk minimization. Vertical diversification aims at reducing the soil loss, high biomass production, high income and employment generation through year round activity and addition of organic matter to soil, organic linkage between agriculture and industry wherein the scope is widened for post-harvest value addition by practicing the enterprises like agroforestry (alley cropping, silviculture, silvipasture, agri-horticulture and agri-silvi-pastoral system), sericulture, rainfed horticulture, olericulture, medicinal aromatic plants, other economic shrubs like dye yielding plants and most importantly animal component for dairy, poultry, apiary, rabbit rearing etc. These complementary enterprises with multiple objectives and advantages in rainfed regions may help for comparative advantage in the present trade regime.

Table 1. Issues and functions provided by diversification in rainfed regions

S. No.	Issues	Functions
1.	Productivity and stability	Increased yield, reduce intra-seasonal variation and improved stability through diverse components, viz. crop, tree, plant and animal
2.	High risk and high cost	Risk and cost minimization through yield and income from annual and perennial mixtures
3.	Unabated land degradation	Minimization of kinds, effect and extent of land degradation by appropriate land care through alternate land use systems
4.	Inadequate employment	Staggered employment round the year
5.	Low profitability	High income generation from various Components
6.	Poor energy management	Energy efficient implements

Table 2. Factors affecting decision for agroforestry in crop production in dryland regions in India

S. No.	Factors	Decision for agroforestry
1.	Short term profit factors	Crop production and quality, forage production level, quality and timing, yields of trees, economic shrubs and forages, input costs, output prices for annuals, perennials and livestock products
2.	Dynamic factors optimum	Short term to medium term: Soil health, tree and forage density, abiotic stresses, water harvesting, tillage
3.	Sustainable factors	Soil degradation, nutrient loss, tree/forage establishment, risk factors. yield variability, price variability,flexibility of the enterprise in response to changed conditions, the farmers attitude to risk.
4.	Whole-farm factors quality and cost the	Total arable area, machinery, total feed requirements, financial support, labour availability, farmer's objectives (profit, risk reduction, sustainability), traditional wisdom.

India is amongst the largest vegetable oil economies in the world next only to USA, China and Brazil. In the agricultural economy of India, oilseeds are important next only to food grains in area, production and value. The diverse agro-ecological conditions in the country are favourable for growing nine oilseeds which include seven edible oilseeds (groundnut, rapeseed-mustard, soybean, sunflower,sesame, safflower and niger) and two non - edible oil seeds (castor and linseed). Oilseed crops have potential for increasing cropping intensity and profitability in wide ranging cropping systems. Oilseeds crops by nature are hardy, mostly grown under rainfed conditions and impart stability of production system under harsh conditions. Oilseeds will have an edge over other crops in price, wider adoptability and relative optimal production under environmental stress conditions.

Introduction of high-yielding varieties in oilseed crops replaced a number of traditional low yielding crops because of their higher efficiency in the utilization of rainfall and moisture, resulting in higher yield and returns. Safflower has comparative economic advantage over other popular crops like rainfed wheat, coriander, chickpea etc. The crop is distinctly remunerative under conditions of limiting moisture in several traditional as well as non-traditional areas viz black soil areas of Karnataka, Maharashtra, Andhra Pradesh and Rajasthan. Sunflower, by virtue of its photo-insensitivity and wide adaptability to soil types, has a greater role to play in contingency cropping plans. It is suitable for late planting in kharif in delayed rains.

It can also be planted whenever the *kharif* planted crop fails due to lack of rains. Sesame has great potential as summer crop under limited irrigation in Telangana region of Andhra Pradesh. Good yields of soybean have been obtained from the crop grown in post-rainy season (*rabi/kharif*) in many states.

Diversification of rice-wheat system with oilseeds: Punjab and Haryana have become major rice and wheat producing states from the last three decades with the dominant rice-wheat cropping system. To reduce the fatigue of this rice-wheat system, alternate crops like oilseeds can be grown without hampering the profitability of the system. At least 0.5-0.6 m.ha of rice area in Punjab could be shifted to soybean in comparatively upland and less irrigated area. The area which goes to late wheat due to delayed harvest of basmati rice can be shifted to sunflower cropping. Coarse rice-potato-sunflower recorded higher returns (Rs 70,262/ha) compared to coarse rice-wheat system (Rs 35,881/ha). In Haryana, the basmati rice area invariably goes for late sown wheat. It is suggested that the basmati area may be put to sunflower cropping in place of wheat. Soybean and pulses like pigeonpea should be promoted by diverting some of the rice area. In Uttar Pradesh, where rice-wheat system is important, it is possible to shift part of rice area to soybean and early pigeonpea. Rabi sunflower could also be promoted in place of wheat.

Diversification of oilseeds in upland rice/rice fallow situation: Wherever water resources are limiting such as tail end area of irrigation command, tank fed area and well irrigated areas, oilseed crops like groundnut, soybean, sunflower, sesame etc. can be profitably cultivated. In rice fallow situations of Cauvery delta (Tamil Nadu) and coastal Andhra Pradesh, soybean and sesame can be profitably grown. In Tungabhadra project areas of Karnataka and Andhra Pradesh, Sriramsagar project area of Andhra Pradesh and Jayakwadi project area of Maharashtra, it is profitable to grow groundnut, sunflower and sesame under rice fallow situations. Groundnut has great potential under residual moisture after the harvest of *kharif* rice in coastal region of Karnataka and Andhra Pradesh. There is great potential of *rabi*/summer groundnut in rice fallows and on residual moisture in flood plains in Assam, West Bengal, Orissa and Chhatisgarh. Mustard has an excellent potential in rice fallows in North-eastern states. As a summer crop in eastern states, sesame is profitably grown after rice. In the upland areas of Orissa, Tamil Nadu, Bihar and Andhra Pradesh, groundnut (*kharif*), soybean and sunflower are more remunerative as compared to upland rice.

Diversification with oilseeds in nontraditional areas and crop substitution

Oilseed crops by virtue of low irrigation requirement and better remunerative price are ideally suited to replace low yielding other crops and become popular even in non-traditional area.

The diversification of traditional crop base with annual oilseed crops is given in table 3. Diversification of oilseeds through intercropping

Groundnut

(a) Inter-cropping with short duration annuals

Groundnut, a long duration crop can be grown along with sunflower so that early season rains benefit sunflower and late rains benefit groundnut, thereby giving some as assurance to the dry land farmers. Advancing sowing of groundnut by 15-30 days prior to sunflower reduces the competition due to shading of sunflower. Research data indicated that a net return of Rs 10,120 and 12,615/ha could be obtained under sunflower groundnut intercropping system in Saurashtra region of Gujarat and West Bengal, respectively. Intercropping groundnut with sesame was also found profitable in this region; the optimum row ratio is 4:1 with net returns to the tune of Rs 9,947-12,292/ha. Groundnut with pearl millet and sorghum is common in red soils of the semi-arid tract of India. The optimum row ratio in this intercropping system is 1:1 with Virginia groundnut and 1:2 with Spanish groundnut. Two rows of sorghum with eight rows of groundnut is one of the best combinations providing 38-53per cent yield advantage over sole crops. The system provides net returns to the tune of Rs 3,000/ha at Hyderabad and Indore. Groundnut is commonly intercropped with maize in Madhya Pradesh and Bihar. In Sundarban area of West Bengal (coastal saline tract) and Dharwad region of Karnataka yield advantage was found quite high when 3 rows of groundnut is alternated with 3 rows of chillies.

(b) Intercropping with long duration annuals

Among the legumes, pigeonpea + groundnut (3:1) were the most prevalent intercropping system in India. At Jalgaon (Maharashtra), this system provided net returns up to Rs 22,338/ha. Raising 2-3 rows of groundnut in between cotton rows spaced 2 m apart is reported to give higher income than raising either of them alone. Castor intercropped with groundnut is better than growing castor alone. This system provided net returns of about Rs 10,000/ha at Junagadh and at Kanpur.

(c) Intercropping with perennial/plantation crops at early stages

Intercropping groundnut under cassava could give yield advantage of 33 to 55%. In Trivandrum and Orissa 12 q/ha of groundnut, in addition to full yield of tapioca was obtained. Growing groundnut between wide spaced rows of banana is common practice in Tamil Nadu, Maharashtra and parts of Karnataka. Groundnut under coconut plantation is common in Kerala.

Soybean

Soybean when introduced as intercrop exerts less competition to companion crop, it does not impose any allelopathic effects on companion crop and it helps in fertilizer economy. Some of the prominent intercropping systems found suitable are: soybean+pigeonpea, soybean + fingermillet, soybean + sugarcane, soybean + sorghum, soybean + groundnut, soybean in coconut/ mango/guava orchards in various soybean growing areas of the country. Soybean offers much scope for intercropping in sugarcane. By growing soybean as intercrop in sugarcane, the N fertilizer requirement could be cut down to some extent. Rapeseed-mustard Diversification is possible by growing mustard as an intercrop with autumn planted sugarcane, potato, wheat, lentil and chickpea in northern India. Sunflower

Sunflower + soybean in Marathwada and Vidarbha region of Maharashtra, sunflower + urdbean in Uttar Pradesh are found to be efficient for increasing productivity and monetary returns.

Additional net returns expected from different intercropping systems are:

Groundnut + The oilseed crops in diversification are explained here.

Groundnut: As a replacement crop for minor millets in Bihar and Orissa. As an irrigated crop in Kosi command and in Tawa command in Bihar and Madhya Pradesh. Substitute rice-groundnut with rice-rice system to prevent build up of pests and diseases.

Soybean: As a replacement crop for minor millets in Bihar and Orissa. Soybean is also as a rotational crop in pest endemic areas of rainfed cotton. In non-traditional areas of Northeastern hills under agripastoral or agri-silvicultural system, diverting some *kharif* cereal areas to soybean in situations of water scarcity and to restore soil-health in North India.

Rapeseed-mustard: As a replacement crop for low yielding rainfed wheat, In diara tract in northern and eastern India and Gujarat.

Sunflower: As a replacement crop for desi wheat, cotton, chickpea, sorghum in black cotton soils in peninsular India. As a spring crop in northern India.

Sesame: As a summer crop in central peninsular and eastern India where only limited irrigation is available.

Castor: As a replacement of cotton in some regions of western Haryana and Rajasthan, As bund crop in all regions.

Table 3. Diversification of traditional crop base with annual oilseed crops

S. No.	Prevailing crop	Crop suggested	Region
1.	Rice fallow	Soybean and sesame	Cauvery deltaic areas of Tamil Nadu and coastal Andhra Pradesh
2.	Upland rice	Groundnut (*kharif*), soybean and sunflower	Upland rice areas of Orissa, Tamil Nadu, Bihar and Andhra Pradesh
3.	Cotton	Safflower and sunflower	Karnataka and adjoining areas of Andhra Pradesh
		Chickpea, dryland safflower (sole crop)	Andhra Pradesh (rabi), Karnataka, Maharashtra
		Wheat and Coriander	Part of Madhya Pradesh (Malwa region)
		Linseed and Barley Safflower (sole crop)	South eastern Rajasthan (Udaipur)
4.	Rice fallow	Groundnut, sunflower and sesame	Most part of rice growing areas

Crop diversification with oilseed crops

Sunflower (Rs 2100 to 9200/ha); Pigeonpea + sunflower (Rs 3500 to 5200/ha) and Soybean + sunflower (Rs 4000 to 10000/ha) in peninsular India. Sunflower + aswagandha (1:6) were found to be highly profitable and merits for diversification in north Karnataka over sole sunflower under rainfed conditions in 2006.

Castor

Castor is usually raised either as a sole crop or intercrop with grain/legumes (pigeonpea, groundnut, mungbean, urdbean and cowpea) and sometimes with horticultural crops like chillies, turmeric, ginger, Dolichos and cucumber. The followings are some of the remunerative viable intercropping systems recommended for Southern states of the country. The additional net returns from castor + pigeonpea (4:1) ranges from Rs 3,700 to 12,400/ha, from castor + groundnut (1:3) between Rs 4,200 and Rs 23,700/ha, from castor+urdbean Rs.1,000 to Rs.2,400/ha, from castor + clusterbean (1:2) Rs 4,000 to Rs 6,000/ha, castor + mungbean

(1:2) Rs.10,000 to Rs.12,000/ha. Studies conducted at Hyderabad revealed that the seed yield of sunflower and castor did not differ significantly in all alley widths (3m × 3m, 4.2 × 3m and 5.4 × 3m) when grown with *Faidherbia albida* in comparison with seed yields of respective sole crops during 1994.

Safflower

Intercropping of safflower with other rainfed rabi crops increases the net returns compared to that of sole cropping. An additional net returns of Rs 2,500 to 4,500/ha could be expected in chickpea + safflower and coriander + safflower depending on row ratio of system and status of soil moisture in Karnataka and Andhra Pradesh. In linseed + safflower and mustard + safflower intercropping system, an additional net returns of Rs 3,500 to 5,000/ha could be realized. Intercropping of safflower in wheat increases the net returns by Rs 1,000 to 1,500/ha over sole of wheat.

Sesame

It can be raised as an intercrop with groundnut grown at wider spacing in Gujarat, with finger millet, groundnut in Andhra Pradesh, Karnataka and Madhya Pradesh.

Linseed

Linseed can be grown as intercrop with wheat, chickpea, lentil, coriander and safflower in major crop growing areas.

Diversification with Oilseeds in Crop Sequences

Oilseeds being high value crops have been given priority for inclusion in the cropping systems mainly in cereal and legume-based crop sequence. The level of fertilizer and water supply required for high-yielding varieties of rapeseed-mustard, groundnut, sesame, sunflower, and soybean can give highest output/unit area and enhanced farm income have been identified for different agro-eco regions.

(a) Need based cropping systems for different agro-climatic conditions

In arid ecosystem, the oilseeds are mostly grown after cereal, whereas in semi-arid, cereal-oilseed, legume-oilseed, oilseed-cereal, and oilseed-legume are prevalent. The details of sequence having oilseed as a component crop clearly indicates the scope for enhancing the production of various oilseed crops in different states/unit of area. Use of quality seed and fertilizer for oilseeds as well as other crops in the sequence enhanced the total production of the system. For instance, the sunflower productivity in arid and semi-arid ecosystems were realized 1.14, 1.18, 1.30, and 1.40 t/ha in various sequences as against the national average productivity of 0.66 t/ha during 1996-97. Similarly the groundnut and soybean yield in cotton-groundnut and rice-soybean sequence was 3.45 t/ha and 2.45 t/ha respectively compared to country's average productivity of 1.15 and 1.0 t/ha in semi-arid ecosystem. The rapeseed-mustard productivity in coastal ecosystem was marked (1.41 t/ha) in rice-mustard-brinjal cropping system. This rate of growth was measured 38.9% higher than average productivity of the nation.

(b) Promising crop sequences involving oilseeds in different states

Directorate for Cropping System Research has identified promising crop sequences for different states wherein oilseed is one of the component crops either as *kharif* or *rabi* in the systems. In Ranchi, (Bihar), rice-linseed is promising crop sequence under limited water supply situations, In Mandi (Himachal Pradesh), inclusion of toria between rice-potato and maize-potato, increased the total farm income to the tune of Rs 41,950/ha and Rs 40,744/ha, respectively by enhancing the system productivity. In Ujjain district of Madhya Pradesh and in Manipur, blackgram-mustard, soybean-wheat, and soybean-gram were found promising crop sequence having oilseeds as one of the component crop. In Maharashtra, soybean was observed prominent oilseed crop associated with cereal, legume and other oil crops in the sequence. The income realized was Rs 20,460, 27,276, and Rs 33,099/ha from soybean-gram, sesame-sunflower-groundnut and sorghum-sunflower-groundnut in the region. Growing of mustard and linseed after rice, urd and maize was a common practice in eastern Uttar Pradesh. Soybean-wheat recorded almost similar net returns as that of rice-wheat system in Punjab with high B: C ratio of 2.61 as against 2.41 in rice-wheat system.

(c) Maximizing productivity and resource use from oilseed crops

The oilseed crops offer excellent opportunity for maximizing productivity under limited moisture availability. Sunflower crop productivity can be increased by more than 60% with limited irrigation at critical stages.

Safflower yields can be doubled by providing two irrigations in Malwa plateau. In Western zone of Uttar Pradesh, rising of rice-mustard-mungbean for 1 year followed by rice-wheat-mungbean in succeeding 2 years registered in saving of water to the tune of 10.5 per cent clearly favouring for the diversification of the rice-wheat system.

In Bihar, rice-potato-sunflower system recorded higher rice equivalent yield, net returns, B: C ratio, land use efficiency, production efficiency than traditional rice-wheat-green manuring. At Jalandhar (Punjab), rice-potato-sunflower registered higher rice equivalent yield, economic efficiency, land use efficiency, irrigation water productivity and nutrient productivity compared to rice-wheat system.

The hybrids available in sunflower, castor and safflower have shown the productivity improvement to the magnitude of 28.6, 18.2 and 11 per cent, respectively over varieties under moisture and nutrient stress situations in the respective major crop growing areas.

Mustard in inter-cropping with potato, sunflower after potato have been found efficient in utilizing the fertilizer applied to potato. The fertilizer can be economized with soybean under intercropping system. In Telangana region of Andhra Pradesh, fertilizing sunflower at 50 per cent recommended dose with *kharif* groundnut as a sequence crop has given sunflower yields comparable at 150% recommended dose of sunflower-sunflower sequence. Hence, when legume like groundnut precedes sunflower; the fertilizer requirement for succeeding sunflower needs suitable modification. By adopting integrated nutrient package about 33% higher yield (116 kg/ha) could be obtained in summer sesame during 1999.

Role of Tree Species in Nutrient Conservation

Trees can increase nutrient inputs to agroforestry systems by retrieval from lower soil horizons and weathering rock. Agroforestry systems involving-nitrogen fixing tree species help in building soil fertility through leaf litter fall, fixation of atmospheric nitrogen and recycling of nutrients from lower soil layer to surface soil. The beneficial association of nitrogen-fixing trees with sorghum and cowpea in semi-arid tropics are well documented.

Increasing nutrient use efficiency

Increasing the nutrient use efficiency through agroforestry can be achieved by

» Increasing the cycling of nutrients from tree litter and pruning via the soil into the crops;

» Reducing losses by leaching and

» Reducing the losses by erosion.

The nutrient transfer from tree residues to crops is through litter and prunings. Perennial-crop combinations can provide 6-20 t/ha/year of leaves and small branches and hedgerows typically provide 5-12 t. The annual hedgerow biomass can be as low as 2 t in SAT and rise to 20 t in humid areas. Planted fallows can produce 12-30 t/ha/year during the tree fallow period. The nutrient content from NFTs is typically N 2.5-4 per cent, P 0.1-0.3 per cent, K 1.0- 2.5 per cent and Ca 1.5-2.0 per cent. Pruning has the advantage over natural litter in that the leafy matter is transferred before loss of nutrients by senescence. The magnitude of addition through 5t tree residues would be N 120-200, P 5-15, K 50-100 and Ca 70-100 kg/ha/year. To these should be added substantial quantities in the decay of fine roots (which contain 50 per cent or more of leaf N). The nutrients in a maize crop with a grain yield of 3 t/ha are N 120-150, P 20-25, K 80-100 and Ca 20-30 kg/year. So, except for P, the tree residues can meet the nutrient requirements of associated crop. In SAT, N supply through *Leucaena* pruning's was 35-74 kg/ha/year. The available N at sowing was higher by 49 and 19 kg respectively in plots where pruning returned and not returned than sole sorghum. A saving of 9.5, 9.4 and 4.3 kg N to groundnut crop was observed when pruning of *Dalbergia*, *Leucaena* and *Albizia* were returned. The organic carbon, P and K were higher (1.02 per cent, 32 kg and 180 kg/ha) in plots with *F.albida* trees (@ 625 trees/ha) than no tree control (0.69 per cent, 16 kg and 162 kg/ha) in SAT.

India being a vast country of continental dimensions presents wide variations in agro-climatic conditions. Such variations have led to the evolution of regional niches for various crops. Historically, regions were often associated with the crops in which they specialize for various agronomic, climatic, hydro-geological, and even, historical reasons. In new challenges and new changes including the achievement of food self-sufficiency, the area shift that tended towards cereals in the immediate aftermath of the green revolution, has started moving in the opposite direction i.e. from cereals to non-cereals like oilseeds and pulses. This review has clearly brought out the advantages of crop diversification with oilseeds in marginal eco-systems to make the cropping enterprise as a profitable venture avoiding risks. Hence, crop diversification with oilseeds could be used as a tool for maximizing productivity, profitability leading to resource conservation.

Crop Diversification in Punjab

1. **Alarming decline in underground water table:** In a rice-wheat cropping system, rice being a water guzzling crop and it utilizes about 3000-5000 litres of water to produce one Kilogram of rice.

2. **Impact on soil environment:** Intensive cropping system lead to depletion of soil macro nutrients (N, P& K), in addition to micro nutrients (Zn , Fe , Mn & B) and secondary nutrients (S, Mg & Ca) from the root zone and making it deficient which results into low

availability of the nutrients to plants and ultimately low crop yield. Continuous cultivation of major crops under intensive cropping system on same land leads to deterioration of chemical, physical and biological properties of the soils and imbalanced & inadequate supply of plant nutrient to the soils which pollute the soil environment. For example the rice-wheat system removed 500 kg/ha nutrients from the soil which is several times more than its addition.

3. **Incidence of weeds, insects and diseases:** Mono-cropping system leads to development of same sort of flora and fauna, therefore increases their incidence. For example, in south-western districts of Punjab where cotton is main *Kharif* crop, incidence of insect- pest like jassid, whitefly, mealy bug, tobacco caterpillar etc increases or may result to crop failure. Similarly, in Rice-Wheat cultivation increased the incidence of *Phalaris minor* (Gulli danda) in wheat and *Echinocloa crusgalli* (Swank) in rice. Similar was the case with diseases.

4. **Development of resistance:** When the incidence of insect-pest and diseases increases then large amount of chemicals are used for their control. The continuous use of a particular chemical develop the resistance in the insect, diseases and weeds which creates a major problem or sometimes leads to crop failure.

5. **Marketing of the produce:** Production of same crop on large scale cause great hindrance in the marketing of produce. Sometimes farmers sold their produce at very low price than the Minimum support price.

6. **Decrease in profitability margins**: From the above said points it was found that imbalanced use of herbicides, insecticides & fungicides, fertilizers, more irrigation water in rice-wheat rotation, increases the cost of production and reduced profit of the farmer For example, There is an excessive use of insecticides to control the attack of mealy bug and herbicides (roundup) for the control of weeds in cotton which has been recommended Generally, the farmers are using the herbicides like roundup for the control of weeds on bunds and channels etc. So as the demand for roundup increases and increases its cost which ultimately increase the cost of production and decrease the margin of profit. Similarly, with decline in water table, farmers shifting from centrifugal to submersible pumps whose installation charges ranging from Rs.80,000 to Rs.1,00,000/- which adds to cost of production and effecting on the economic condition of the farmers.

It is clear from the above that there is need to diversify the existing cropping system to sustain the agriculture in Punjab. The Punjab Agricultural University has given the few substitute options of crop rotations in place of rice-wheat cropping system.

1. **Different crop rotations:** Instead of Rice-wheat rotation farmers are advised to follow the following rotations which are promising in environment and net return basis. These rotations give more net returns as compare to Rice-Wheat and also their requirement for nutrients and water is much less than Rice-Wheat rotation.

   a. **Basmati rice-mentha/bersem (fodder and seed)/celery-bajra fodder**

   Generally, Punjab farmers are following Rice-Wheat cropping sequence in which sowing of wheat get delayed which results to reduction in productivity. Both rice and wheat, being cereals, exhaust the soil fertility. Inclusion of short duration leguminous crops in the sequence can rejuvenate the soil health besides improving the productivity. It was found that basmati rice-mentha, basmati rice- berseem (seed & fodder) and basmati rice-celery- bajra fodder were more remunerative to fetch more net returns of Rs. 61,700, 53,600 and 51,200/ha, respectively, as compared to Rs. 40,700/ha in rice-wheat system.

   b. **Maize(June)-potato-mentha**

   This cropping sequence provides almost double crop productivity in terms of Maize equivalent yield (325.5 q/ha) than rice – wheat system (91.49 q/ha). This system is more remunerative as maximum net returns of Rs. 91,400/ha was obtained as compared to Rs. 41,100/ha from rice – wheat system.

   c. **Maize(August)- mentha/wheat & bajra fodder/celery and bajra fodder**

   These sequences are highly stipendiary than rice- wheat system. Maize(August) – mentha, Maize(August) – wheat – bajra (fodder) and Maize(August) – celery – bajra (fodder) gave net returns of Rs. 60,600, 49600 and 47,900/ha as compared to 41,100 in rice – wheat system. Similarly, the rotations given below are equally profitable as the above as rice-wheat cropping system.

   d. **Basmati rice - Celery- Bajra (fodder)**

   This system is more remunerative and productive than the existing basmati rice – wheat system with sizeable saving of irrigation water. Transplant basmati rice in mid july which will vacate the field in mid November. Then grow celery which vacate the field in first fortnight of May and after this grow bajra crop for fodder.

   e. **Maize/Rice–Gobi Sarson –Summer Mungbean**

   These systems produced more yield and economic returns than the maize – wheat and rice – wheat system. Therefore, the maize/rice should be sown in the first fortnight of June, Gobi Sarson from 10-30th October and summer mungbean in the first fortnight of April. The summer mungbean can be sown without tillage after applying pre-sowing irrigation that saves water and reduce the cost of production.

f. Maize-potato-mentha

This cropping system is doubly profitable than rice – wheat system and provides considerable saving of irrigation water. In this system, sow early maturing varieties of maize with zero tillage (var. Paras) in mid June which will vacate the field in 2nd fortnight of September. Then grow potato (Kufri Chandramukhi) in the first week of October that vacate the field in mid January and after then grow mentha crop in the second fortnight of January. The soil fertility in relation to organic carbon, available P and K also improves over time.

2. **Inclusion of pulses in the cropping system:** Inclusion of pulses in Rice-wheat crop rotation not only improves and maintains the soil fertility but it also increases the net returns per unit land. As the pulses like mungbean and urdbean in *Kharif* and Chickpea (gram), lentil and field pea in rabi season were able to fix atmospheric nitrogen with symbiotic relationships with *Rhizobium* bacteria present in their roots. By this they not only fulfill their own requirement but also fix nitrogen in soil which will be fully available to the succeeding crop. Pulses can fix nitrogen ranging from 30-68 kg N/ha. They also improve the physical property of soil as their root system go deep in soil which helps to reduce erosion, increase water holding capacity of soil. Root exudates of some pulses enhance the availability of phosphorous and other nutrients. By shedding their leaves and roots in soil before harvest they adds large amount of organic carbon.

3. **Intercropping:** Intercropping of cereals with leguminous crops is good option to increase the productivity as well as farm income. It will also help to maintain the soil physical and chemical properties. These crops will also increase C: N ratio of the soil when buried in the soil.

Further Thrust Areas of Research

- » In the present scenario of resource conservation followed in the Indo-Gangetic plains where cereal-based cropping system is predominant; the potentiality of leguminous oilseed like soybean may be exploited.

- » The development of cost effective or high yield-low cost technology can change the scenario for oilseeds production in the country.

- » Evaluation of cultivars of various oilseeds crops to suit to the requirement of non-traditional areas and seasons.
- » Efforts should be diverted to develop the potential oilseeds growing areas based on the agro-eco region basis and involving oilseeds of tree origin.
- » Both technological and price factors are important variables influencing supply response of edible oil-seeds.
- » Effective crop insurance policies
- » Adequate Minimum Support Price (MSP) of crops other than rice & wheat
- » Easy Availability and good quality inputs at nominal prices
- » Adequate marketing infrastructure
- » Strengthening of research and extension by involving private sectors

Concept, Significance And Sustainability Of Cropping System

CROPPING SYSTEM

A system consists of several components, which are closely related and interacting among them. In agriculture, cultivated practices are usually developed for individual crops. However, the different crops are grown in different seasons based on their adaptability to a particular area, domestic needs, productivity and profitability. Therefore, production technologies should be formulated as the various crops grown in a year or more than one year. Such package of practices of these crops leads to efficient use of costly inputs. Keeping in view, residual effect of applied manures and fertilisers to one after other crop, growing of legume crops in sequence and as green manuring crops in between the main crops help to fix the atmospheric nitrogen can considerably bring down the use of fertilizers which resulted in reduced production cost. Therefore, the importance of growing of crops in a system is increasing.

Concept and Definition

The term cropping system refers to the crops, crop sequences and management techniques used on a particular agricultural field over a period of years. It includes all spatial and temporal aspects of managing an agricultural system. Historically, cropping systems have been designed to maximise yield, but modern agriculture is increasingly concerned with promoting environmental sustainability in cropping systems. System means an arrangement of components, which process inputs into outputs. Each system consists of boundaries, components, interactions between components, inputs and outputs. Crop system is an arrangement of crop populations that transform solar energy, nutrients, water and other inputs into useful biomass. The crop can be of different species and variety but they only constitute one crop system if they are managed as a single unit. The crop system is a subsystem of a cropping system. In the maize crop system, maize is the dominant crop, which is grown in association with other crops.

A **cropping system** is the sum total of all *crops* and the practices used to grow those *crops* on a field or farm. Example as one variety grown each year in the same field with nutrients provided as fertilizer to replace nutrients sold off the farm with the *crop*.

Sequence cropping is defined as growing of two or more crops in sequence on the same piece of land in a farming year.

A cropping system refers to the type and sequence of crops grown and practices used for growing them. It encompasses all cropping sequences practiced over space and time based on the available technologies of crop production. Cropping systems have been traditionally structured to maximize crop yields.

Cropping system is a land-use unit comprising soils, crop, weed, pathogen and insect subsystems that transform solar energy, water nutrients, labour and other inputs into food, feed, fuel and fibre. The cropping system is a subsystem of the farming system.

The cropping patterns used on a farm and their interaction with farm resources, other farm enterprises, available technology and environment (physical, biological and socio-economic) which determine their makeup, constitute the cropping system.

Efficient cropping zone

The zone, where both relative yield and spread indices are maximum

$$\text{Relative yield index (RYI) of a crop} = \frac{\text{Average yield of state/zone}}{\text{Average yield of country}} \times 100$$

$$\text{Relative spread index (RSI) of a crop} = \frac{\text{Area in state/zone}}{\text{Area in country}} \times 100$$

Criteria efficient cropping zone

» Most efficient cropping zone when RYI and RSI both more than 100

» Efficient cropping zone when RYI is less than 100 and RSI more than 100

» Not efficient cropping zone when RYI is more than 100 and RSI less than 100

» Highly inefficient cropping zone when RYI and RSI both less than 100

Cropping System and Pattern

Cropping system varies widely from the simplest system of two crops a year in sequence to complex intercropping with many crops. Multiple cropped lands can be broadly grouped into lowlands, irrigated uplands, and rainfed uplands. The yearly sequence and spatial arrangement of crops or of crops and fallow are on a given area.

Cropping pattern' and 'cropping system' are two terms used interchangeably; however these are two different concepts. While cropping pattern refers to the yearly sequence and spatial arrangement of crops or of crops and fallow in a particular land area; Cropping system refers to cropping pattern as well as its interaction with resources; technology, environment etc. Thus, a cropping system comprises cropping pattern plus all components required for the production of a particular crop and the interrelationships between them and environment. Cropping system is a critical aspect in developing an effective ecological farming system to manage and organize crops so that they best utilize the available resources.(soil, air, sunlight, water, labour, equipments). It represents cropping patterns used on a farm and their interaction with farm resources and farm enterprises and available technology which determine their makeup. It is executed in the field level.

Multiple cropping is the practice of growing two or more crops in the same piece of land in same growing season instead of one crop. It is a form of polyculture. The most important aspect of the multiple cropping is the practice of intensification of cropping system in time & space dimension that is more no. of crops within a year & more no. crops in a same piece of land. Some additional terms are also used as agroforestry, mixed and intercropping etc.

Types of Multiple Cropping Systems

1) Sequential cropping: Growing two or more crops in sequence on the same field per year. The succeeding crop is planted after the preceding one has been harvested.

- » Crop intensification is only in the time dimension.
- » There is no intercrop competition.
- » Farmers manage only one crop at a time.

Types of sequential cropping

1. Double Cropping: growing of 2 crops in a year. Eg: cowpea- bajra green gram- jowar
2. Triple Cropping: growing of 3 crops in a year. Eg: Rice- potato-groundnut Cowpea- mustard- jute
3. Quadruple cropping: growing of 4 crops in a year. eg: maize – toria – potato - wheat green gram - maize – toria – wheat
4. Ratoon cropping/ stubble cultivation: cultivation stubble re-growth after the harvest of the crop. eg: sugar cane mulberry

2) Inter cropping: Growing two or more crops simultaneously on the same field per year. • Crop intensification is in both time and space dimensions.

- » There is intercrop competition during all or part of crop growth.
- » Farmers manage more than one crop at a time in the same field.

Types of Inter cropping

1. Mixed intercropping: Growing two or more crops simultaneously with no distinct row arrangement as wheat + mustard fodder maize + fodder cowpea
2. Row intercropping: growing two or more crops simultaneously with one or more crops planted in rows as cotton + chilly ground nut + maize
3. Multi storied cropping: Growing plants of different height in the same field at the same time is termed as multistoried cropping.
   - » It is the practice of different crops of varying heights, rooting pattern and duration to cultivate together.
   - » It is mostly practiced in orchards and plantation crops.

4. Strip intercropping: Growing two or more crops simultaneously in different strips wide enough to permit independent cultivation but narrow enough for the crops to interact agronomically as corn + alfalfa ragi+ groundnut

5. Relay intercropping/over lapping cropping: Growing two or more crops simultaneously during part of each one's life cycle. A second crop is planted after the first crop has reached its reproductive stage of growth, but before it is ready for harvest as rice + pulses potato + wheat

6. Alley Cropping: is planting rows of trees at wide spacing with a companion crop grown in the alleyways between the rows. Alley cropping can diversify farm income, improve crop production and provide protection and conservation benefits to crops as silver oak+ ragi

7. Ley farming is a system of rotating crops with legume or grass pastures to improve soil structure and fertility and to disrupt pest and disease lifecycles. It has been practiced in many parts of the world for centuries as grass + stylosanthu

Advantages multiple cropping

- » With multiple cropping the risk of total loss from drought, pests and diseases is reduced. Some of the crops can survive and produce a yield.

- » It gives maximum production from small plots. This can help farmers cope with land shortages.

- » Including legumes in the cropping pattern helps maintain soil fertility by fixing nitrogen in the soil.

- » Different types of crops can be produced, thereby providing a balanced diet for the family because of high planting density weeds are suppressed.

- » Efficient uses of resources available.

Disadvantages multiple cropping

- » Because of year-long crop some pests can shift from one crop to another.

- » The large number of different crops in the field makes it difficult to weed.

- » New technologies such as row planting, modern weeding tools and improved varieties may be difficult to introduce.

Shifting cultivation

Shifting cultivation is commonly known as land rotation and jhumming cultivation. Shifting cultivation is an agricultural system in which plots of land are cultivated temporarily, then abandoned and allowed to revert to their natural vegetation while the cultivator moves on to another plot. Swidden agriculture, also known as *shifting cultivation*, refers to a technique of rotational farming in which land is cleared for *cultivation* (normally by fire) and then left to regenerate after a few years.

In case of shifting cultivation, forest land is cleared and cultivated. Due to cultivation of the same crop on the same cleared forest land year after year, soil productivity is lost and the crop is shifted to other slashed and burnt land. Here same crop is grown year after year. In this case land is rotated but crop is fixed. Therefore, it may also be called land rotation. Shifting of land hence called shifting cultivation, also called Jhum cultivation. It causes soil erosion. Practised in north-eastern states of India, Chhotanagpur Plateau of Jharkhand, M.P. and in hilly areas.

Crop Rotation: Crop rotation is the reverse of land rotation. Here land is fixed but crop is rotated year after year.

Sustainable agriculture

Agriculture has been the basic source of subsistence for man over thousands of years. It provides a livelihood to half of the world's population even today. Modern day world, and especially developing societies, faces numerous challenges; among which the progressive increase in the population and thus the increase of food consumption and the expansion of industrial production. These factors require shifting the progress of agriculture toward sustainability. According to the Food and Agricultural Organisation (FAO), people in the developing world where the population increase is very rapid, may face hunger if the global food production does not rise by 50-60 per cent.

Indian agriculture has made impressive progress and so is more resilient to the vagaries of the monsoon, although the country's population increased from 361 million in 1951 to more than one billion in 2005. During this period, the size of farm holdings and the per capita availability of agricultural land have also been decreasing and they are expected to be around 1.4 and .14 hectares, respectively, by the turn of this century. With competing demands on land for others sectors of development, this decline is likely to aggravate further.

Agriculture is considered the biggest profession on earth; it provides the world with its need of daily nutrition, which may reach up to 7.3 billion tons of dairy products every year, and around 2.25 billion cups of coffee every day. Plus, agriculture occupies 40 per cent of the Land on which we are living, and consumes 70 per cent of water resources and 30 per cent of green reserves around the globe. World population today is about more than 6 billion. It is projected to become over 8 billion by 2025 and nearly 10.5 billion by the end of next century. In simple terms, the basic food production must double to maintain the status quo. The hunger must be banished from the surface of earth, as a first responsibility of any civilised society to provide sufficient food for the people who are below poverty line.

Factors for agricultural sustainability

- Firstly, taking into regard ensuring financial income, in other word the profitability, whereas agricultural activity shall be considered as an investment.

- Secondly, the creation of a new social order benefiting from these agricultural activities, as well as providing work, development, and training for targeted individuals.

- Thirdly, the preservation of the environment and ensuring its diversity.

Such factors could be fulfilled through exploiting each and every available resource, taking into account to protect and develop such resource in order to provide safe nutrition and agricultural produce over the long term. In fact such precautions are able in one hand to achieve sufficiency to agricultural societies that are engaged in these activities, as well as to non agricultural societies and in the other hand to manage profitability for investors. Over the long term sustainable agriculture could preserve the surrounding environment and ensure the bio diversity. If agriculture would not conserve the environment and its diversity therefore there will not be any sustainability or continuity of agricultural investment, and thus there will not be provided any safe nutrition to consume anymore.

Steps to achieve sustainable agriculture

- Developing bio diversity and preserving non harmful environment

- Preserving land and soil quality

- Smartly managing water resources and consumption

- Planning and creating new orders and rural societies, and advancing their health and social conditions

- » Increasing both quantity and quality of agricultural production
- » Smartly exploiting the land
- » Adequately and rationally consuming the energy
- » Considering climate changes.

Concept of Sustainable Agriculture

The concept of sustainable agriculture has come up because yield from modem farming technique (modem commercial agriculture) reaching a plateau and the environmental poblems due to excessive use of chemicals, fertilizers and pesticides in food chain.

Types of agriculture

1. **Subsistence agriculture:** has a low level equilibrium with the objectives to sustain the life and family needs with the use of low inputs and also has low output.

2. **Commercial farming:** has a high level equilibrium with the objectives to obtain high income with the use of high inputs for high output.

3. **Sustainable agriculture:** has natural/ecological equilibrium with the objectives for ecological balance with the use of low inputs for high output.

Sustainable agriculture is that form of farming which produces sufficient food to meet the needs of the present generation without eroding the ecological assets and productivity of life supporting systems of future generations. Natural farming is an excellent illustration of sustainable agriculture. It is also known as ecological farming/ eco-farming or organic farming or perm culture. It is called eco forming because ecological balance is given importance and organic farming because organic matter is the main source for nutrient management It is a system of cultivation with use of manures, crop rotation and minimal ullage. Sustainable agriculture also involves agro-forestry and multi-level cultivation and integrated animal husbandry. The term sustainability denotes the characteristic of a process that can be maintained indefinitely and sustainable use of the eco system refers to making use of the system without impairing its capacity for renewal or regeneration.

Definition of sustainable agriculture

It is the practice of farming using principles of ecology, the study of relationships between organisms and their environment. It has been defined as "an integrated system of plant and animal production practices having a site-specific application that will last over the long term.

Technical Advisory Committee or the Consultative group on International Agricultural Research (TAC/CGIAR) States: Sustainable agriculture is the successful management of resources for agriculture to satisfy human changing needs, while maintaining or enhancing the quality of the environments and conserving natural resources.

A sustainable agriculture is a system of agriculture that is committed to maintain and preserve the agriculture base of soil, water and atmosphere ensuring future generations the capacity to feed them with an adequate supply of safe and wholesome food.

A Sustainable agriculture system is one that can indefinitely meet demands for food and fibre at socially acceptable, economic and environment cost. Sustainable Agriculture refers to an agricultural production and distribution system that:

» Achieves the integration of natural biological cycles

» Protects and renews soil fertility and the natural resource base

» Reduces the use of non renewable resources and purchased (external or off-farm) production inputs

» Optimizes the management and use of on-farm inputs

» Provides an adequate and dependable farm income

» Promotes opportunity in family farming and farm communities

» Minimizes adverse impacts on health, safety, wildlife, water quality and the environment.

Sustainable agriculture is the successful management of resources for agriculture to satisfy changing human needs while maintaining or enhancing the quality of environment and conserving natural resources.

Sustainable agriculture is also known as eco-farming or organic farming or sometimes as natural farming or permaculture. Some other designating, it as regenerative agriculture or alternative farming. Sustainable agriculture is a food and fibre production and distribution system that:

- » Supports profitable production
- » Protects environmental quality
- » Uses natural resources efficiently
- » Provides consumers with affordable, high quality products
- » Decreases dependency on non renewable resources
- » Enhances the quality of life for farmers and rural communities and
- » Will last for generations to come.

Indian Agriculture before Green Revolution

Our traditional farming systems were characterised mainly by small and marginal farmers producing food and basic animal products for their families and local village communities. Farming was highly decentralised with individual farmers deciding on the types of crops to grow depending on climate and soil conditions. Even Alexander Walker, resident of Baroda in Gujarat, wrote in 1820 that green fodder was being grown throughout the year; intercropping, crop rotation, fallowing, composting and manuring were practised; all these allowed continued farming on the same land for more than 2000 years without drop in yield. One of the reasons for the decline in sustainable system of the agriculture was the land revenue collected by British. Even sacred groves, which were preserved since time immemorial, were turned into coffee, tea, teak wood and sugarcane plantations. Hence, from 1865 through 1900 India experienced the most severe series of protracted famines in its entire history.

Green revolution

After the green revolution was launched in India, substantial increase in the production of food grains was achieved through the use of improved crop varieties and higher levels of inputs of fertilizers and plant protection chemicals.
The ills of green revolution are stated to be:

- » Reduction in natural fertility of soil
- » Destruction of soil structure, aeration and water holding capacity
- » Susceptibility of soil erosion by water and wind
- » Diminishing returns on inputs

- » Breeding more virulent and resistant species of insects
- » Reducing genetic diversity of plant species
- » Pollution with toxic chemicals from agrochemicals
- » Health of farmers
- » Cash crops displacing nutritious food crops
- » Chemical changing natural taste of food
- » High cost inputs
- » Depleting fossil fuel resources
- » Lowering drought tolerance of crops
- » Appearance of problematic and difficult weeds
- » Throwing financial institutions into disarray
- » Agricultural and economic problems sparking off social and political turmoil resulting in violence.

Sustainability

It works on the principle that we must meet the needs of present without compromising the ability of future generation to meet their own needs. Organic agriculture claims to be sustainable in the context of agriculture. Sustainability refers to the successful management of resources of agriculture to satisfy human needs while at the same time maintaining or enhancing the quality of the environment and conserving natural resources. Sustainability in organic farming must therefore be seen in a holistic sense, which includes ecological, economical and social aspects.

Ecological sustainability

- » Recycling the nutrients instead of applying external inputs
- » No chemical pollution of soil, air and water
- » Promote biological diversity
- » Improve soil fertility and build up humus

- » Prevent soil erosion and compaction
- » Animal friendly husbandry
- » Using renewable energies

Social sustainability

- » Sufficient production of subsistence and income
- » A safe nutrition of the family with healthy food
- » Good working conditions for both men and women
- » Building a local knowledge and traditions

Economic sustainability

- » Satisfactory and reliable yields
- » Low costs on external inputs and investments
- » Crop diversification to improve income safety
- » Value addition through quality improvement and on-farm processing
- » High efficiency to improve Competitiveness

Thus only if the above three dimensions are fulfilled an agricultural system can be called sustainable.

Concept of sustainability in cropping system

The concept of sustainability applied to agriculture developed mainly as a result of growing awareness of negative impacts of intensive farming systems on the environment and the quality of life of rural and neighbouring communities.

- » Protecting the natural resources
- » More efficient use of arable lands and water supply
- » The sustainability concept has promoted the need to propose major adjustments in

conventional agriculture to make it more environmentally, socially and economically viable and compatible. The concept of sustainability is useful

» It captures a set of concerns

» Several possible solutions to the environmental problems.

» The main focus lies on the reduction or elimination of agrochemical inputs.

Basic principles of sustainable agriculture

» Based on both biological potential and biological diversity, land can be classified into conservation, restoration and sustainable intensification areas.

» Effectiveness in water saving, equity in water sharing and efficiency in water delivery and use are important for sustainable management of available surface and groundwater resources.

» An integrated system of energy management involving the use of renewable and non-renewable resources of energy in an appropriate manner is essential for achieving desired yield levels.

» Soils in India are often not only thirsty but also hungry. There is need for reduction in the use of market purchased inputs and not of inputs *Pre se*. It is in this context integrated systems of nutrient supply assume importance.

» Genetic diversity and location specific varieties are essential for achieving sustainable advances in productivity.

» The control of weeds, insect pests and pathogens is one of the most challenging jobs in agriculture.

» Whole plant utilization methods and preparation of value added products from the available agricultural biomass are important both for enhancing income and for ensuring good nutritional and consumer acceptance properties.

» Recycling of crop waste and livestock management.

» Growing legume crops

» Genetically Engineering crops (GM crops)

The aim of sustainable agriculture

The goal of sustainable agriculture is to maintain production at levels necessary to meet the increasing demand of an expanding world population without degrading the environment and that sustainability implies concern for the i) generation of income ii) the promotion of appropriate policies iii) the conservation of natural resources iv) enhance efficiency of use of input v) minimize adverse environmental impacts on adjacent and downstream environments vi) minimize the magnitude and rate of soil degradation and to enhance soil quality and resilience so that the crop productivity can be sustained with minimum adverse impact on soils and environment vii) enhance compatibility with social and political conditions.

However, many people use a wider definition judging agriculture to be sustainable if it is:

- » Ecologically sound: quality of natural resources is maintained
- » Economically viable: farmers can produce enough for self-sufficiency
- » Socially just: resources and power are distributed in such a way that basic needs of all members of society are met and their rights to land use, adequate capital and technical assistance and market opportunities.
- » Humane: all forms of life (plant, animal and human) are respected.
- » Adaptable: rural communities are capable of adjusting to the constantly changing for farming, population growth, policies, market demand etc.

Objectives of sustainability in cropping system

1. Satisfy human food and fiber needs.
2. Enhance environmental quality and the natural resource base on which the agricultural economy depends.
3. Make the most efficient use of non-renewable resources and on-farm resources.
4. Sustain the economic viability of farm operations.
5. Enhance the quality of life for farmers and society as a whole.
6. Use renewable resources at a rate less than the natural rate of generation.
7. Maintain wastes from production at a level below the assimilative capacity of the environment.

8. Ensure that the reduction of stock resources is compensated by increases in renewable resources.

9. Depletion of stock resources should occur with an increased standard of living.

10. Protect and renew soil fertility and the natural resource base.

11. Minimizes adverse impacts on health, safety, wildlife, water quality and environment.

12. Sustain the economic viability of farm operations i.e. the benefit cost ratio should be high.

13. Save natural and available resources for future generation:
    - Make best use of the resources available
    - Minimize use of non-renewable resources
    - Protect the health and safety of farm workers, local communities and society
    - Protect and enhance the environment and natural resources
    - Protect the economic viability of farming operations
    - Provide sufficient financial reward to the farmer to enable continued production and contribute to the well-being of the community
    - Produce sufficient high-quality and safe food
    - Build on available technology, knowledge and skills in ways that suit local conditions and capacity.

Need of sustainability in cropping system

- Monoculture a method of growing only one crop at a time in a given field is a very widespread practice, but there are questions about its sustainability, especially if the same crop is grown every year.

- If yields can be maintained over time with increased applications of chemical fertilizers, the cost of these additional units of fertilizer over time ultimately decreases profits. Thus, this "conventional" farming system illustrates a pattern of decreasing profitability over the long period.

- Continuous fall in soil fertility is major problem in many parts of India. Sustainable agriculture improves soil fertility and also prevents soil erosion. It also increases

» The way in which the agriculture is practised it contributes significantly to global climate change. There are more energy intensive activities such as production of artificial fertilizers by adopting sustainable agriculture it can reduce very significant.

» The introduction of agricultural practices in all biomes covered with natural vegetation profoundly affected microbial biomass-C (MB-C) with an overall decrease of 31%. Annual crops most severely reduce microbial biomass and soil organic C, with an average decrease of 53 per cent in the MB-C.

Ways to bring sustainability in cropping system

1. Increase in the soil organic matter by retaining more residues on the soil is important. Promoting existing biological cycle and soil biological activity, maintaining environmental resources and using them more carefully and efficiently and reusing residues as much as possible can help sustaining the rice-wheat cropping system. Thus, minimizing only pollution both on-site and off-site is an important feature of reducing soil degradation.

2. Organic amendments reduced chemical N input by 56 per cent, P input by 60 per cent, and K input by 72 per cent in the treatments with double dose green manure or green manure plus rice straw, but the mean yields were less than 3 per cent than NPK treatment but the differences were not significant. Substitution of 30-70 per cent of inorganic fertilizers with amendment of organic materials that maintained or even increased the rice yields in subtropical China, this positive effects were mainly attributed due to the increased soil organic soil organic carbon and improved soil nutrient status.

3. The okra + cowpea intercropping system at 60 cm by 45 cm recorded the highest okra equivalent yield, lower weight of weeds from the inter space and highest gross return during both the seasons. The results revealed the scope of above combination as an economically viable, biologically suitable and it was sustainable cropping system to increase the productivity of vegetables.

4. Crop rotation can improve the profitability and sustainability of crop production. Crop rotations optimizing crop water use, nutrient N use and market prices and reducing adverse effects of plant diseases. The Chickpea-wheat rotation resulted in the greatest gross margins and profitability of water use (water use efficiency 7.7 kg/ha-mm) and higher grain yield (2.72 t/ha) as compare to wheat-wheat system in which crops such as faba bean (legume crop) and canola and summer crops such as grain sorghum and mung

bean was also profitable compared with continuous winter cereal. The grain legumes chickpea and faba bean also provided soil nitrogen fertilizer requirements.

5. Cropping system also includes growing of perennials in mixtures because perennial roots hold the soil and provide a diversity of species to thwart insect or pathogen attacks. Similarly, the potential for restoring unproductive cropland to natural habitats has been demonstrated.

6. A long term study (2006-2009) revealed that double no-till practice in rice-based system is cost-effective, restored soil organic carbon (70.75%), favoured biological activity(46.7%), conserved water and produced yield (49%) higher than conventional tillage. Therefore, conservation tillage practised in terrace upland, valley upland and low-land situations ensured double cropping, improved farm income and livelihood in rain fed north-east India.

7. The water resources are under great stress. The real water saving will come by obtaining more crop production from same amount of water. Bed planting and laser levelling can help to reduce this stress.

8. Puddling of alkali soils further degrades the soil structure, and can facilitate formation of subsurface plough pan further restricting the percolation of water through soil profile. Reduced infiltration slows down the process of reclamation; therefore, puddling should be avoided. In another studies have shown that puddling can be avoided in transplanted rice.

9. The existing practices like straw burning leads to pollution, which is spread of, from smoke of burnt straw. Farmers do not bear the cost of such pollution, which is publicly unacceptable. Zero tillage can effectively serve as an opportunity to evolve residue management technologies because management of surface residue is easier than incorporation.

10. Legumes are important in maintaining soil organic matter and increasing soil N reserves. In addition, they protect the soil from run off water, wind erosion and improve infilteration; agro-forestry systems use leguminous and other trees to provide alternative crops.

11. Plant biodiversity plays an important role in pest, disease and weed management. Crop rotations are effective in controlling pests, diseases and weeds. Living mulches control weeds and minimize the need for herbicides.

12. Increases in structural diversity within the crop canopy leads to greater diversity in insects and less damage from insect pests.

Practices of Sustainable Agriculture

Crop rotation

Crop rotation is the practice of growing a series of dissimilar types of crops in the same area in sequential seasons for various benefits such as to avoid the buildup of pathogens and pests that often occurs when one species is continuously cropped. Crop rotation also seeks to balance the fertility demands of various crops to avoid excessive depletion of soil nutrients. It can also improve soil structure and fertility by growing dissimilar types of crop. Crop rotation avoids a decrease in soil fertility, as growing the same crop repeatedly in the same place eventually depletes the soil of various nutrients.

A crop that leaches the soil of one kind of nutrient is followed during the next growing season by a dissimilar crop that returns that nutrient to the soil or draws a different ratio of nutrients, for example, Paddy followed by cotton.

By crop rotation farmers can keep their fields under continuous production, without the need to let them lay fallow, and reducing the need for artificial fertilizers, both of which can be expensive. Rotating crops add nutrients to the soils.

Selection of crops in rotations

- » Adaptable to the local climatic conditions.
- » Have demand in the market.
- » Short duration.
- » Timing of input requirements should differ.
- » Maturity of crop should match with the market demand.
- » Crop demand should match with the available resources.
- » Resistant to pest and diseases
    a. Soil amendment
    b. Low External Input Sustainable Agriculture or Low Input Sustainable Agriculture (LEISA/LISA)
    c. Sustainable agro ecosystem
    d. Organic farming

Organic farming

Food safety and quality issues are receiving a great deal of attention today. Organic farming is such a method of cultivation that does not involve the use of artificial inputs like chemicals to either enrich the soil, fight pests or increase productivity. This method of farming provides for food produce that is healthy for man's consumptions at the same time protects the soil and the environment. India has been consistent increase in areas where new farmers are adopting the organic practices every year. According to the apex body for organic products production and exports in India-Agricultural and Processed Food Products Export Development Authority (APEDA), the area under certified organic farming in India is estimated at 2.5 million hectares including 2.4 million hectares in the forest areas of M.P. and U.P.

Components of organic farming

Thus organic agriculture is comparatively free from the complex problems identified with modern agriculture. It is basically a farming system, devoid of chemical inputs, in which the biological potential of the soil and the underground water resources are conserved and protected from the natural and human induced degradation or depletion by adopting suitable cropping models including agro-forestry and methods of organic replenishment, besides natural and biological means of pest and disease management, by which both the soil life and beneficial interactions are also stimulated and sustained so that the system achieves self regulation and stability as well as capacity to produce agricultural outputs at levels which are profitable, enduring over time and consistent with the carrying capacity of the managed agro-ecosystem. Crop production and health in organic farming systems is attained through a combination of structural factors and tactical management components to ensure products of sufficient quality and quantity for human and livestock consumption.

Soil fertility management: The aim of nutrient or soil fertility management within organic farming systems is to work, as far as possible, with in a closed system .Organic farming aims to manage soil fertility through use of organic manures (FYM & farm compost, vermicompost), recycling of crop residues such as straw, plant residues, grasses etc., dung and urine from domesticated animals and wastes from slaughter houses, human excreta & sewage, biomass of weeds, organic wastes from fruit and vegetable production & processing units and household wastes, sugarcane trash, oil cakes, press mud and fly ash from thermal power plant. Biological nitrogen fixation through blue green algae, azolla for rice, rhizobium for legumes, azatobactor & azospirillum for other crops, green manuring & green-leaf manuring, manure form biogas plants, legumes in crop rotations & intercropping systems.

Weed control: Organic farmers often identify weeds as their key problem. Within organic systems an integrated approach to weed control using a combination of cultural and direct techniques is necessary. Appropriate soil cultivation viz., deep ploughing in summer, harrowing, inter-cultivation using mechanical hoes and harrows, and the timing of field operations and good crop establishment are vital for successful control of weeds. Mulching the soil surface can physically suppress weed seedling emergence. Soil solarization, to heat field soil under plastic sheeting to temperatures high enough to kill weed seeds (>65 °C) can also be used for weed, control in some parts of India. Good seedbed preparation, timely sowing, line sowing, crop rotation, smoother crops & intercropping systems etc., suppress the weed growth and favour normal growth and development of crops in organic systems.

Natural pest and disease control: One of the important features of organic farming is the exclusion of plant protection chemicals for pest and disease control. The system relies on the on-farm diversity, improved health of the soil and crops, protective influence of beneficial soil organisms against soil borne pathogens and use of plant based insecticides and biological control measures. The population of naturally occurring beneficial insects and other organisms which act as bio control agents multiplies making natural control of pests possible when the system is free from the indiscriminate use of chemicals.

Few examples are:

- Manipulation of crop rotations, to minimize survival of crop specific pests (in the form of, for example insect eggs, fungi) which can infest the next crop.
- Strip cropping, to moderate spreading of pests over large areas.
- Manipulation of the moisture level or pH level of the soil (in irrigated areas).
- Manipulation of planting dates, to plant at a time most optimal for the crop, or least beneficial for the pest.
- Adjustment of seeding rate, to achieve an optimal density given the need to check weeds or avoid insects.
- Use of appropriate plant varieties for local conditions.
- Biological control methods, to encourage natural enemies of pests by providing habitat or by breeding and releasing them in areas where they are required.

- *Bacillus thuringensis* against caterpillars of *Heliothis, Earias, Spodoptera* etc.
- *Pseudomonas fluoroscenes* against *Pythium spp., Rhizoctonia spp., Fusarium spp.*
- Nematodes like Green commandoes and Soil commandoes against caterpillars & grubs
- Nuclear Polyhedrosis virus (NPV) against caterpillars
- *Trichoderma virdi* against many common diseases of vegetables and spices
- Weevils *Neochitina eichorniae* & *N. bruchi* against water hyacinth
- Beetle *Zygogramma biocolorata* against parthenium

» Trapping insects, possibly with the use of lures such as pheromones

» Use of domesticated birds

» Biological pesticides (for example neem oil, nicotine) of which the active ingredient is short-lasting, and which may be produced locally.

Integrated nutrient management

Integrated nutrient management system envisages conjunctive use of organic manures, crop residues, biofertilizers, legumes in crop rotation and green manuring. It combines traditional and improved technologies to gain from the symbiosis and synergy of crop- soil-environment bio-interactions. The concept is for optimization of all available sources of plant nutrients to improve soil fertility availing nature's gifts. Development of INM system involving and appropriate mix of organics, biological N fixation, phosphate solubilising microbes, and need based chemical fertilizers would be crucial for sustainability of production and soil as a resource base for it.

Bulky organic manures: In India, the estimate production of rural compost is about 226 million tons and urban compost of 6.6 million tons annually. Aggregate stability, decrease in pH, resistance to compaction, infiltration and water holding capacity. Proper methods of preparation of FYM/Compost therefore have to be popularized.

Recycling of organic wastes: Substantial quantity of crop residue (about 350 million tons) is produced in India. Crop residues in combination with organics have been shown to improve availability of plant nutrients, soil organic matter, aggregate stability, infiltration rate, microbial population etc.

Bio-fertilizers: Bio-fertilizers such as *rhizobium* culture is an effective source of N supply to leguminous corps. *Azotobacter* and *Azospirillium* help in N fixation and supply to crops like rice, wheat, sorghum, maize, cotton, sugarcane, fruit corps and vegetables. Phosphate solubilising bacteria viz., *Bacillus aspergillus* helps in making available soil P to the crops and increases the solubility of indigenous sources of P like rock phosphate. Blue green algae and *Azolla* have shown promise in low land rice. These are renewable and environment friendly supplementary sources of nutrients and are presently being used in quantities between 8-10 tons per year. *Vesicular arbuscular mycorrhiza* (VAM) has beneficial effect on plant growth, particularly in P deficient soils. Improved uptake of water, production of plant hormones and microbial activity are the prime benefits of *mycorrhizal* inoculations.

Determinants of crop diversification for sustainability

Number of factors governs nature and speed of crop diversification:

- » Resource endowments
- » Agro climatic conditions
- » Soil
- » Labour
- » Facility of irrigation
- » Technological factors
- » House hold factors
- » Institutional and infrastructural factors
- » Price factors

In Punjab, sustainability in agriculture is possible and essential to save the crumbling agriculture economy and environment of Punjab. However the process and strategies of making it happen are not as easy as said. People who are actual players in the field have a definite mindset and conditioned behaviour. Conducive conditions are another aspect. Suggestion alone is not the solution to the problem. There is need to think of practical and workable strategies. First of all, it must be understood that diversification is a dynamic phenomenon and can be multidimensional. In the present context, there may be two-pronged diversification:

Crop-wise diversification is related to crops outside the normal cycle of paddy and wheat and also to the shift from one variety of rice and/or wheat to some other variety that can be more useful and relevant.

Area-wise diversification is that certain areas may be identified for one set of crops while other areas for another. An added advantage of this type of diversification may be in the form of marketing management.

The following suggestions may be considered for area-wise diversification:

- There should be a survey by soil testing in various parts of the state because over fertilisation and pesticides have disturbed the macro/micro nutrients of the soil. In accordance with the results of these tests and the agro-climatic conditions, an action plan for area-wise diversification, incentives and marketing may be prepared.

- Diversification may be between two varieties of rice, for example a shift from paddy rice to superior quality basmati rice and between two crops, or to a shift from paddy rice cycle to pulses, oil seeds, floriculture, sugarcane and horticulture.

- The Kandi belt and Shivalik foothills of Doaba, the Malwa belt, Pathankot, Hoshiarpur and Ropar can be encouraged to grow lichi, mango and citrusfruits.

- Certain pockets in Patiala district are ideal for guava, ber and mango. Faridkot and Abohar are suitable for guava, citrus fruits like kinnow, red blood malta, etc.

- The Barnala area is good for grapes and citrus fruit. Bathinda, Abohar and Sangrur are good for cotton and groundnut.

- The Majha area, in particular the Ravi belt, can opt for superior quality basmati rice for the domestic market as well as export because of the agroclimatic conditions-Gurdaspur district and pockets in Amritsar and Kapurthala districts are ideal for this purpose.

- The rest of the area can be used to grow sugarcane, barley, oil seeds, pulses, groundnut, soyabean, maize, sunflower etc.

- It is also suggested that farmers may be encouraged to set up fishery, poultry, piggery, dairy, etc.

- The state government should initiate a process of agro-based industrialisation with the help NRI/private entrepreneurs.

» The required infrastructure facilities to facilitate sorting, grading, packaging along with cold storage, dehydration units and movement of produce from the producer to the consumer in the shortest time may be developed. The government should identify one centre each in Pathankot, Hoshiarpur, Rajpura and the Sangrur area for these activities.

» Before setting up a medium to large sized food processing units, the entrepreneurs must expose themselves to the international and national market. They must focus on the market share rather than on strategies to pay less to the farmers; only then the farmers at the grassroots level may be ready to diversify.

» The government has to come forward for the supply of quality seeds/sapling.(www.ggssc.net/files/pdf/crop_diversification).

In Punjab, Agricultural production in Punjab has been characterised by a sharp decline in diversity in the cropping pattern and the emergence of wheat-rice specialisation over the past few decades. This declining diversity has serious repercussions in terms of overuse of natural resources, ecological problems and growing income risk. As diversity in the production pattern declines, variability in the gross value of production also increases.

Crop diversification has been perceived as the only alternative to end the problems arising due to paddy- wheat rotation. The present study was aimed at quantifying the shift in area between the year 2002-2006 from paddy and wheat to other crops in general and to vegetables in particular. The data collected personally on a structured interview schedule from six agro-climatic zones of Punjab reveals that majority of the vegetable growers were growing vegetables in less than 3.5 acres of the land and the share of vegetable crops was maximum in the area under diversification. However, area under vegetable crops increased by 0.08 per cent and 2.66 per cent during and *Rabi* season respectively. With vegetable cultivation emerging as major alternative preferred by farmers, the extension personnel and the policy makers should address the problem areas in vegetable cultivation and its marketing.

Scope of Sustainability in Cropping System

Saving of soil resource

When farmers grow and harvest crops, they remove nutrients from the soil, without replenishment, land suffers from nutrient depletion and becomes either unusable or suffers from reduced yields. . The farmer who reduces purchased chemical fertilizer use by relying more heavily on green-manure crops in a rotation to improve soil fertility will likely experience some change in the pattern of profitability over time. Under sustainable agriculture possible sources of nutrients that would be available indefinitely, include:

- » Recycling crop waste and livestock manure (FYM, Compost, and Vermicompost).
- » Growing legume crops and forages such as peanuts or alfalfa that form symbioses with nitrogen-fixing bacteria called *rhizobium*.
- » Crops that require high levels of soil nutrients can be cultivated in a more sustainable manner with certain fertilizer management practices.
- » Soil erosion can also be control by no-tillage, wind-breaks to hold the soil, incorporate all nutrients back to soil.
- » Growing of cover crops such as cowpea, sweet potato and stop using chemical fertilizer which have sufficient amount of salts.

Common green manure crops and their N content

Crop species	Biomass	N (%)
Sunnhemp	21.2	0.43
Dhaincha	20.0	0.43
Pillipersara	18.3	1.10
Mungbean	8.0	0.53
Cowpea	15.0	0.49
Guar	20.0	0.34
Senji	28.6	0.57
Khesari	12.3	0.54

Source: Abrol and palaniappan (1987)

It is generally recognized that any type of agricultural land use will result in a significant loss in topsoil. Even idle land in grass steadily loses topsoil. An environmental goal of a sustainable cropping system might not be to actually increase the quantity of topsoil on the farm, but rather to employ a cropping system that minimizes the amount of topsoil loss over time.

Saving of water resources

Considering the fact that 80 to 90% of the developed water resources are diverted towards irrigated agriculture, its efficient management is of crucial importance. For irrigation systems to be sustainable they require proper management and must not use more water from their source than is natural replenished, otherwise the water source becomes a non-renewable resource. Improvements in water well drilling technology and submersible pumps combined with the development of drip irrigation and sprinkler system have made it possible to regularly achieve high yield crops as well as by conserving water resources. Sustainable recommendations have been evolved for efficient use of irrigation water.

- » On-farm water management technology such as land shaping, land levelling and efficient design and layout of irrigation methods.
- » Furrow/alternate furrow/skip furrow irrigation for all row crops.
- » Adopting proper cropping systems in place of continuous lowland rice.
- » Conjunctive use of surface and ground water and canal and tank command area for increasing the productivity of crops.
- » Introduction of sprinkler and drip irrigation systems in water scarcity areas.

It is a better strategy to increase irrigated area than to increase irrigation intensity. A close coordination between irrigation engineers and agronomists will result in improvements.

Management of rain water

There is wide variability in rainfall occurrence over space and time. About 72% of the cultivated area of the country depends entirely upon rainfall. Hence, efficient rain water management is of prime importance for insuring stability and sustainability in agricultural production. Options for sustainable management are:

- » Agro-climatic analysis that gives an index of probability of rainfall.
- » Contingent crop planning for aberrant weather conditions.
- » In-situ soil moisture conservation practices.
- » Runoff harvesting and recycling

Rainwater Management is focused on providing the right storm water solution for your project. With a range of technologies that provide storm water filtration, separation, screening, absorption, recycling, infiltration and detention, Rainwater Management can provide you with an engineered solution for many of your needs.

Water is a precious resource that most of us take for granted. We utilize it in everything that we do. Our increasing population is putting larger demands on our water resources. Rainwater Management is committed to offering solutions to help lessen our negative impact on our waterways.

Managing rainwater and drainage

We are focusing on making the most of rain when it falls and on managing the flood risk from heavy rainfall. Putting rainwater is to good use. Using rainwater rather than let it all go down the drain can help to relieve pressures on the drainage system – reducing flood risk and the demand for fresh water.

Encourage the practices

» Green roofs – for example, plant-covered roofs that make good use of the rain that falls on them

» Rainwater harvesting - collecting rainwater to water your garden, for example

» Grey water recycling - for example, using wastewater from baths to flush loos

» Sustainable drainage – which helps reduce the volume and speed of water flowing into sewers

Making sure heavy rainfall doesn't result in flooding

London is vulnerable to surface water flooding. Heavy rainfall can swiftly overwhelm the drainage network, leading to flooding of low-lying areas. As climate change increases the frequency and intensity of heavy rainfall and London's growth puts added pressure on the drainage network – the risks of floods rise ever higher as do the consequences.

Storm water (Rainwater) management plans

Another important aspect of water and watershed planning is the management of rainwater or storm water. All communities in British Columbia are subject to rainfall events and need to provide adequate drainage to prevent localized flooding. It is widely recognized that urban development can increase impervious surfaces, impacting the hydrology of local streams. Rural resource activities such as forestry can also influence land cover, and thus impact, hydrology.

Projections of increased storm water runoff as a result of climate change and urban development have led some municipalities to seek additional funds for infrastructure upgrades, while other communities are examining alternative approaches to managing rainwater. In addition to, or instead of, building bigger pipes, ditches and pumps to convey water from bigger storms, some communities are taking the approach of "store it, spread it, and sink it". Rainwater and storm water can be viewed as a valuable resource that can be stored in wetlands and detention ponds and infiltrated into the ground to recharge water tables. This can reduce or avoid many of the economic and environmental costs associated with the traditional model of conveying rainwater as quickly as possible from roads, rooftops and parking lots into storm sewers, drainage ditches and streams. Urban storm water runoff can adversely impact stream hydrology, fish habitat and watershed health by eroding stream banks, by causing water turbidity and siltation of spawning and rearing habitat, and by introducing other pollutants, such as oil from roads, into the system.

Under the *Local Government Act*, municipalities are responsible for the provision of drainage, and in some cases, Regional Districts may also have associated responsibilities. How we handle storm water has a huge impact on aquatic ecosystems. Integrated storm water management planning is a proactive process that utilizes land use planning tools to protect property and aquatic habitat from storm water flows, while at the same time accommodating urban growth. Ideally, the aim of these plans is to ensure that storm water runoff resembles natural runoff patterns (i.e., volume and timing of surface water runoff), and does not transport pollutants or sediment from the land into watercourses. Storm water management plans are typically completed at the local government level, with guidance from the Province.

BC's Storm water Planning Guidebook, released in 2002, is premised on the idea that land development and watershed protection can be compatible. It assumes that municipalities exert control over runoff volume through their land development and infrastructure policies, practices and actions.

Integrated Storm water Management Plans (ISMPs) are required to be developed and implemented by Metro Vancouver's member municipalities in accordance with the Integrated Liquid Waste and Resource Management Plan. To help facilitate this, Metro Vancouver developed a comprehensive ISMP Template (consistent with the provincial guide) to guide member municipalities with this process. The updated LWMP is pending approval from the Minister of Environment.

Outdated approaches to storm water management, which fail to respect natural systems and water cycles, have been accused of being one of the largest water pollution challenges in the province. In recent years, rainwater management has emerged as a new way of thinking about the precipitation that falls on the land. Instead of viewing storm water as a site-specific problem that is best solved by piping water away from properties into streams, rainwater management considers the dynamics of the entire watershed and identifies how development can use "green infrastructure" to maintain natural systems and protect buildings. Green infrastructure is a concept that emphasizes the importance of the natural environment in decisions about land use planning. For example, the installation of permeable pavements, rain gardens, bioretention ponds and constructed wetlands help reduce the volume of runoff that enters sewer systems and increase absorption. Instead of relying heavily on pipes and concrete, green infrastructure takes advantage of the natural absorption, storage, evaporation and filtration services that nature provides. As opposed to the quick, high-impact flush that comes with traditional approaches to storm water management, lower impact green development seeks to mimic the natural water cycle by allowing water to infiltrate down through the ground and slowly release into the watershed.

Climate change and in particular the potential for more frequent and more intense extreme precipitation events is an important consideration in this type of planning. Urban flooding is now the leading cause of home insurance claims in Canada, and is a priority of the insurance industry. There is limited science on which to base projections for a specific location; however, some jurisdictions have identified voluntary planning contingencies for extreme precipitation based on the best science available. For example, on Vancouver Island, the Capital Regional District has identified a planning contingency of a 15 per cent increase in frequency and duration of winter storms for the next 100 years. These contingencies should inform decisions about which storm water management techniques to adopt and what the management goals should be in terms of rate of runoff.

Characteristics, benefits and applications

Under the *Environmental Management Act* (Municipal Sewer Regulation), storm water management planning is a formal requirement of Liquid Waste Management Plans, which is one of several mechanisms used by the Province to regulate storm water in BC. Although municipalities are generally not required to have Liquid Waste Management Plans, they often opt to complete such plans because the plans allow a suitable length of time to develop and implement effective and affordable solutions.

The outcome of integrated storm water planning includes regional or watershed level objectives and priorities, integration of these objectives into community planning, and implementation of on-site practices that reduce volume and rate of run-off and improve water quality. Storm water management plans need to be integrated with Official Community Plans and zoning bylaws that regulate the location of development and density of use.

Some benefits of undertaking storm water management planning and implementing completed plans include:

» Protection of the aquatic environment, including water quality and stream flow, and protection from flooding

» Protection of community assets and infrastructure from localized flooding

» Protection of water supply (e.g., groundwater recharge areas)

» Management of erosion and sedimentation processes and

» Protection of aesthetic values and recreational uses of water.

Key elements and steps

The Storm water Planning Guidebook provides a detailed explanation of the various stages of preparing a plan. It outlines three key steps and associated methods that work towards integrated storm water management solutions, and five guiding principles to uphold in the process, as outlined below. They include the following:

Storm water planning steps

1. Identify at-risk catchments,
2. Set preliminary performance targets and,
3. Select appropriate storm water management site design solutions.

Guiding principles of Integrated Storm water Management (ADAPT)

1. Agree that storm water is a resource,
2. Design for a complete spectrum of rainfall events,
3. Act on a priority basis in at-risk catchments,

4. Plan at four scales (regional, watershed, neighborhood and site),

5. Test solutions and reduce costs by adaptive management.

Storm water is any water running off a land surface before it reaches a natural water body. It occurs when the rate of precipitation is greater than it can infiltrate, or soak, into the soil. Runoff also occurs when the soil is saturated. Runoff remains on the surface and flows into streams, rivers, and eventually large bodies such as lakes or the ocean. Movement of this storm water across the soil causes erosion. It can also carry and deposit untreated pollutants, such as sediment, nutrients and pesticides, into surface-water bodies (as seen on the left). Impervious surfaces such as driveways, sidewalks, and streets block rainfall and other precipitation from infiltrating naturally into the ground, leading to even more storm water and potential pollutant runoff. Rainwater harvesting offers a small-scale best management practice to reduce storm water runoff and the problems associated with it. By harvesting the rainfall and storing it, you can slowly release the water back into the soil, either through irrigation or direct application. The water then moves into groundwater table, providing a steady supply of water to local streams and rivers.

Rain Gardens: One method of storm water control that also adds an aesthetic benefit to your landscape is the rain garden. Please follow this link to go the rain garden page.

Storm water Management: Pollutants in storm water discharges remain a significant source of environmental impacts to water quality. The Clean Water Act regulates certain discharges of storm water. Learn which sources are regulated and what can be done to control storm water.

Rainwater Harvesting: A soil storage and infiltration system collects rainfall runoff from the roofs of buildings and directs it underground where it infiltrates the soil. Such a system conserves water and protects it from surface pollution.

Conservation of environment

» Minimizing the use of non-renewable resources such as natural gas (used in converting atmospheric nitrogen into synthetic fertilizer), or mineral ores (phosphate).

» Judicial use of chemical fertilizers and pesticide which are responsible for the problem of nitrate accumulation in ground water (permissible limit 45 mg/litre nitrate but in Muktsar 600 mg/litre nitrate found).

» Residue management by incorporating them in soil instead of burning because

- burning of residues creates air pollution deteriorates fertility status of soil and kills beneficial organisms present in soil.
- Recycling of crops wastes and animal dung's as manure and treating sewage water for irrigation purposes are sustainable approaches toward conserving the environment.

Pest management

Pest management in cropping system is carried out by:

- A number of cultivars resistant to pests and diseases have been developed. Selection of such cultivars can bring down the pest and disease incidence considerably.
- Ploughing brings about unfavourable conditions for multiplication of pests. Summer ploughing reduces the reserve of several pests.
- Early sowing in the season considerably reduces the pest and disease incidence.
- Mechanical or manual inter cultivation suppresses the pests similar to that of preparatory tillage. Microclimate conducive to pest and disease build up will be disturbed by inter cultivation.
- Many pests prefer feeding on a particular plant to others. Groundnut as component crop in cotton reduces pest load on cotton. Based on the pests infesting a particular crop, companion or border crops can be grown to minimise the incidence of pests.

Research priorities in sustainable agriculture

India's National Agricultural Policy accords high priority to the sustainability of agriculture. ICAR and the State Agricultural Universities, which comprise the National Agricultural Research System (NARS), also emphasize the importance of incorporating the sustainability perspective into their research and education programmes. But this requires an analytical framework for sustainable agriculture that can guide a transition from research and education directed towards productivity goals to research and education that addresses productivity issues keeping sustainability concerns in sight. Identifies the challenges posed to the existing agricultural research and education systems in India in the transition towards sustainable agriculture.

The National Agricultural Policy of the Government of India aims at agricultural growth (4% annually to 2020) with sustainability, by a path that will be determined by three important factors: technologies, globalization, and markets. Agricultural research and education of the future must therefore address two related challenges: increasing agricultural productivity and profitability to keep pace with demand, and ensuring long-term sustainability of production. The National Agricultural Research System (NARS) deals with the first challenge. Development of short-duration, high-yielding cultivars, irrigation, and intensive use of fertilizers and other agro-chemicals provided the technological basis for increasing agricultural production and the green revolution. Central to the adoption of green revolution technologies were the micro or farm economics - which governed the use of inputs such as land, cultivar, labour, machinery, and chemicals balanced against profits from crop yields - and the macro economics that ensured better access to inputs and markets. The research and education systems have evolved within this framework with a commodity/ productivity focus.

Sustainability as a goal of agricultural research and development is a relatively recent concept. In recent years, national and international research organizations have responded to the increasing importance of sustainability in agricultural development. But integrating the concept of sustainability into the institutional strategy and design of research and education programmes is proving difficult because sustainability requires dealing with interactions between technology, society, and environment and therefore with multiple stakeholders. It broadens the both scope and scale of agriculture from farm production and profitability to agribusiness issues that encompass regional and global development and environmental concerns. To address such concerns, a transparent and integrative analytical framework is needed through which the stakeholders can understand the interactions and discuss, define structure, formulate, and measure concepts, issues, and criteria related to agricultural sustainability and its assessment. This process is necessary for the effective redesign of agricultural research and education systems for sustainability.

Any definition of sustainability must recognize its multiple dimensions: physical, economic, ecological, social, cultural and ethical. Sustainability can be defined only in the boundaries of a system's framework, that is, after specification of what is to be sustained. Choosing the boundary is difficult because agricultural systems operate at multiple levels: soil-plant system, cropping system or farming system, agro-ecosystem and so on to higher regional, national, and global levels. The level chosen thus also defines the spatial scale of operation for the definition. Decisions at the farm level have impacts at the agro-ecosystem and higher levels and vice versa. The linkages between agricultural systems at different levels of hierarchy (spatial scales) are important. By this definition, an agricultural production system is sustainable if, over the long term, it enhances or maintains the productivity and profitability of farming in the region, conserves or enhances the integrity and diversity of both the agricultural production system and the surrounding natural ecosystem, and also enhances health, safety, and aesthetic satisfaction of both consumers and producers. Reduced use of synthetic chemical inputs, biological pest control, use of organic manures, soil and water conservation practices, crop rotations, biological nitrogen fixation, etc., are all relevant and important technological components of sustainable agriculture. But central to the concept of sustainability is the integration of these components in systems 3framework at specified levels and to meet specified objectives. The above definition may be considered an acceptable starting point for the Indian NARS as well.

Sustainable agriculture is the practice of farming using principles of ecology, the study of relationships between organisms and their environment. It has been defined as "an integrated system of plant and animal production practices having a site-specific application that will last over the long term:

» Satisfy human food and fibres need

» Enhance environmental quality and the natural resource base upon which the agricultural economy depends

» Make the most efficient use of non-renewable resources and on-farm resources and integrate, where appropriate, natural biological cycles and controls

» Sustain the economic viability of farm operations

» Enhance the quality of life for farmers and society as a whole.

Sustainable agriculture in the United States was addressed by the 1990 farm bill. More recently, as consumer and retail demand for sustainable products has risen, organizations such as Food Alliance and Protected Harvest have started to provide measurement standards and certification programs for what constitutes a sustainably grown crop.

Implications for research

The above framework helps define a vision for agricultural research, target it effectively to prioritize investments and set production and productivity goals at various levels that match national goals. But, it calls for a major paradigm shift in agricultural research and education from the current commodity and input- based approach to management of agricultural resources, to and approach that emphasizes a systems framework and process-knowledge based management to increase production. The new emphasis is on alternatives to agrochemical use and increasing the rates of existing biological processes to control nutrient cycling and pests. The concept of economic discounting of future value of natural resources is also altered. All of these will place far greater demands on research capacity and farmer knowledge (Lynam and Herdt, 1989). They will also require agricultural research to become more grounded in theory than it has been so far. Other major issues for agricultural research policy and design are described below:

- **Characterization of systems:** Research designs for sustainability will require clear characterization of production systems, agro ecosystems and their boundaries, the marketing systems, and the linkages between them. Setting objectives: An appropriate balance is required between commodity focused research based on intensive use of agrochemicals (which formed the research paradigm up to now and which was responsible for the green revolution) and the resource management focused research (which forms the backbone of research within the sustainability paradigm).

- **Research prioritization:** Higher priority would be needed for research on systems which currently are tending towards unsustainability. And to problems which are contributing most to the degradation of the system.

- **Externalities and measurement of sustainability:** Research will have to be initiated on identification and measurement of externalities and tradeoffs to develop sustainability indicators for agricultural systems at different levels. This research will be interdisciplinary and will need interactions with economics and ecology, and between theory and experiment.

- **Farmer response:** The demands on farmer knowledge and responses will be much higher for sustainable agriculture than for traditional agriculture.

- **The organizational challenge:** Incorporating the sustainability perspective into research policy, design, and management will require important organizational changes in the Indian NARS. Major changes in all the three vital organizational components, namely, its structure, systems, and skills will be required.

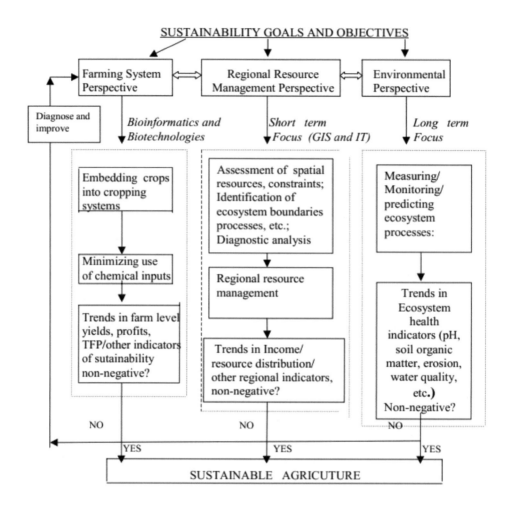

Fig. 1: Analytical framework for sustainable agriculture (adopted from Barnett *et al*, 1995)

Areas of Concern Where Change is Most Needed

Food and agricultural policy: Existing federal, state and local government policies often impede the goals of sustainable agriculture. New policies are needed to simultaneously promote environmental health, economic profitability, and social and economic equity. For example, commodity and price support programs could be restructured to allow farmers to realize the full benefits of the productivity gains made possible through alternative practices. Tax and credit policies could be modified to encourage a diverse and decentralized system of family farms rather than corporate concentration and absentee ownership. Government and land grant university research policies could be modified to emphasize the development of sustainable alternatives. Marketing orders and cosmetic standards could be amended to encourage reduced pesticide use. Coalitions must be created to address these policy concerns at the local, regional, and national level.

Land use: Conversion of agricultural land to urban uses is a particular concern in California, as rapid growth and escalating land values threaten farming on prime soils. Existing farmland conversion patterns often discourage farmers from adopting sustainable practices and a long-term perspective on the value of land. At the same time, the close proximity of newly developed residential areas to farms is increasing the public demand for environmentally safe farming practices. Comprehensive new policies to protect prime soils and regulate development are needed, particularly in California's Central Valley. By helping farmers to adopt practices that reduce chemical use and conserve scarce resources, sustainable agriculture research and education can play a key role in building public support for agricultural land preservation. Educating land use planners and decision-makers about sustainable agriculture is an important priority.

Labour: In California, the conditions of agricultural labour are generally far below accepted social standards and legal protections in other forms of employment. Policies and programs are needed to address this problem, working toward socially just and safe employment that provides adequate wages, working conditions, health benefits, and chances for economic stability. The needs of migrant labour for year-around employment and adequate housing are a particularly crucial problem needing immediate attention. To be more sustainable over the long-term, labour must be acknowledged and supported by government policies, recognized as important constituents of land grant universities, and carefully considered when assessing the impacts of new technologies and practices.

Rural community development: Rural communities in California are currently characterized by economic and environmental deterioration. Many are among the poorest locations in the nation. The reasons for the decline are complex, but changes in farm structure have played a significant role. Sustainable agriculture presents an opportunity to rethink the importance of family farms and rural communities. Economic development policies are needed that encourage more diversified agricultural production on family farms as a foundation for healthy economies in rural communities. In combination with other strategies, sustainable agriculture practices and policies can help foster community institutions that meet employment, educational, health, cultural and spiritual needs.

Consumers and the food system: Consumers can play a critical role in creating a sustainable food system. Through their purchases, they send strong messages to producers, retailers and others in the system about what they think is important. Food cost and nutritional quality have always influenced consumer choices. The challenge now is to find strategies that broaden consumer perspectives, so that environmental quality, resource use, and social equity issues are also considered in shopping decisions. At the same time, new policies and institutions must be created to enable producers using sustainable practices to market their goods to a wider public. Coalitions organized around improving the food system are one specific method of creating a dialogue among consumers, retailers, producers and others. These coalitions or other public forums can be important vehicles for clarifying issues, suggesting new policies, increasing mutual trust, and encouraging a long-term view of food production, distribution and consumption.

Indices of sustainability

Quantification of sustainability is essential to objectively assess the impact of management systems on actual and potential productivity and on environment. One can assess sustainability or several indices. Indices may be simple involving one parameter or complex involving several parameters. Although general principle may be the same, there indices must be fine tuned and adapted under local environments. Some indices of sustainability include the following:

a. Productivity **(P):** Production per unit of resource used can be assessed by

$$P = P/R$$

Where,
    P is productivity
    P is total production
    R is resource used

b. Total Factor Productivity **(TFP):** It is defined as productivity per unit cost of all factors involved.

c. Coefficient of sustainability **(CS):** It is measure of change in soil properties in relation to production under specific management system.

$$Cs = F(Oi, Od, Om)t$$

Where,
    Cs is coefficient of sustainability

O_i is output per unit that maximizes per capita productivity or profit
O_d is output per unit decline in the most limiting or non-renewable resource
O_m is the minimum assured output
t is time

The time scale is important and must be carefully selected.

1. **Index of sustainability (Is):** It is a measure of sustainability relating productivity to change in soil and environmental characteristics

$$I_s = f(P_i * S_i * W_i * C_i)t$$

Where,
   I_s is index of sustainability
   S_i is alteration in soil properties
   W_i is change in water resources and quality
   C_i is modification in climatic factor and t is time

2. **Agricultural Sustainability (Ag):** It is a broad based on several parameters associated with agricultural production.

$$A_s = d(P_t * S_p * W_t * C_t)dt$$

Where,

   A_s is agricultural sustainability
   P_t is productivity per unit time of the limited or non renewable resources
   S_p is critical property of rooting depth, soil organic matter content
   W_t is available water capacity including water quality
   C_t is climatic factor such as gaseous flux from agricultural activity
   T is time

Crop productivity as an indicator of sustainability

A measure of crop productivity is a good integrator of all soil, water, climatic and biotic factors. It is important to assess potential vis-a-vis actual productivity. In a science based management system, actual production exceeds potential production in soils of low inherent fertility and in harsh environments. The potential productivity, soils productive potential within a biome, can be estimated by several models e.g., CERES and Tropical soil productivity are Land use

Factor (L), Land Equivalent Ratio (LER) and Area Time Equivalent Ratio (ATER) etc.
Sustainability coefficient (SC): It is complex and a multipurpose index based on a range of parameters and is similar to as it is defined as:

$$Sc + F (Pt*Pd*Pm) t$$

$$Sc = d (Pd*Wt*Ct) dt$$

Where,

Pt is productivity per unit input of limited resource
Pd is productivity per unit decline in soil property
Sc is critical level of soil property
Wt is soil water regime and quality
Ct is climatic factor
t is time

The Land use factor (L) is defined as the ratio of cropping period C plus fallow period F to cropping period C.

$$L = C + F/C$$

The factor L is generally high for low intensity systems e.g., shifting cultivation

Input management for sustainable agricultural systems

1. **Optimizing nutrient availability:** A very important condition for good plant growth and health and, indirectly for good animal and human health is the timely provision of sufficient and balanced quantities of nutrients that can be taken up by the plant roots. Nutrient deficiencies and imbalances are main constraints to crop production, especially in regions with poor or alkaline soils. There is a constant flow of nutrients through the farm. For example, it has been estimated that in Africa nutrient losses through soil erosions and other processes exceed application of artificial fertilizers.

2. **Micronutrient deficiencies:** Due to intensive cropping the micronutrients are removed to a considerable extent, which control various aspects of plant growth. A tudy of Ranchi, India revealed that applying looks NPK (10:25:25) per ha. Led to depletion of Zinc by 0.619/ha and copper by 0.49/ha, this can depress yields by up to 4 t/ha in rice,

2 t/ha in wheat and 3.4 t/ha in maize. Also iron is a limiting factor in rice production in the new rice-wheat rotation evolved in the non-traditional rice growing areas of Punjab. At Punjab Agricultural University, Ludhiana field experimental results proved that application of poultry manure, pig manure and farmyard manure were effective in meeting zinc requirements in a maize-wheat rotation.

3. **Limiting nutrient losses:** Nutrient losses can be limited by:

    » Recycling organic wastes by returning them to the field, either directly or treated (composted, fermented etc.)

    » Applying organic and artificial fertilizers in such a way that nutrients are not leached by excessive rain or volatilized by high temperature or solar radiation.

    » Reducing losses due to run-off and soil erosion.

    » Minimizing nutrient losses due to biomass bussing.

    » Reducing volatilization of nitrogen by denitrification under wet soil conditions.

    » Avoiding leaching by using organic and artificial fertilizers, which release nutrients slowly maintaining high humus content in the soil and intercropping plant species with different rooting depth.

    » Limiting nutrient export in products by producing crops with relatively high economic value relative to nutrient content.

4. **Use of chemical fertilizers:** The use of chemical fertilizers is essential for obtaining high crop yields. However, many small landholders and resource poor farmers cannot offer costly fertilizers. Most soils in the tropics are so deficient in primary nutrients that it is imperative that strategies be developed for adding them from outside the ecosystem. There is some potential for enhancing N supply by biological N fixation. Additional fertilizer N and other nutrients must be supplied. The requirements for chemical fertilizers however can be reduced considerably by decreasing losses, recycling nutrients and through biological N fixation.

5. **Nutrient recycling:** Nutrient recycling or regime is an important strategy for sustainable crop production. It involves returning nutrients removed by crops to the soil for further use. In addition, soil fauna (e.g., earth worms, termites) also play an important role in recycling of plant nutrients. Growing deep-rooted crops is important in order to recycle nutrients from the sub soil by returning them through crop residue to the surface where

the succeeding shallow rooted crops can use them. Use of mulches, incorporation of crop residues and animal waste, growing legumes as intercrops in cereals etc., can substantially reduce chemical fertilizer requirements.

6. **Use of crop residues:** Crop residue contains substantial quantities of plant nutrients. The beneficial effects of returning crop residues as mulch on crop yields are well known. These benefits are not only to the recycling of plant nutrients but also to improvements in soil moisture and temperature, enhancement of soil structure and soil erosion control.

7. **Biological nitrogen fixation:** Augmenting the nitrogen supply to crops through biological nitrogen fixation is a viable officer for resource poor farmers of the tropics. The amount of N fixed by legumes can range from 20-250 kg/ha/yr depending on the species, soil type, climate and agro-eco-region.

8. **Use of biofertilizers:** Biofertilizers have been recognized as important inputs in integrated plant nutrition systems. The use of legume green manure, blue green algae and Azolla for rice: Azotobacter and Azospirillum for wheat, millets and vegetable crops; Rhizobium for pulses and oil legume crops, Phosphate Solubilizers Vesicular Arbuscular Mycorrhizae) for various crops is well reported, on an average these biofertilizers can minimize the use of inorganic N by 25-50 kg/ha.

9. **Green manuring:** Green manuring is the practice of ploughing or turning into the soil undecomposed green plant tissues for the purpose of improving physical structure as well as fertility of soil. The plant material used in this way is called a green manure (GM). The short duration legume crops are grown and buried in the same site when they attain the age of 60-80 days after sowing. This system of on-site nutrient resource generation is most prevalent in northern and southern parts of India where rice is the major crop in the existing cropping systems. Green leaves and tender plant parts of the plants are collected from shrubs and trees growing on bunds, degraded lands or nearby forest and they are turned down or mixed into the soil. It is done 15-30 days before sowing of the crops depending on the tenderness of the foliage or plant parts.

There are now unique opportunities for launching a food-for- sustainable development initiative, in the form of a "grain for green" movement.

Advantages of green manuring

- » Improves the soil fertility by the increasing the humus content of the soil
- » Adds nutrients and organic matters

- » Improves the soil structure
- » Improves soil aeration
- » Increases the water holding capacity of the soil
- » Increases the nutrient holding capacity of the soil
- » Helps control the soil born insects/ mite (neem, Vitex, etc), nematodes (marigold) and diseases (crucifers)
- » Used as a soil binder in the sloppy areas
- » Conserves and recycles plant nutrients
- » Protects the soil from erosion
- » Promotes habitat for natural enemies
- » Serves as a good food for the earthworms

Increases soil biodiversity of beneficial microbes by stimulating their growth.

Approach, Objectives And Components Of Farming System

FARMING SYSTEM APPROACH

It has now been realized that component, commodity and discipline oriented research approach has reached its limits. Any kind of future approach must be holistic involving biophysical and socio-economic settings targeting at interdisciplinary and inter-institutional synergies. The basic idea should be to exploit growth potential of the components with rational and efficient use of natural and other resources with particular reference to rainfed regions. The two issues of paramount importance ire:

» Orientation from cropping system research to farming system research focusing on multiple use of inputs and recycling for reduction in cost and resource conservation.

» Integrated farming systems of simultaneous or sequential production of crops, trees, livestock and fisheries after their proper assessment and evaluation that is expected to expand livelihood basket. Assessment of farming systems has to be carried out at micro level for their adoption by the farmers with proper testing and monitoring of the suggested farming system. There is also an urgent need of developing a sustainable farming system model for marginal, small, medium and large farm situations in all regions of the country.

Farming System

Farming systems research is considered a powerful tool for natural and human resource management in developing countries such as India. This is a multidisciplinary whole-farm approach and very effective in solving the problems of small and marginal farmers. The approach aims at increasing income and employment from small-holdings by integrating various farm enterprises and recycling crop residues and by-products within the farm itself.

The basic aim of integrated farming system (IFS) is to derive a set of resource development and utilization practices, which lead to substantial and sustained increase in agricultural production. There exists a chain of interactions among the components within the farming systems and it becomes difficult to deal with such inter-linking complex systems. This is one of the reasons for slow and inadequate progress in the field of farming systems research in the country. This problem can be overcome by construction and application of suitable whole farm models. However, it should be mentioned that inadequacy of available data from the whole farm perspective currently constrains the development of whole farm models. Integrated farming systems are often less risky, if managed efficiently, they benefit from synergisms among enterprises, diversity in produce, and environmental soundness.

Definition

» Farming system is a complex inter-related matrix of soils, plants, animals, implements, labour and capital, inter-dependent farming enterprises.

» The farm is viewed in a holistic manner (multi-disciplinary approach)

The Farming System, as a concept, takes into account the components of soil, water, crops, livestock, labour, capital, energy and other resources with the farm family at the centre managing agricultural and related activities. The farm family functions within the limitations of its capability and resources, the socio-cultural setting, and the interaction of these components with the physical, biological and economic factors.
Farming system focuses on:

» The interdependencies between components under the control of household.

» How these components interact with the physical, biological and socio-economic factors, which is not under the control of household.

» Farm household is the basic unit of farming system and interdependent farming enterprises carried out on the farm.

» Farmers are subjected to many socio-economic, bio-physical, institutional, administrative and technological constraints.

» The operator of the farming system is farmer or the farming family.

Objectives of farming system approach (FSA)

» To develop farm - household systems of rural communities on a sustainable basis.

» To improve efficiency in farm production

» To raise farm and family income

» To increase welfare of farm families and satisfy basic needs

	TARGET GROUP FARMER'S CIRCUMSTANCES	
NATURAL	ECONOMIC	SOCIAL AND CULTURE
Climate Soil Biology	Market opportunities Input distribution Credit, Technology, Institution	Tenure Religion Caste/Trie
FAMILIES PRIORITIES		FAMILIES RESOURCES
Food Security Income Social obligations		Land Labour Technical know how Cash
	FARMER'S DECISION ON RESOURCES ALLOCATION AND MANAGEMENT STRATEGIES	
CROPS	LIVESTOCK	OFF-FARM
Rice Maize Cotto Pulses	Cattle Sheep/Goats Poultry	Handicrafts Casual labour Non-farm enterprises
	TARGET GROUP FARMER'S ACTIVITIES	

Source: Collinson, M.P. 1987.

Components of Farming Systems

The potential enterprises which are important in farming system in the way of making a significant impact of farm by generating adequate income and employment and providing livelihood security are as follows:

1. Cropping system

Cropping system is an important farming practice adopted invariably by every farmer. It is an integral part of farm activities in the country. The cropping system should provide enough food for the family, fodder to the cattle and generate sufficient cash income for domestic and cultivation expenses. These objectives could be achieved by adopting intensive cropping. Types of cropping systems include monocropping, multiple cropping. Alteration of crop geometry may help to accommodate intercrops without loosing the base crop population.

i. **Monocropping:** It refers to growing of only one crop on a piece of land year after year. Under rainfed conditions, groundnut or cotton or sorghum are grown year after year due to limitation of rainfall. In canal irrigated areas, under waterlogged condition, rice crop is grown as it is not possible to grow any other crop.

ii. **Multiple cropping:** Growing two or more crops on the same piece of land in one calender year. It is the intensification of cropping in time and space dimensions. It includes intercropping, sequence cropping and multier cropping.

a. **Intercropping**: intercropping is growing two or more crops on the same piece of land with a definite row pattern. Thus cropping intensity in space dimensions is achieved. Based on the per cent of plant population used for each crop in intercropping system it is divided into two types:

   » **Additive series**: one crop is sown with 100 percent of its recommended population in pure stand which as known as base crop. Another crop is grown into the base crop by adjusting or changing crop geometry.

   » **Replacement series**: both the crops are called component crops. By scarifying certain proportion of population of one component, another component is introduced.

b. **Sequential Cropping Systems:** In sequential cropping, two or more crops in sequence on the same piece of land in a farming year. Preceding crop has considerable influence on the succeeding crop. This includes the complementary effects such as release of N from the residues of the previous crop, particularly legume, to the following crops and carries over effects of fertilizer applied to preceding crops. The adverse effects include allelopathy, temporary immobilization of N due to wide C/N ratio of the residues and carry over effect of pest and diseases.

In India, food crop is predominantly grown in most suitable seasons and thus particular food crop is basic to the cropping system followed by the farmers. Accordingly the cropping systems are usually referred to as:

i. Rice-based cropping system

ii. Sorghum-based cropping system

iii. Pearl millet-based cropping system

iv. Wheat and gram-based cropping system

c. **Multi-tier Cropping:** The practice of growing different crops of varying height, rooting pattern and duration is called 'multi-tier cropping' or multi-storied cropping. Multi-storied cropping is mostly prevalent in plantation crops like coconut and areca nut. There is scope for intercropping in coconut garden up to the age of 8 years and after 25 years. The objective of this system of cropping is to utilize the vertical space more effectively. In this system, the leaf canopies of intercrop components occupy different vertical layers. The tallest components have foliage tolerant of strong light and high evaporative demand and the shorter component(s) with foliage requiring shade and on relatively high humidity e.g. coconut + black pepper + cocoa + pineapple.

2. Dairy farming

Dairy farming is an important source of income to farmers. Besides producing milk and/or draft power, the dairy animals are also good source of farm yard manure, which is good source of organic matter for improving soil fertility. The farm byproducts in turn are gainfully utilized for feeding the animals. Though the total milk production in the country as per current estimates have crossed 90 million tones/annum marks, the per capita availability of milk is still about 220g/day against the minimum requirement of 250g/day as recommended by Indian Council of Medical Research.

The dairy sector in India is characterized by very large number of and very low productivity. Around 70 per cent of Indian cows and 60 per cent of buffaloes have very low productivity. This sector is highly livelihood intensive and provides supplementary incomes to over 70 per cent of all rural and quite a few urban households. The sector is highly gender sensitive and over 90 per cent of the households dairy enterprise is managed by family's women folk.

3. Goat and sheep rearing

The system of sheep and goat rearing in India is different from that adopted in the developed countries. In general, smaller units are mostly maintained as against large scale in fenced areas in the developed countries.

i. **Goat Rearing**: In India, activity of goat rearing is sustained in different kinds of environments, including dry, hot, wet and cold, high mountains or low lying plains. The activity is also associated with different systems such as crop or animal-based, pastoral or sedentary, single animal or mixed herd, small or large scale. Goat is mainly reared for meat, milk, hide and skin. Goat meat is the preferred meat in the country. A goat on hoof (live goat) fetches a better price than a sheep on hoof.

ii. **Sheep Rearing:** Sheep are well adapted to many areas. They are excellent gleaners and make use of much of the waste feed. They consume large quantities of roughage, converting a relatively cheap food into a good cash product. Housing need not be elaborate or expensive. However, to protect the flock from predatory animals, the height of the fencing should be raised to two meters.

4. Poultry

Poultry is one of the fastest growing food industries in the world. Poultry meat accounts for about 27 per cent of the total meat consumed worldwide and its consumption is growing at an average of 5 per cent annually. Poultry industry in India is relatively a new agricultural industry. Till 1950, it was considered a back yard profession in India. In the sixties, the growth rate of egg production was about 10 per cent and it increased to 25 per cent in the seventies. The growth rate came down to 7-8 per cent by 1990 due to price-rise in poultry feed. By 2000, the total egg production may reach up to 5000 crores. Broiler production is increasing at the rate of 15 per cent per year. It was 31 million in 1981 and increased up to 300 million in 1995. Nearly 330 thousand tonnes of broiler meat are currently produced. The average global consumption is 120 eggs per person per year and in India, it is only 32-33 eggs per capita year. As per the nutritional recommendation, the per capita consumption is estimated at 180 eggs/year and 9 kg meat/year.

5. Duck rearing

Ducks account for about 7% of the poultry population in India. They are popular in states like West Bengal, Orissa, Andhra Pradesh, Tamil Nadu, Kerala, Tripura and Jammu and Kashmir. Ducks are predominantly of indigenous type and reared for egg production on natural foraging. They have a production potential of about 130-140 eggs/bird/year. Ducks are quite hardy, more easily brooded and resistant to common avian diseases. In places like marshy riverside, wetland and barren moors where chicken or any other type of stock do not flourish, duck rearing can be better alternative.

6. Apiculture

Apiculture is the science and culture of honeybees and their management. Apiculture is a subsidiary occupation and it is a additional source of income for farm families. It requires low investments and so can be taken up by small, marginal and landless farmers and educated unemployed youth.

7. Fishery

Ponds serve various useful purposes, viz., domestic requirement of water, supplementary irrigation source to adjoining crop fields and pisciculture. With the traditional management, farmers obtain hardly 300-400 kg of wild and culture fish per ha annually. However, composite-fish culture with the stocking density of 5000-7500 fingerlings/ ha and supplementary feeding can boost the total biomass production.

8. Sericulture

Sericulture is defined as a practice of combining mulberry cultivation, silkworm rearing and silk reeling. Sericulture is a recognized practice in India. India occupies second position among silk producing countries in the world, next to China. The total area under mulberry is 188 thousand ha in the country. It plays an important role in socio-economic development of rural poor in some areas. In India more than 98% of mulberry-silk is produced from five traditional sericultural states, viz., Karnataka, Andhra Pradesh, West Bengal, Tamil Nadu, and Jammu and Kashmir.

9. Mushroom cultivation

Mushroom is an edible fungus with great diversity in shape, size and colour. Essentially mushroom is a vegetable that is cultivated in protected farms in a highly sanitized atmosphere.

Just like other vegetables, mushroom contains 90 per cent moisture with high in quality protein. Mushrooms are fairly good source of vitamin C and B complex. The protein have 60-70 per cent digestibility and contain all essential amino acids. It is also rich source of minerals like Ca, P, K and Cu. They contain less fat and CHO and are considered good for diabetic and blood pressure patients.

10. Agroforestry

Agroforestry is a collective name for land use systems and technologies, in which woody perennials (trees, shrubs, palms, bamboos etc) are deliberately combined on the same land-management unit as agricultural crops and/or animals, either in some form of spatial arrangement or in a temporal sequence.

The different commonly followed agro-forestry systems in India are: (1) Agri-silviculture (crops + trees), which is popularly known as farm forestry (2) Agri-horticulture (crops + fruit trees); (3) Silvi-pasture (Trees + pasture + animals); (4) Agri-horti-silviculture (crops + fruit trees + MPTS + pasture); (5) Horti-silvi-pasture (fruit trees + MPTs+ Pasture); (6) Agri-silvi-pasture (crops + trees + Pasture); (7) Homestead agroforestry (multiple combination of various components); (8) Silvi-apiculture (trees + honey bees); (9) Agri-pisci-silviculture (crops + fish + MPTS); (10) Pisci-silviculture (Fish + MPTs) etc.

11. Biogas

A biogas unit is an asset to a farming family. It produces good manure and clean fuel and improves sanitation. Biogas is a clean, unpolluted and cheap source of energy, which can be obtained by a simple mechanism and little investment. The gas is generated from the cow dung during anaerobic decomposition. Biogas generation is a complex biochemical process. The celluloitic material is broken down to methane and carbondioxide by different groups of microorganisms. It can be used for cooking purpose, burning lamps, running pumps etc.

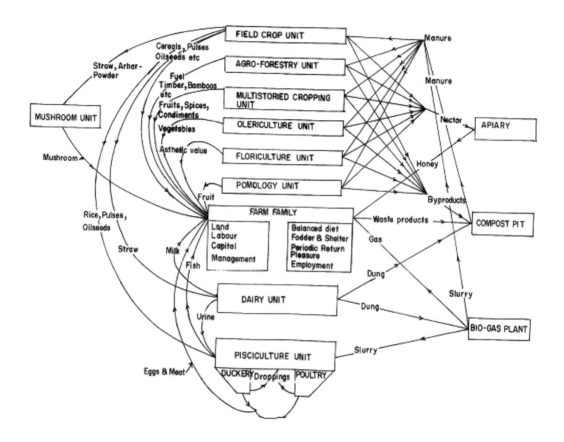

Methodology

1. **Analysis of existing farming system**

    » Identification of major socio-economic situations

    » Understanding dominant enterprises and most common farming systems

    » Analysis of economic viability of existing farming systems

    » Understanding relationship between different enterprises within the system

    » Analysis of linkages between different farming systems

2. **Understanding the modifications made in existing farming system by innovative farmers**

    » Understanding the changing scenario in rural areas and its impact on existing farming system.

» Identification of new market opportunities and its impact and relevance to socio-economic situation

» Suitable modification made by innovative farm families in existing farming system

» Type of modification made (diversification or intensification of the enterprises)

3. **New options recommended by the researchers**

   » Identification of new suggested options by researchers around each dominant enterprise

   » Understanding the technological details about new options

4. **Economic analysis of recommended options**

   » Analysis of relative profitability of recommended options as compared to existing farming system

   » Understanding of implications of each options with regard to reallocation of resource

5. **Testing of recommended options**

   » Selecting representative village / farm families

   » Training about technical skills

   » Testing the effectiveness of recommended options

Integrated Farming System (IFS), a component of farming systems introduces a change in the farming techniques for maximum production in the cropping pattern and takes care of optimal resource utilization.

Unlike the specialized farming system (SFS), integrated farming systems activity is focused round a few selected, interdependent, interrelated and often interlinking production systems based on a few crops, animals and related subsidiary professions.

The on-station study involving enterprises such as crop, fishery, poultry, duckery, apiary and mushroom production revealed that there is chain of interactions among these enterprises. The byproduct of one enterprise may effectively utilize for the other enterprise, thus ensuring higher and efficient resource use efficiency.

Per capita availability of land in India has declined from 0.5 ha in the year 1950-51 to 0.15 ha in the year 2000-01. Due to conversion of valuable irrigated agricultural lands for non-agricultural purposes viz. residential houses, industrial and business establishments and subdivision and fragmentation of holdings, the per capita availability of land is declining day by day. Therefore, no single farm enterprise is able to meet the growing demands of food and other necessities of the small and marginal farmers. Agriculture is in the hands of 125 million farm families of which 75% are the marginal farmers (<1 ha holding). World prices of wheat and rice have declined substantially. The odds are pitted against small and marginal farmers. Burden on marginal farmers are becoming unbearable. Hence, there is necessity of adoption of "Farming Systems approach" by these vulnerable sections of farming community.

13 Degradation And Management Of Land Use Resource

Land use is the human use of land. Land use involves the management and modification of natural environment or wilderness in to built environment such as settlements and semi-natural habitats such as arable fields, pastures, and managed woods. Land management practices describe the way that land is managed - the means by which a land use outcome is achieved. It also has been defined as "the arrangements, activities and inputs people undertake in a certain land cover type to produce, change or maintain it".

Land-use (or Land Resources) Planning is a systematic and iterative procedure carried out in order to create an enabling environment for sustainable development of land resources which meets people's needs and demands. It assesses the physical, socio-economic, institutional and legal potentials and constraints with respect to an optimal and sustainable use of land resources, and empowers people to make decisions about how to allocate those resources.

These are matched through a multiple goal analysis and assessment of the intrinsic value of the various environmental and natural resources of the land unit. The result is an indication of a preferred future land use, or combination of uses. Through a negotiation process with all stakeholders, the outcome is improved, agreed decisions on the concrete allocation of land for specific uses (or non-uses) through legal and administrative measures, which will lead eventually to implementation of the plan.

Soil and Land-use Management is not only important for protecting the still undervalued resource soil but also critical to balance food production and food security with biomass production for energy and carbon sequestration and should be considered a prime control option for water management. In many regions of the world, biomass production and water security is at risk due to the overuse of water resources and soil degradation. Climate change will further exacerbate these risks, particularly in water-limited environments and in marginal regions. It is sure that population growth will increase the demand for food, feed, fiber and wood. At the same time, the accelerated urbanization will not only reduce the amount of land available for biomass production but also increase the risk of soil degradation through waste disposal and air pollution. Soil resources of good quality are limited. Population growth and accelerating urbanization induces a significant stress on soils with good quality. Enhancing and maintaining soil quality is therefore one of the key issues for sustaining livelihoods, human well-being and socio-economic development.

The Soil and Land-use Management section at UNU-FLORES promotes interdisciplinary and transdisciplinary research to make sure that society can benefit from an improved understanding of the nexus of soil, water and waste. Advancing the nexus, the Soil and Land-use Management section develops site and region specific management options for developing and emerging countries which may help to protect and develop soils and water resources, advance the food security and water security, enhance the water use efficiency and the water productivity, maintain and improve soil and water related ecosystem services, and enable the production of biomass under an uncertain climate and degrading and dwindling soils. The work of the Soil and Land-use Management unit is based on the assumption that a sound understanding of the interactions between atmosphere, biosphere, hydrosphere, lithosphere, and pedosphere allows the development and successful implementation of management options, and a reliable, model-based assessment of how these adaptive management concepts prevent a further decline in soils and in water resources.

Definitions

Land use means the purpose to which the land cover is committed. Some land uses, such as agriculture, have a characteristic land cover pattern. These usually appear in land cover classifications. Other land uses, such as nature conservation, are not readily discriminated by a characteristic land cover pattern. For example, where the land cover is woodland, the land use may be timber production, grazing or nature conservation.

Land management practice means the approach taken to achieve a land use outcome - the 'how' of land use (eg cultivation practices, such as minimum tillage and direct drilling). Some land management practices, such as stubble disposal practices and tillage rotation systems, may be discriminated by characteristic land cover patterns and linked to particular issues.

Land cover refers to the physical surface of the earth, including various combinations of vegetation types, soils, exposed rocks and water bodies as well as anthropogenic elements, such as agriculture and built environments. Land cover classes can usually be discriminated by characteristic patterns using remote sensing.

Fractional cover is the fraction of an area (usually a pixel for the purposes of remote sensing) that is covered by a specific cover type such as green or photosynthetic vegetation, non-photosynthetic vegetation (i.e. stubble, senescent herbage, leaf litter) or bare soil/rock. Areas that have been burnt resulting in ash/blackened soil are considered as a 'bare soil' cover type.

Ground cover is a sub-component of land cover and can be used to infer land management practices. Ground cover is defined as the vegetation (living and dead), biological crusts and stone that is in contact with the soil surface. The non-woody ground cover such as crops, grass forbs and chenopod-type shrubs may change monthly rather than annually making this component a good indicator of land management performance. Ground cover is a sub-component of land cover and (from a remote sensing perspective) is the fractional cover of the non-woody understory.

Regulation

Land use practices vary considerably across the world. The United Nations' Food and Agriculture Organization Water Development Division explains that "Land use concerns the products and/or benefits obtained from use of the land as well as the land management actions (activities) carried out by humans to produce those products and benefits." As of the early 1990s, about 13 per cent of the Earth was considered arable land, with 26 per cent in pasture, 32 per cent forests and woodland, and 1.5 per cent urban areas.

As Albert Guttenberg (1959) wrote many years ago, «›Land use› is a key term in the language of city planning.» Commonly, political jurisdictions will undertake land use planning and regulate the use of land in an attempt to avoid land use conflicts. Land use plans are implemented through land division and use ordinances and regulations, such as zoning regulations. Management consulting firms and Non-governmental organizations will frequently seek to influence these regulations before they are codified.

In colonial America, few regulations existed to control the use of land, due to the seemingly endless amounts of it. As society shifted from rural to urban, public land regulation became important, especially to city governments trying to control industry, commerce, and housing within their boundaries. The first zoning ordinance was passed in New York City in

1916, and, by the 1930s, most states had adopted zoning laws. In the 1970s, concerns about the environment and historic preservation led to further regulation.

Today, federal, state, and local governments regulate growth and development through statutory law. The majority of controls on land, however, stem from the actions of private developers and individuals. Three typical situations bringing such private entities into the court system are: suits brought by one neighbor against another; suits brought by a public official against a neighboring landowner on behalf of the public; and suits involving individuals who share ownership of a particular parcel of land. In these situations, judicial decisions and enforcement of private land-use arrangements can reinforce public regulation, and achieve forms and levels of control that regulatory zoning cannot.

Land use and land management practices have a major impact on natural resources including water, soil, nutrients, plants and animals. Land use information can be used to develop solutions for natural resource management issues such as salinity and water quality. For instance, water bodies in a region that has been deforested or having erosion will have different water quality than those in areas that are forested. Forest gardening, a plant-based food production system, is believed to be the oldest form of land use in the world.

The major effect of land use on land cover since 1750 has been deforestation of temperate regions.[More recent significant effects of land use include urban sprawl, soil erosion, soil degradation, salinization, and desertification. Land-use changes, together with use of fossil fuels, are the major anthropogenic sources of carbon dioxide, a dominant greenhouse gas.

According to a report by the United Nations› Food and Agriculture Organisation, land degradation has been exacerbated where there has been an absence of any land use planning, or of its orderly execution, or the existence of financial or legal incentives that have led to the wrong land use decisions, or one-sided central planning leading to over-utilization of the land resources - for instance for immediate production at all costs. As a consequence the result has often been misery for large segments of the local population and destruction of valuable ecosystems. Such narrow approaches should be replaced by a technique for the planning and management of land resources that is integrated and holistic and where land users are central. This will ensure the long-term quality of the land for human use, the prevention or resolution of social conflicts related to land use, and the conservation of ecosystems of high biodiversity value.

The urban growth boundary is one form of land-use regulation. For example, Portland,

Oregon is required to have an urban growth boundary which contains at least 20,000 acres (81 km^2) of vacant land. Additionally, Oregon restricts the development of farmland. The regulations are controversial, but an economic analysis concluded that farmland appreciated similarly to the other land.

Natural Resource Conservation

According to various definitions of wastelands proposed by authors and National wasteland Development board it is estimated the wastelands in India vary from as low as 53 mha to 175 mha. It is found out that over cultivation, uncontrolled grazing and irrational irrigation are the major factors of land degradation. With the help of National Remote Sensing Agency (NRSA) and Survey of India the various types of wastelands have been classified. After understanding the necessity of wasteland reclamation Govt. of India enunciated the land use policy in 1983. Since then National Land Resources Conservation and Development Commission (NLRCDC), the National Land Board (NLB), National Land Use and Wastelands Development council (NLWC), The National wasteland Development Board (NWDB) etc. and many wasteland development schemes has been launched. At present IWDP (Integrated wasteland development Project) is under implementation. The focus and objective are high and it has been successful in region like Sivganga district of Tamilnadu. Though many schemes implemented the wasteland development action programme so far proved ineffective in dealing with the problem. Lessons learnt from review suggest that current strategy need to be modified .It should include diversification of the programme into multidisciplinary effort based on integrated planning done in consultation with local community. The management system of programme would need a greater element of public participation so that its outcome is beneficial to the people and the programme eventually develops into a people's movement.

Wasteland Development

For the policy makers, agriculture scientists and the people who follow development sector the term wasteland is not a new term at all. Wasteland is precisely defined as "those lands (a) which are ecologically unstable, (b) Whose top soil has been nearly completely lost, and (c) which have developed toxicity in root zones for growth of most plants, both annual crop and trees." The National Wasteland Development Board (NWDB) has defined wasteland as "degraded land which can be brought under vegetative cover with reasonable effort and which is currently under utilized and land which is deteriorating for lack of appropriate water and soil management or on account of natural causes". In a sense the two definition of wasteland complement each other and describe most of its features.

Causes of Land Degradation

Just we have studied that wasteland is degraded land. The four main reasons of land degradation are (a) over cultivation (b) deforestation (c) overgrazing and (d) improper irrigation. Let us discuss in some details.

a. **Over cultivation-** The proportion of landless and marginal farmers in India is high. Because of scarcity of land, farming of ecologically vulnerable areas is taken up resulting in erosion and associated land degradation problems.

b. **Deforestation** – Tree are among the most effective preservers of land; however, at present, trees are being cut for various reasons. Let us briefly discuss those.

   » **Fuel wood:** Deforestation for fire wood is estimated at around 5000 hectares annually; the actual figure in fact, could be much higher. In most part of the country there is no cheaper alternative of fire wood. Relative costs of fire wood and its substitute have not been worked out for all the regions, for the simple reason that fire wood is free, just collected from the forests, orchards, canal side plantations or trees along roads and railways. According to planning commission if all the current and projected planting plans are implemented it can only produce about 49 million ton against the requirement of 133 million tons.

   » **Shifting cultivation:** It is a traditional practice locally known as jhum in north eastern region, podu in Orissa and dhya in MP. Due to increase in population and land shortages have resulted in a shortening of the forest fallow period. The jhum cycle, which was 30 to 40 years a few decades back, varies at present between 1 to 17 years. The result is poor crop yields necessitating the shifting cultivators to go in for fresh clearing of forests causing extensive land degradation.

   » **Commercial timber exploitation:** Timber exploitation to meet the ever increasing need of industries and urban areas has been a major cause of extensive deforestation. It is the commercial demand not the local people who cause large scale forest destruction.

   » **Clearing for permanent non forestry purpose:** Activities such as cultivation, human settlement, setting up of industries etc. often bares the soil to ravages of rain and wind, with all the subsequent dangers of flooding, silting and drought when rainfall is markedly seasonal, especially if inappropriate methods of farming are used on erosion sensitive soils.

c. **Over grazing-** Over grazing is as destructive as deforestation but its effects are not immediately noticeable. According to an FAO estimate, one buffalo eats seven tones of leaves (by fresh weight) per year, and a cow two and a half tones –all these leafy material coming from forests adjoining the villages. Thus uncontrolled grazing has contributed to destruction of forests and grasslands thereby exposing the soil to wind and water erosion and consequent land degradation.

d. **Improper Irrigation-** The farmer is totally oblivious of the cost that improper irrigation, viz, over use of water and non provision of adequate drainage, imposes on others. In most cases , especially in large plains, the water table, and capillary arising from the higher ground water level increases accumulation of salts both in ground water and the soil near the surface thus subsequently lowering soil productivity, and in extreme cases, making it unfit for crop production.

Types of wastelands

The National wasteland development Board (NWDB) of the ministry of Environment and forests, in addition to requesting the state governments to conduct detailed surveys and document all categories of land suffering from physical and other deficiencies or under sub optimal use, prepared maps of wastelands in respect of 146 districts in 19 states with the help of National Remote Sensing Agency (NRSA) and Survey of India. Wastelands in the following 13 categories were identified and mapped:

1. Gullies and/or ravines
2. Upland with or without scrub
3. Waterlogged and marshy land
4. Land affected by salinity/alkalinity in coastal and inland areas
5. Land under shifting cultivation
6. Under utilized/degraded notified forest land
7. Degraded pasture/grazing land
8. Degraded land under plantation crops
9. Shifting sands- inland/coastal
10. Mining /industrial wastelands

11. Barren rocky/stony waste/sheet rock areas

12. Steep sloping area

13. Snow covered and/or glacial area.

Structures and Policies

The govt. of India in the early 1970s urged the states to set up state land use board (SLUB) under the chairmanship of the chief minister for providing policy directions and for coordinating activities of the Department concerned with soil and land resources. In 1983, a two tier body, viz. National Land Resources Conservation and Development Commission (NLRCDC) and the National Land Board (NLB) were set up. In 1983 NLB as reconstituted as National Land Use and Wastelands Development council (NLWC) with prime minister as chairman. It emphasized the twin objectives of proper land use and development of watersheds. Under NLWC two separate bodies were created in 1985. These are:

» The National Land Use and conservation Board (NLCB) in place of NLRCDC. The deputy chairman Planning commission is the chairman of the Board. Its members include secretaries of Department concerned of the central GOVT. and representatives from five states selected on rotational basis.

» The National wasteland Development Board (NWDB) with the minister of Environment and forest as chairman. The principal aim of the NWDB was to bring about a qualitative change in the programme of reclamation of wastelands in the country through a massive programme of afforestation with the participation of the people.

The country has paid a heavy price for its neglect of land use policies, especially those relating to the uncultivated land. Land policy is fairly well defined for private agricultural lands but it is not true for uncultivated lands- forest land, grazing lands, barren and unculturable lands which are fast becoming depleted lands. The basic issue is to ensure optimal management of land. The National Land Use policy enunciated by the Govt. of India in 1983 emphasizes the following action points directly relevant to wasteland improvement:

» The problem of wasteland must be tackled on an emergency basis. Programmes must be drawn up for providing vegetative cover to 40 million hectares of degraded forests.

» A massive campaign for increasing the land under productive use for fuel and fodder species needs to be launched.

- » Voluntary effort should be generally preferred in supporting these objectives. Efforts by farmers' cooperatives and voluntary organizations should be fully recognized and assisted.

- » Mining of land for house building material such as bricks can be reduced by developing alternative building materials.

Wasteland Development Schemes

Till the sixth five year plan, no specific programme of wasteland development was taken up. It is only in1985 with establishment of NWDB that the problem of wasteland development received a new thrust. With the setting up of NWDB, a number of new schemes were initiated to secure people's participation, besides continuation of ongoing afforestation schemes. These are:

- » Grants-in-aid to voluntary agencies * Ariel seeding programme
- » Decentralized People's nurseries. * Plantation of minor forest produce
- » Silvipasture farms * Margin money schemes
- » Seed development * Rural employment scheme
- » Area oriented fuel wood and fodder projects

Major ongoing project

IWDP (Integrated wasteland development Project) Scheme: This scheme is under implementation since 1989-90, and has come to this Department along with the National Wastelands Development Board. The development of non-forest wastelands is taken up under this Scheme. The scheme provides for the development of an entire micro watershed in an holistic manner rather than piecemeal treatment in sporadic patches.. The thrust of the scheme continues to be on development of wastelands.

Objectives: The basic objective of this scheme is an integrated wastelands development based on village/micro watershed plans. These plans are prepared after taking into consideration the land capability, site condition and local needs of the people. The scheme also aims at rural employment besides enhancing the contents of people's participation in the wastelands development programmes at all stages, which is ensured by providing modalities for equitable and sustainable sharing of benefits and usufructs arising from such projects.

Activities: The major activities taken up under the scheme are:

- In situ soil and moisture conservation measures like terracing, bunding, trenching, vegetative barriers and drainage line treatment.
- Planting and sowing of multi-purpose trees, shrubs, grasses, legumes and pasture land development.
- Encouraging natural regeneration.
- Promotion of agro-forestry & horticulture.
- Wood substitution and fuel wood conservation measures.
- Awareness raising, training & extension.
- Encouraging people's participation through community organization and capacity building.
- Drainage Line treatment by vegetative and engineering structures
- Development of small water Harvesting Structures.
- Afforestation of degraded forest and non forest wasteland.
- Development and conservation of common Property Resources.

The regions under consideration in India

S. No.	Category of wastelands in India	Area (in sq.Kms.)
1	Snow Covered/Glacial	55788.49
2	Barren Rocky/Sheet Rock	64584.77
3	Sands-inland/coastal	50021.65
4	Land affected by salinity/alkalinity	20477.38
5	Gullied/or ravenous land	20553.35
6	Upland with or without scrub	194014.29
7	Water logged & Marshy	16568.45
8	Steep sloping area	7656.29
9	Shifting cultivation land	35142.20

10	Mining/Industrial Wastelands	1252.13
11	Degraded/pastures/grazing land	25978.91
12	Under utilized/degraded notified forest land	140652.31
13	Degraded land under plantation crop	5828.09
Total		638518.31

Success story: A major success story of wasteland development is the Sivganga district of Tamilnadu. Where extensive jatropha plantation has been carried out on wasteland which is very much required for biofertilizer and it is beneficial also. Due to integrated development approach the fate of the people as well the optimum use of land resources has been done. The success story can be seen on the district website.

Modified strategies and suggestions

Keeping in view the lesson learnt the following suggestions need to be considered while revising strategy:

1. Govt. departments must view afforestation as a definite support to agriculture, e.g. shelterbelts, agro forestry, mixed plantation etc.

2. People's involvement can be mobilized by understanding the community structure and their needs.

3. Sufficient funds should be earmarked and made available to finance projects aimed at integrated rural resource management which are multi disciplinary in nature at every stages.

4. Banks like NABARD should establish a separate line of credit for afforestation projects.

5. Suitable action plan should be taken up for integrated development of wastelands for ecological restoration and to meet essential needs of fuel wood, fodder and timber for local community.

6. The lab to land programme should seek to extensively popularize transfer of available technologies.

7. Sufficient advance planning should be done for raising nursery stock of the required species and quality to avoid planting of poor quality seedling stock and consequent failure of plantation.

8. All development projects e.g. mining, road, irrigation and power etc. which by their very nature either create wastelands or degraded local environment should earmark a budget provision in the project estimates for reclaiming such wastelands or regenerating natural vegetation so damaged.

The Natural Resources Conservation Service (NRCS)

The Natural Resources Conservation Service (NRCS), formerly known as the Soil Conservation Service (SCS), is an agency of the United States Department of Agriculture (USDA) that provides technical assistance to farmers and other private landowners and managers.

Its name was changed in 1994 during the presidency of Bill Clinton to reflect its broader mission. It is a relatively small agency, currently comprising about 12,000 employees. Its mission is to improve, protect, and conserve natural resources on private lands through a cooperative partnership with state and local agencies. While its primary focus has been agricultural lands, it has made many technical contributions to soil surveying, classification and water quality improvement. One example is the Conservation Effects Assessment Project (CEAP), set up to quantify the benefits of agricultural conservation efforts promoted and supported by programs in the Farm Security and Rural Investment Act of 2002 (2002 Farm Bill). NRCS is the leading agency in this project.

Farm bill

In 1994 a lot of farmers did not make use of the farm bill and it was then dropped (Curdy). The purpose of this program is to work with land owners to purchase development rights to current farm and ranch land, in order to keep said land from being developed for other uses. The program matches funds from property owners, and can be applied whether potential buyers are private or from state or local government, as well as Native American tribes. In order to receive funds from the program, the land must be privately owned, and have an offer for sale pending. The land must be large enough to support substantial agricultural yield, and be surrounded by land with a similar nature. If there is a danger for soil erosion, a conservation plan must be included.

Grasslands reserve program (GRP)

Volunteer programme increases animal, plant biodiversity and to protect grasslands. Participants limit use of grassland for commercial and agricultural development. The land may still be grazed or seeded, with the exception of the nesting seasons of bird species that are protected under law. A grazing management plan must be submitted for participation.

Healthy forests reserve program

(HFRP) Landowners volunteer to restore and protect forests in 30 or 10 year contracts. This program hands assisting funds to participants. The objectives of HFRP are to:

1. Promote the recovery of endangered and threatened species under the Endangered Species Act (ESA)
2. Improve plant and animal biodiversity
3. Enhance carbon sequestration

Wetlands reserve program

(WRP) Volunteer program for landowners to protect or restore wetlands on properties they own. The program offers both financial and technological support to these landowners in order to help cultivate long term wetland health with optimal biodiversity per acre.

NRCS National Ag Water management team

(AGWAM) serves 10 states in the Midwest United States in helping to reduce Nitrate levels in soil due to runoff from fertilized farmland. The project began in 2010 and initially focused on the Mississippi Basin area. The main goal of the project is to implement better methods of managing water drainage from agricultural uses, in place of letting the water drain naturally as it had done in the past. In October 2011, the National "Managing Water, Harvesting Results" Summit was held to promote the drainage techniques used in hopes of people adopting them nationwide.

Snow survey and water supply forecasting

Snow survey and water supply forecasting includes water supply forecasts, reservoirs, and the Surface Water Supply Index (SWSI) for Alaska and other Western states. NRCS agents collect data from snowpack and mountain sites to predict spring runoff and summer stream flow amounts. These predictions are used in decision making for agriculture, wildlife management, construction and development, and several other areas. These predictions are available within the first 5 days of each month from January to June.

Wildlife habitat incentive program

Wild habitat incentive program (WHIP) is a volunteer program for improving habitats for wildlife on farmlands, private lands, and Indian land. The program was renewed in 2008 under the Food, Conservation, and Energy Act. WHIP provides up to 75 percent cost share and technical assistance, and includes both terrestrial and freshwater aquatic habitats. The program aims to protect seven threatened species in particular:

- » Lesser Prairie Chicken
- » New England Cottontail
- » Southwestern Willow Flycatcher
- » Greater Sage-Grouse
- » Gopher Tortoise
- » Bog Turtle
- » Golden-winged Warbler

Conservation technical assistance program

Conservation technical assistance program (CTA) is a blanket program which involves conservation efforts on soil and water conservation, as well as management of agricultural wastes, erosion, and general long-term sustainability. NRCS and related agencies work with landowners, communities, or developers to protect the environment. Also serve to guide people to comply with acts such as the Highly Erodible Land, Wetland (Swampbuster), and Conservation Compliance Provisions acts. The CTA can also cover projects by state, local, and federal governments.

International programs

The NRCS (formerly SCS) has been involved in soil and other conservation issues internationally since the 1930s. The main bulk of international programs focused on preventing soil erosion by sharing techniques known to the United States with other areas. NRCS sends staff to countries worldwide to conferences to improve knowledge of soil conservation. There is also international technical assistance programs similar to programs implemented in the United States. There are long term technical assistance programs in effect with one or more NRCS staff residing in the country for a minimum of one year. There are currently long term assistance programs on every continent. Short term technical assistance is also available on a two week basis.

These programs are to encourage local landowners and organizations to participate in the conservation of natural resources on their land, and lastly landscape planning has a goal to solve problems dealing with natural resource conservation with the help of the community in order to reach a desired future outcome.

Technical Resources

Water

Pollution of water due to a number of different pollutants has driven the NRCS to take action. Not only do they offer financial assistance but they also provide the equipment needed for private land owners to protect our water resources. Water gets polluted by nitrogen and phosphorus which causes algae to grow proliferously causing the oxygen concentrations to decline rapidly, life is no longer supported in this habitat. Excessive sedimentation is also another concern along with pathogens threats that can find their way into water systems and cause detrimental effects. NRCS works in a way to help both the land owner and the water systems that need prevention or restoration.

Water management

Water management strives to manage and control the flow of water in a way that is efficient while causing the least amount of damage to life and property. This helps provide protection in high risk areas from flooding. Irrigation water management is the most efficient way to use and recycle water resources for land owners and farmers. Drainage management is the manipulation of sub surface drainage networks in order to properly disperse the water to the correct geographical areas. The NRCS engineering division is constantly making improvements to irrigation systems in a way that incorporates every aspect of water restoration.

Water quality

A team of highly trained experts on every aspect of water is employed by the NRCS to analyze water from different sources. They work in many areas such as: hydrology and hydraulics, stream restoration, wetlands, agriculture, agronomy, animal waste management, pest control, salinity, irrigation, and nutrients in water.

Watershed program

Under watershed programs the NRCS works with states, local governments, and tribes by providing funding and resources in order to help restore and also benefit from the programs.

They would like to provide: watershed protection, flood mitigation, water quality improvement, soil erosion reduction, irrigation, sediment control, fish and wildlife enhancement, wetland and wetland function creation and restoration, groundwater recharge, easements, wetland and floodplain conservation easements, hydropower, watershed dam rehabilitation.

Plants and animals

Plants and animals play a huge role in the health of our ecosystems. A delicate balance exists between relationships of plants and animals. If an animal is introduced to an ecosystem that is not native to the region that it could destroy plants or animals that should not have to protect itself from this particular threat. As well as if a plant ends up in a specific area where it should not be it could have adverse effects on the wildlife that try to eat it. NRCS protects the plants and animals because they provide us with food, materials for shelter, fuel to keep us warm, and air to breathe. Without functioning ecosystems we would have none of the things mentioned above. NRCS provides guidance to assist conservationists and landowners with enhancing plant and animal populations as well as helping them deal with invasive species.

Fish and wildlife

NRCS for years has been working toward restoration, creation, enhancement, and maintenance for aquatic life on the nearly 70 per cent of land that is privately owned in order to keep the habitats and wildlife protected. NRCS with a science based approach provides equipment to wildlife and fish management. They also do this for landowners who qualify to benefit from these technologies.

Insects and pollinators

Pollination via insects plays a huge role in the production of food crop and flowering plants. Without pollinators searching for nectar and pollen for food the plants would not produce a seed that will create another plant. NRCS sees the importance of this process so they are taking measures to increase the declining number of pollinators. There are many resources provided from the NRCS that will help any individual do their part in conservation of these important insects. Such as Backyard Conservation, this tells an individual exactly how to help by just creating a small habitat in minutes. There are many others such as: Plants for pollinators, pollinators habitat in pastures, pollinator value of NRCS plant releases in conservation planting, plant materials publications relating to insects and pollinators, PLANTS database: NRCS pollinator documents. All of these are valuable resources that any individual can take advantage of.

Invasive species and pests

Many adverse effects are present due to invasive species. Plants and animals both inhabit areas that they are not intended to be. The kudzu vine for example covers miles of foliage. These invasive species cause America's reduction in economic productivity and ecological decline. Humans are unknowingly transporting these invasive species via ships, planes, boats, and their own bodies. NRCS works in collaboration with the plant materials centers scattered throughout the country in order to get a handle on the invasive species of plants. These centers scout out the plants and take measures to control and eradicate them from the particular area. Invasive animals such as feral hog, European gypsy moth, and the sirex woodwasp pose a significant threat to America's wildlife as well as to the health of human beings. The hog was introduced as a food source for humans, but now the swine pandemic is a serious threat to humans. They gypsy moth destroys natural forests that are habitat to many beneficial species. The Woodwasp feeds on pine trees as well as providing a means of transportation for a fungus that kills pine trees.

Livestock

Livestock management is an area of interest for the NRCS because if not maintained valuable resources such as food, wools, and leather would not be available. The proper maintenance of livestock can also improve soil and water resources by providing a waste management system so that run off and erosion is not a problem. The NRCS provides financial assistance to land owners with grazing land and range land that is used by livestock in order to control the run off of waste into fresh water systems and prevent soil erosion.

Plants

Plants are a huge benefit to the health of ecosystems. NRCS offers significant amounts of resources to individuals interested in conserving plants. From databases full of information to financial assistance the NRCS works hard to provide the means needed to do so. The plant materials program, Plant materials centers, Plant materials specialists, PLANTS database, National Plant Data Team (NPDT) are all used together to keep our ecosystems as healthy as possible. This includes getting rid of unwanted species and buildings up species that have been killed off that are beneficial to the environment. The NRCS utilizes a very wide range of interdisciplinary resources.

The NRCS also utilizes the following disciplines in order to maximize efficiency:

- » Agronomy
- » Erosion

- Air quality and atmospheric change
- Animal feeding operations and confined animal feeding operations
- Biology
- Conservation innovation grants
- Conservation practices
- Cultural resources
- Economics
- Energy
- Engineering
- Environmental compliance
- Field office technical guide
- Forestry
- Maps
- Data and analysis
- Nutrient management
- Pest management
- Range and pasture
- Social sciences
- Soils and water resources

These Science-Based technologies are all used together in order to provide the best conservation of natural resources possible.

Land Use Planning System in India

Land includes benefits to arise out of land, and things attached to the earth or permanently fastened to anything attached to the earth.

- » Land is the most important component of the life support system.
- » It is the most important natural resource which embodies soil and water, and associated flora and fauna involving the ecosystem on which all man's activities are based.
- » Land is a finite resource.
- » Land availability is only about 20% of the earth's surface. Land is crucial for all developmental activities, for natural resources, ecosystem services and for agriculture.
- » Growing population, growing needs and demands for economic development, clean water, food and other products from natural resources, as well as degradation of land and negative environmental impacts are posing increasing pressures to the land resources in many countries of the world.
- » For India, though the seventh largest country in the world, land resource management is becoming very important.
- » India has over 17% of world's population living on 2.4% of the world's geographical area.

Infrastructure across the country must expand rapidly

- » Industrialization, especially in the manufacturing sector, is inevitable and will accelerate.
- » Urbanization is on drastic rise.
- » Agricultural lands are becoming important as livelihood of a significant amount of the country's population is dependent on it.
- » For all these, land is an essential requirement.
- » In addition, the Government also requires land from time to time for a variety of public purposes.
- » The developmental targets of India on one hand and the social, cultural and environmental aspects on the other hand demand land.
- » These demands for land could be competing by different sectors for the same land or even leading to conflicting land uses once put to a use by a sector.
- » In the recent years, there has been tremendously increasing pressure on land in India posing challenges for sustainable development.

Need for optimal utilization of land resources

» The country can no longer afford to neglect land, the most important natural resource, so as to ensure sustainability and avoid adverse land use conflicts.

» There is a need to cater land for industrialization and for development of essential infrastructure facilities and for urbanization.

» While at the same time, there is a need to ensure high quality delivery of services of ecosystems that come from natural resource base and to cater to the needs of the farmers that enable food security, both of which are of vital significance for the whole nation.

» Also, there is a need for preservation of the country's natural, cultural and historic heritage areas.

» In every case, there is a need for optimal utilisation of land resources.

Constitution explains as below:

Article 39 of the Constitution explains

» The ownership and control of the material resources of the country should be so distributed as best to serve the common good; and

» The operation of the economic system should not result in a concentration of wealth or a means to production to the common detriment.

Article 243ZD(1) of the Constitution explains:

"There shall be constituted in every State at the district level a District Planning Committee to consolidate the plans prepared by the Panchayats and the Municipalities in the district and to prepare a draft development plan for the district as a whole". While the Constitution provided for spatial planning, the National level activities currently are focused to evolving policies, guidelines and model laws for adoption by the States, disbursing and monitoring assistance/grants/funding, and formulating development plans and policies for Union Territories.

» In the area of land utilisation, there is no single approach currently being followed across the country.

» Various sectors at central level such as urban, rural, industrial, transport, mining, agriculture etc. follow their own approaches. For example, in the case of rural sector, since nearly 50% of India's population is dependent on agriculture, the sector lays focus on reforms on land acquisition and resettlement & rehabilitation, watershed management and modernisation of land records, and there is no yet an approach in place for planning and management of land resources in rural areas.

Proper planning of land use

Proper planning of land and its resources allows for rational and sustainable use of land catering to various needs including social, economic, developmental and environmental needs.

» Proper land use planning based on sound scientific, and technical procedures, and land utilization strategies, supported by participatory approaches empowers people to make decisions on how to appropriately allocate and utilize land and its resources comprehensively and consistently catering to the present and future demands.

» There is a need for scientific, aesthetic and orderly disposition of land resources, facilities and services with a view to securing the physical, economic and social efficiency, health and well- being of communities.

» There is a need for an integrated land use planning which inter-alia includes agriculture, industry, commerce, forests, mining, housing infrastructure and urban area settlements, transportation infrastructure etc. to settle claims/counter-claims of these sectors.

The National Commission on Agriculture (1976) emphasized on scientific land use planning for achieving food security, self reliance and enhanced livelihood security.

The National Policy for Farmers (2007) has recommended revival of existing Land Use Boards and their linkage to district-level land-use Committees, so that they can provide quality and proactive advice to farmers on land use.

The Committee on "State Agrarian Relations and the Unfinished Task in Land Reforms" (2009) has also emphasized the need for land use planning in the country. The Sustainable Development strategy of Agenda 21, a non-binding and voluntarily implemented action plan of the United Nations (UN) that was ratified by more than 170 countries at the United Nations Conference on Environment and Development (UNCED) held in Rio de Janeiro, Brazil, in 1992, advocates achievement of sustainable development through appropriate land use planning and management.

Definition of "land use planning" by the United Nation's Food and Agriculture Organization and the United Nations Environment Programme published in 1999 reflects consensus among the international organizations. Land use planning is understood as a systematic and iterative procedure carried out in order to create an enabling environment for sustainable development of land resources which meets people's needs and demands. It assesses the physical, socio- economic, institutional and legal potentials and constraints with respect to optimal and sustainable use of natural resources and land and empowers people to make decisions about how to allocate those resources.

Another definition of "land use planning" is the process of evaluating land and alternative patterns of land use and other physical, social and economic conditions for the purposes of selecting and adopting those kinds of land use and course of action best suited to achieve specified objectives. Land use planning may be at national, regional, state, district, watershed, city, village or other local levels.

Drawbacks of land use planning

There is lack of comprehensive and integrated land use planning in the country, which enables rationale and optimal land utilization.

- » The current land use planning in the country is inadequate and does not cover all the levels of local, regional as well as at state levels.

- » There is a need for a systematic and scientifically based land use planning.

- » The Constitution (Seventy-fourth Amendment) Act, 1992 provides for District Planning and Metropolitan Area Planning that consolidates plans of both panchayats and municipalities having regard to spatial (land use) planning.

The District Plans prepared currently, in general, do not cover spatial (land), environmental as well as urban concerns.

- » District level spatial land use plans do not generally exist, the regional development triggered by urbanization or industrialization, or the regional development that needs to be regulated due the presence of large eco sensitive zone calls for initiating land use planning of such (urban or industrial or eco sensitive) areas so as to ensure sustainable development.

- » If no immediate actions are taken, the unplanned and haphazard development has potential to cause adverse impacts, including land use conflicts with neighboring land uses, particularly with agricultural areas, rural uses, natural resource areas and environmentally sensitive and fragile ecosystem, as well as of losses of productive land and ecosystem services.

Competing and conflicting land uses are a major concern.

a. "Competing land uses" are those that compete for the same parcel of land for their location.

b. Such land uses competing in rural areas with agriculture could be, for example, cash crops or food crops; industrial or agro-industrial uses; Special Economic Zones (SEZs) or Special Investment Regions; highways; peri-urban development or outgrowth, integrated townships or theme cities, and mega projects (e.g. industrial corridors/power plants/ports).

Conflicting land uses are those that are in conflict with the existing land use.

» Certain land uses cause impacts on other land uses nearby. Such conflicting land uses include, for example, agriculture in forest areas, mining in forest areas, highways in eco- sensitive areas, polluting industries in rural or eco-sensitive areas, urban/industrial development on agriculture lands, agriculture encroaching into forest lands, and urban waste disposal in peri-urban areas.

» The basic concern is the negative impact that such land uses cause on other land uses. For example, an industrial area can cause impacts on the neighboring areas due to air pollution, or an urban expansion can lead to destruction of the ecosystem service of natural drainage thereby impacting the existing lakes and water bodies.

» Indiscriminate land use changes in eco-sensitive zones could directly affect wildlife habitat and thereby impacts local and global biodiversity.

» Both, competing and conflicting land uses frequently are also the reason for social conflicts between the local population and the authorities and prospective investors.

Important land use changes

Between 1950/51 and 2007/08, land utilisation in India underwent significant changes.
» While the lands under net sown area, forests and non-agricultural uses have increased, the lands under "other areas" uses have almost halved from 40.7% to 22.6% meaning that for future land demands, the forest lands and agricultural lands may have to be used.

- » Also, the per capita amount of agricultural land has reduced by 67% from 1951 (0.48 Ha) to 2007/08 (0.16 Ha). Degradation of soils and land due to soil erosion and other degradation processes is a severe problem in many regions in India.

- » As per available estimates (2010), total degraded land in the country is about 120.40 million ha. Land degradation leads to decline in soil fertility, creates problems of alkalinity/salinity/acidity and water logging etc.

- » The degraded soils are often used by marginal farmers and tribal population.

- » According to studies, the economic losses of reduced productivity of these lands count for approx. Rs. 285,000 million, which is about 12 per cent loss of total value of productivity of these lands.

- » Land use changes, such as conversion of forest lands to agriculture or industrial use can be a factor in increasing CO_2 (carbon dioxide) (a dominant greenhouse gas) atmospheric concentrations, thereby contributing to climate change.

- » Due to growing urbanization and industrialization, as well as usage of chemicals for agriculture, there are threats of pollution and disasters.

- » There are potential impacts from handling, storage and transportation of hazardous chemicals/materials and wastes, emission of pollutants including toxic emissions, discharge of effluents, especially those that are not easily biodegradable and toxic, pollution of ground water, streams, rivers, lakes, oceans, or other bodies of water, industrial disaster risks and natural disaster risks.

- » There are also growing risks to biodiversity with loss of species and threat to ecosystem services.

Water resources projects are frequently being planned and implemented in a fragmented manner without giving due consideration to optimum utilization of water resources, environmental sustainability and holistic benefit to the people.

- » The natural water bodies and drainage channels are being encroached upon, and diverted for other purposes.

- » The groundwater recharge zones are often blocked.

- » There is growing pollution of water sources, especially through industrial effluents, which is affecting the availability of safe water besides causing environmental and health hazards. In many parts of the country, large stretches of rivers are both heavily polluted and devoid of adequate flows to support self-purification, aquatic ecology, cultural needs and aesthetics.

» The characteristics of catchment areas of streams, rivers and recharge zones of aquifers are changing as a consequence of land use and land cover changes thereby affecting water resource availability and quality.

» Due to climate change, there are increasing temperatures and issues of drought and flooding.

Land use policy

There is a need for a policy framework to be formulated at the national level incorporating concerns of various sectors and stakeholders so as to ensure optimal utilization of land resource through appropriate land use planning and management. Such a policy should provide guiding framework for the States to adopt and formulate their own policies incorporating their state specific concerns. The States should develop land use policies by consulting all stakeholders and ensuring appropriate legal backing. Further, detailed land use strategies and plans should be developed in accordance with these policies so as to achieve sustainable development.

A "National Land Use Policy Guideline and Action Points" (1988) was prepared by the Government of India, Ministry of Agriculture after intensive deliberations. In the said policy, framing of suitable legislation and its sincere enforcement were stressed by imposing penalties, of violation thereof.

The said policy guidelines were placed before the 'National Land Use and Wasteland Development Council' under the chairmanship of Prime Minister and its first meeting was held on 6th February, 1986. The Council agreed to the adoption of policy and circulated the same throughout the country for adoption after suitable considerations at the State level. However, the policy did not make the desired impact. The proposed policy framework at national level is hereinafter referred to as the "National Land Utilization Policy". The policy seeks to order and regulate land use in an efficient and rational way, thus taking care of the needs of the community while safeguarding natural resources and minimizing land use conflicts. The details of the need for such policy, the challenges, the guiding principles etc. are detailed in the following sections.

Indian Territory

India's territory includes 3.287 million sq. km. (328.73 million ha) with west to east extent of approx. 3,000 km and north to south extent of approx. 3,200 km. Numerous developmental activities demand land and in the process of progression of development, land use changes take place with time. If not regulated such changes can detrimental in the long run for the sustainable development of India. During the period 1950-51 to 2007-087, the net sown areas in the country have increased from 41.8 to 46.1 per cent. the forest areas have increased from 14.2 to 22.8 per cent, and the areas under non- agriculture uses, which include industrial complexes, transport network, mining, heritage sites, water bodies and urban and rural settlements has increased from 3.3 to 8.5 per cent. These increases of land use as above have lead to reduction of land use elsewhere. During the same period (1950-51 to 2007-08), the "other areas" that include barren and un- cultural land, other uncultivated land excluding fallow land and fallow lands have drastically decreased by nearly half from 40.7 to 22.6 per cent.

Mining area

The mining areas are about 0.17 per cent of total land of India, the urban areas are about 2.35 per cent and the industrial areas are much less than 1 per cent. However, with rapid industrialization and urbanization, the associated infrastructure development, the lands under these uses will further increase. These increases of demands of land will require land to taken away from other uses. So far, the land under "other areas" was being used. However, these lands may no further be usable as they maybe under steep hills or other such terrains or uses that constrain their use for developmental purposes. In such cases, the demands for additional lands will be resorted to from agricultural uses or forests uses which would be detrimental. There is a need to strategies utilisation of land and its management so that the land use changes are not detrimental to sustainable development of India.

Population in India

Due to growing population in India, the per capita availability of land has reduced from 0.89 Ha in 1995 to 0.27 Ha in 2007/08. It is estimated that by 2030, India will become the most populated country on earth with 17.9% of world's total population. With this, the per capita land availability will further reduce. Such reducing per capita land availability will have a direct bearing on the land requirements for various developmental purposes and community development. The concerns become severe when the land availability is reduced directly in the areas that support human life or natural resources such as water or ecosystems including flora and fauna, and agricultural areas.

As per the 2011 Census, 68.84 per cent of the country's population lives in 6,40,867 villages and the remaining 31.16 per cent lives in 7,935 urban centres. Although agriculture presently accounts for only about 14 per cent of the Gross Domestic Product (GDP), it is still the main source of livelihood for the majority of the rural population, and provides the basis of food security for the nation. Therefore, fertile agriculture land and clean water resources need to be protected effectively for providing and ensuring livelihood to rural population and food security for the nation. Currently, India produces about 245 million tons of food grains while for 2020 it is estimated that the demand for food grains shall rise by 25 per cent to 307 million tons. The agricultural productivity is currently half of what it is in many other countries. The solution for food productivity and security may not lie in stopping diversion of agriculture land in all circumstances, but also in increasing food production through higher productivity. However, the increasing use of soil can cause threat to its productivity, as it was experienced in several other countries. Hence, the stark question is whether soils will be productive enough to sustain a population of one billion by the end of the century at higher standards of living than those prevailing now.

There is a need for long term plans to meet the food security as well as livelihood issues. For this purpose, reasonable restrictions on acquisition and conversion of at least certain types of agricultural land should be introduced. As per the National Policy for Farmers, 2007 (NPF 2007), prime farmland must be conserved for agricultural use and except under exceptional circumstances the use should not be altered. There is a need to protect agricultural areas that are essential for food security including the prime agricultural lands, command areas of irrigation projects, and double cropped land. There is also a need to protect agricultural lands that are essential for livelihood of rural and tribal populations.

Climatic regions

India comprises seven climate regions in three groups:

» Tropical wet-humid group with tropical wet (humid or monsoon climate, and tropical wet and dry or savannah climate.

» Dry climate group with tropical semi-arid (steppe) climate, sub-tropical arid (desert) climate and sub-tropical semi-arid (steppe) climate.

» Sub-tropical humid climate group with sub-tropical humid (wet) with dry winters climate and the mountain, or highland, or alpine climate.

India comprises of nine bio-geographic regions, i.e. the Trans-Himalayan Region (1), The Himalayas (2), The semi-arid areas (3), The Western Ghats (4), The North-West Desert Regions (5), The Deccan Plateau (6), The Gangetic Plain (7), North-East India (8), The islands (9), and the coasts (10).

India has extraordinarily rich biodiversity and several environmentally sensitive and fragile zones. India is one of the twelve mega-biodiversity countries in the world, comprising over 91,000 animal and 45,500 plant species. Nearly 6,500 native plants are still used prominently in indigenous healthcare. Furthermore, India is recognized as one of the eight so called 'Vavilovian Centres of Origin and Diversity of Crop Plants', having more than 300 wild ancestors and close relatives of cultivated plants still growing and evolving under natural conditions. India has a wide range of soils, classified into 27 broad soil classes. Alluvial soils, black cotton soils, and red soils covering a total of approx. 56 per cent of the total land area, which are considered suitable for a wide range of crops. Laterite and lateritic soils and desert soils covering another 15 per cent of the land are not suitable for agriculture. India's water resources are limited and scarce. The country has only 4 per cent of the world's renewable water resources at its disposal. Furthermore, these limited resources are distributed unevenly over time and space. In addition, there are challenges of frequent floods and droughts in one or the other part of the country.

The total area of the recorded forests in the country (2003) is 77.47 million ha, or 23.57 per cent of the country's geographic area. Of the forest areas, 51.6 per cent are notified Reserved Forests, 30.8 per cent are notified Protected Forests and the remaining 17.6 per cent are un-classed forests. The National Forest Policy of 1988, the Indian Forest Act and various other State legislations on the matters pertaining to forests provide for the guiding principles, and ways and procedures through which legally declared forests are to be utilized and administered. These legislations not only affect the way forest lands are to be utilized, these have profound impact on the utilization of non-forest lands as well. India is rich in ample resources of a number of minerals and has the geological environment for many others13. India produces 89 minerals out of which 4 are fuel minerals, 11 metallic, 52 non- metallic and 22 minor minerals.

There are several ecosystem services being offered by the natural environment and its resources. These include:

» Provisioning, such as the production of food, water, pharmaceuticals, industrial products, wind/wave/hyrdo-power and biomass.

» Supporting, such as purification of water and air; nutrient cycles; crop pollination; seed dispersal; disease control.

» Cultural, such as spiritual and recreational benefits (e.g., ecotourism).

» Preserving, genetic and species diversity for future use; accounting for uncertainty.

» Regulating, such as the carbon sequestration; climate regulation.

There is a need for protection of the natural resource areas including biodiversity areas, forests areas etc. and ecosystem service areas.

» While areas such as National Parks, Biosphere Reserves etc. are clearly demarcated due to existing legal provisions, the "Eco Sensitive Zones" being identified around them are at present at the discretion of the agencies involved.

» It has to be ensured that reasonable extents of areas around the environmentally sensitive/fragile areas are demarcated and land use is planned properly.

» There is also a need to regulate and control land uses in such 'Eco Sensitive Zones' so as to avoid conflicts or negative environmental impacts.

Urbanization

Level of urbanization in India has increased from 17 per cent in 1951 to 31 per cent in 2011. According to the world population prospects by the United Nations, 55 per cent population of India will be urban by the year 2050. With this pattern of urbanization, the urban population of 377 million as in 2011 will be 915 million by the year 2050. During the decade 2001-2011, the number of towns in the country has increased from 5,161 to 7,935. The number of urban agglomerations, having a population of more than one million has increased from 5 in 1951 to 53 in 2011. Most of the cities are traditionally located along the major rivers, around lakes and along the coastline, the agriculturally productive belt and environmentally sensitive areas. The urban land is about 7.74 million hectares, which is only 2.35 per cent of the country's total land area. However, several land use conflicts and environmental problems originate from urban area.

» The mega cities are mostly spilling over to rural-agricultural belt (peri-urban areas) due to abnormally high land price in the cities as compared to household income of the average citizens.

» The peri-urban areas or fringes of such agglomerations are under fast transformation resulting into haphazard growth of slums, unauthorized colonies, piecemeal commercial development, intermixes of conforming and non-conforming uses of land coupled with inadequate infrastructures, services and facilities.

- » Cities and towns are emerging as centers of domestic and international investments where most of the commercial activities take place. As the economy grows, towns and cities expand in size and volume and the contribution of the urban sector to the national economy increases.

- » In days to come, the urban sector will play a critical role in the structural transformation of the Indian economy and in sustaining the high rates of economic growth.

- » Ensuring high quality public services for all in the cities and towns will also facilitate the full realization of India's economic potential.

- » The demand for non-farm land use will increase further in future.

Management strategy for urbanization

There is a need for appropriate land utilization and management strategy and land use planning to cater to the growing urbanization needs.

There is scope for re-densification of urban areas by augmenting the existing infrastructure.

Large chunks of institutional land within big cities are lying vacant or under utilized or not available for urban development which can be put to optimum use by way proper planning and land management system.

Large extents of agricultural land still exist within the municipal boundaries of small & medium size towns due to slow pace of physical growth and such lands should preferably be retained as such, particularly if the soils are of high quality.

There is need for proper planning of urban areas and the regions around.

Industrial development

Industrial development, apart from urbanization, is the major driver of economic growth in India. The 12th Five Year Plan provides that the country needs to reach an annual economic growth rate of at least 8 per cent in five years (2012 to 2017) in order to significantly increase the quality of life of its citizens, reduce poverty and foster environmentally sustainable development. The 12th Five Year Plan proposes growth of economy at 9 to 9.5 per cent during the Plan period (2012 to 2017) and has ambitious targets for various sectors. For example, the growth of manufacturing sector is proposed at 9.8-11.5 per cent and the mining and quarrying at 8 - 8.5 per cent. This will bring in demand for additional land and put pressure on existing land uses for their conversion. The National Manufacturing Policy 2011 of the Government of India has set industrial growth rate of 12 to 14 per cent in the medium run and contribution of industrial sector to national GDP by 25 per cent.

The industrial development that is seen in the form of industrial estates, special economic zones, specialized industrial parks, investment zones, NMIZs, special investment regions, PCPIRs (petroleum, chemicals and petro chemical investment regions) and industrial corridors occupies a lot of land. The industrial development is associated with supportive development, viz, housing areas, transport, trade and commerce areas, wasteland, and waste water treatment and disposal areas etc., which also require considerable amounts of land. The Delhi Mumbai Industrial Corridor covering an overall length of 1,483 km and passing through the States of Uttar Pradesh, Haryana, Rajasthan, Madhya Pradesh, Gujarat and Maharashtra and the National Capital Region of Delhi, will have 24 identified Industrial Areas and Investment Regions requiring large quantity of land for development not only for industrial areas but also for supporting population arising out of 3 million jobs that would be created. A similar Chennai-Bengaluru Industrial Corridor is proposed. Similar other mega industrial infrastructure projects will come up in the future in the country. The creations of large number of SEZs as per the Special Economic Zone Act of 2005, involving large extent of fertile agricultural land have added substantially to already aggravate land relations in India.

An SEZ may bring far reaching changes in the local economy. However, people lose access to farmlands, grazing grounds water bodies and other common resources. The agrarian protests against the SEZs are prevalent everywhere in India. It also aims to create 100 million additional jobs by 2022. It has proposed the development of National Manufacturing and Investment Zones (NMIZ) in the form of Industrial Clusters and Integrated Townships each having an area of 5,000 hectares. Paradigm shift in the government policy like, putting a cap of 5,000 ha of land for each SEZ and an important decision that governments will not invoke its powers under 'eminent domain' and 'public purpose' to acquire land for SEZs. However, these policy changes may not prove adequate to address the core issues unless there is accompanying land use planning strategy, particularly because of large requirements of land. The requirements of lands for SEZs which are given 'in–principle approval' stands at 2,00,000 Ha, which, as per the estimates of the Committee on State Agrarian Relations and Unfinished Task in Land Reforms, is capable of producing around 1 million tons of food grains. There is a need for appropriate land utilization and management strategy and land use planning to cater to the growing industrialization needs.

Mineral resources

Land has hidden treasure of vast resources of different kinds of minerals. India is rich in mineral resources such as bauxite, iron, copper, zinc, gold, diamonds etc.

- » Minerals are basic raw materials for many industries and play a key role in the evolution of human society and the development of economies.
- » The wide availability of the minerals in the form of abundant rich reserves/resources makes it very conducive for the growth and development of the mining sector in India.
- » Presently, utilisation of land by mineral sector (excluding atomic, fuel and minor minerals) is about 0.17 per cent of India's total land (as in 2010-11) and contributes about 2.72 per cent to the GDP of India.
- » The sector provides employment to over 5 lakh people directly. The needs of economic development of India make the extraction of the nation's mineral resources an important priority.
- » Minerals are valuable, finite and largely non-renewable natural resources, but are site specific.
- » The State Governments are the owners of minerals located within their respective boundaries.
- » The Central Government is the owner of the minerals underlying the ocean within the territorial waters or the EEZ of India.

Extraction of minerals involves use of land for undertaking mining.

- » Mining industry, unlike other industries, is site specific and degradation of the land and other associated natural resources becomes inevitable.
- » Mining areas are closely linked with forestry and environment issues.
- » A significant part of the nation's known reserves of some important minerals are in areas which are under forest cover.
- » Further, mining activity has potential to disturb the ecological balance of an area.
- » For ensuring sustainable development, there is a need to properly plan and manage mining areas.

Land use by the transport sector

The major land users in the transport sector are: railways (railway tracks, stations, workshops, godowns etc.), roadways (roads, fuel pump stations, toll plazas, utilities etc.), airways (airports, runways, workshops etc.), waterways (ports, workshops, godowns etc.). The total road network in India is 4.69 million km in length. In the case of roadways, under the National Highways Act, 1956, the Central Government has power to acquire land for National Highways. The transport networks require considerable amounts of land and their proper planning is very important, as otherwise the transport networks can trigger land use conflicts due to the development sectors and communities that depend on them.

Displacement of population

The developmental activities require land and they have potentials to displace people, exploit natural resources and cause negative environmental impacts as well as other land use conflicts. There is a need to support various sectors to achieve their development targets, such as those of urban development, industrial development, mining, and infrastructure development (transportation, ports, harbours, airports etc.) through properly guided development in a sustainable and harmonized manner so as not to have land use conflicts or negative impacts.

India has considerable amount of vulnerable populations in the rural, tribal and backward areas, many of whom do not have adequate access to basic amenities and proper livelihood. There are disadvantaged and vulnerable communities including tribal populations, economically weaker sections of people and backward communities. There are issues of livelihood, poverty eradication, inclusiveness and gender. There is a need to support social development addressing these issues. Land plays an important role in all these matters. There is a need to prevent or at least minimize social conflicts arising from acquisition of lands or development of such activities that pose conflicts. Land use planning should be undertaken giving due considerations for social aspects.

Cultural and historic heritage

India has rich cultural and historic heritage. There are several scenic beauty areas and tourism areas. All these areas including religious places of importance, scenic areas, heritage areas, archaeological sites etc. need to be protected from negative impacts of development and land use changes. Depending on the developmental activities coming up in the vicinity of these areas, there could be potential impacts. Though proper land use planning and management such impacts could be prevented and the heritage areas secured.

Lack of systematic data for planning purpose

There is a severe lack of systematic, orderly and up to date spatial data base in the country that is readily available for land use planning purposes. Also, due to lack of systematic database, there would also be difficulties initially in making projections and forecasting of prospective needs for land uses by various sectors. However, the country is quite advanced in the applications of Geographic Information Systems (GIS) and remote sensing, which come handy for generating spatial database. The Government of India is already working on setting up National Spatial Data Infrastructure. Systematically, such spatial databases could be built-up over a period of time. Also, the existing database on land use in the country is highly inadequate. There is no mechanism to monitor land use changes taking place and their impacts.

Introduction of systematic and integrated land use planning at national, state and regional levels is going to be a major challenge. There have to be supportive instruments (mapping, spatial information, planning processes, tools, methods, procedures, standards etc.) for land use planning and management which also take into account inclusiveness, poverty, gender and climate change aspects. Another aspect is the availability of guidelines for uniform land use planning. Except for urban sector, where urban development plan formulation and implementation (UDPFI) guidelines exist; the other sectors such as industry, environment, transport, mining, agriculture etc. do not have similar guidelines in place. For ensuring proper land use planning, there is a need for development of detailed guidelines for following integrated approaches catering to all the sectors. There is also lack of adequate institutional structures at national, state, regional/district and local levels for planning and management of land resource.

Guidelines by Urban Development Plans Formulation and Implementation (UDPFI)

The Urban Development Plans Formulation and Implementation (UDPFI) Guidelines15 (1996) recommended urban development planning system, which consists of a set of the following four inter-related plans:

» Perspective Plan: A long term (20-25 years) policy plan of spatio-economic development of the settlement.

» Development Plan: Conceived within the framework of the approved Perspective Plan, it is a medium-term (generally five years co-terminus with the term of the local (authority) comprehensive plan of spatio-economic development of the urban centre.

» Annual Plan: Conceived within the framework of Development Plan, it is a plan containing the physical and fiscal details of new and ongoing projects that the local authority intends to implement during the respective financial year.

Plans of Projects/Schemes: Conceived within the framework of approved Development Plan/Annual Plan, these are detailed working layouts for execution by a public or private agency. Master Plans and Development Plans are prepared for urban areas, metropolitan areas and sometime Regional Plans such as for Delhi National Capital Region. The Master Plans or Development Plans are prepared by the urban local bodies or the Town Planning Departments or the Development Authorities. One of the main issues is, in the absence of Regional Plans, the urban sprawl forces itself into farmlands and rural areas. If all the urban areas in the country are properly planned, this would bring about 2.25 per cent of the country's land under planned development.

The National Manufacturing Policy (Nov., 2011) of the Government of India, promotes integrated industrial townships, known as the National Investment and Manufacturing Zones (NIMZs) with at least 5,000 ha area and calls for preparation of environment friendly Development Plans.

» Major environmental aspects are required to be taken care of in the NIMZ in the beginning itself by having proper zoning during Master Planning.

» The state level/local level authorities such as industrial development corporations and infrastructure development boards are identifying locations for industrial estates, special economic zones, investment zones/regions and industrial corridors and preparing development/master plans for such areas. For example, the Gujarat Infrastructure Development Board is preparing an elaborate Development Plan for Dholera Special Investment Region, which is a part of the Delhi-Mumbai Industrial Corridor. Such plans guide future land use.

» If all the industrial areas in the country are properly planned, this would bring about 1% of the country's land under planned development.

Under the Environment (Protection) Act, 1986, the Ministry of Environment & Forests, GoI is notifying "Eco Sensitive Zones", which require preparation of Zonal Master Plans or Zonal Development Plans that guide further development in the area.

» "Eco Sensitive Zones" may be defined as areas which contain natural features with identified environmental resources having 'incomparable values' (water resource, flora & fauna etc.) requiring special attention for their conservation.

- » The Eco Sensitive Areas will include protected areas such as National Parks, Wildlife Sanctuaries, Conservation Reserves and Community Reserves (total number: 659), which cover about 4.79% of the total geographic area of the country.

- » The areas other than protected areas such as landscape areas, areas with historical value also are covered under Eco Sensitive Zones.

- » The purpose of declaring Eco Sensitive Zones is to create a kind of 'shock absorber' for the specialized ecosystem that needs to be protected.

- » The Eco Sensitive Zones would act as transition zone from areas of high protection to areas involving lesser protection.

- » These areas are of regulatory nature rather than prohibitive nature, unless or otherwise so required.

Objectives of Eco Sensitive Zones

The objectives of declaring Eco Sensitive Zones are:

- » To maintain the response level of an ecosystem within the permissible limits with respect to environmental parameters.

- » To take care of special protection needs because of its landscape, wildlife, historical value etc. and to ensure that the new activities allowed are within the carrying capacity of that area.

- » To ensure protection and conservation of 'Entities of Incomparable Values' of these zones and regulate development activities based on a scientific basis and based on adequate participation in the decision making by the local communities.

- » To ensure compliance to the provisions contained in the approved Zonal Development Plan/Master Plan/Management Plan through the constitution of high level monitoring committees.

- » The State Governments identify these Eco Sensitive Zones and the Ministry of Environment & Forests, Goal finalizes the same and notifies under the Environment (Protection) Act, 1986.

» Accordingly, the Zonal Development Plans are prepared and implemented for regulating further development or land uses in the areas.

» If all the Eco Sensitive Zones in the country are notified and planned, this would bring about 5% of the country's land under planned development.

Steps taken in mining areas

In the case of mining areas, the steps being currently taken include allotment of mineral concession (i.e., mining lease) for extraction of mineral, seeking consent from land owner or the agency having surface right for the area covered under lease, operating the allotted mine with complying all existing applicable laws including implementation of mine closure plan as approved by the competent authority.

No mining lease would be granted to any party, private or public, without a proper mining plan including the environmental management plan approved and enforced by statutory authorities. However, elaborate land use planning is not undertaken. If this process is initiated, about 0.17 per cent of the country's land will come under planned development.

The Government of India initiated a number of centrally sponsored schemes like Integrated Wasteland Development Programme, Drought-Prone Area Programme and Desert Development Programme for assisting states to increase productivity of marginalized land. Later in 2009, all these programmes were merged under single integrated scheme called Integrated Watershed Management Programme, covering not only the marginal lands but also the area under rain-fed agriculture. This was based on the realization that irrigated area under Green Revolution has already reached its productivity limits and the increase in productivity of vast extent of rain-fed area is the main plank to address the looming food security issues of the country. The watershed areas, if associated with land use planning could serve planned development.

Coastal environment plays a vital role in nation's economy by virtue of the resources, productive habitats and rich biodiversity. India has a coastline of about 7,500 km of which the mainland accounts for 5,400 km, Lakshadweep coasts extend to 132 km and Andaman & Nicobar Islands have a coastline of about 1,900 km. Nearly 250 million people live within a distance of 50 km from the coast. The coastal zone is also endowed with a very wide range of coastal ecosystems like mangroves, coral reefs, sea grasses, salt marshes, sand dunes, estuaries, lagoons, etc., which are characterized by distinct biotic and abiotic processes and ecosystem services. The coastal areas are assuming greater importance in recent years, owing to increasing human population, urbanization and accelerated developmental activities. These anthropogenic activities have put tremendous pressure on the fragile coastal environment. There has been significant degradation of coastal resources and ecosystem services in recent years due to poorly planned developmental activities and overexploitation of natural resources.

For the purpose of protecting and conserving the coastal environment, the Ministry of Environment & Forests, GoaI issued the Coastal Regulation Zone Notification dated 19.2.1991 under Environment (Protection) Act, 1986. This notification regulates all developmental activities in the Coastal Regulation Zone area. This notification imposed formidable restrictions on the land use in the coastal region. The Government of India has initiated, with the support of the World Bank, the Integrated Coastal Zone Management (ICZM) Project for building national capacity for implementation of comprehensive coastal management approach in the country, and piloting the integrated coastal zone management approach in states of Gujarat, Orissa and West Bengal. The project has an important element of preparation of an Integrated Coastal Zone Management Plan.

During late 1980's, the Government of India launched centrally sponsored programme Computerization of Land Records (CLR) and Strengthening of Revenue Administration & Updating of Land Records (SRA&ULR) to improve revenue administration and the sordid state of land records in the country. Various States achieved differently in these programmes. The Government of India again took initiative to revitalize the land administration agenda by merging the earlier two programmes into a single integrated programme called the 'National Land Records Modernization Programme (NLRMP)' which aims at ushering in a system of updated land records, automated and automatic mutation of land transactions, integration between textual and spatial land records, inter-connectivity between revenue and registration systems, and finally replacing the present deeds registration and presumptive title system with that of conclusive titling with title guarantee.

During 1970's, all the States established 'State Land Use Boards' under the Chairmanship of respective Chief Minister of the State. These Boards were meant to provide policy directions and coordinate the activities of different departments dealing with soil and land resources. These Boards never functioned the way they were meant to be, and they felt in disuse overtime and nearly all of them have been abolished. At present, the States do not have any mechanism at their disposal to deal with land policy issues in a coherent manner. Hence, the responses of the States to land issues are impulsive and ad hoc without consistency. No rationale and scientific considerations appear to be guiding the decisions on land use.

There are several existing policies relating to land use. These include:

- The National Water Policy 2013
- The National Land Use Policy Outlines 1988
- The National Forest Policy 1988
- The Policy Statement of Abatement of Pollution 199
- The National Livestock Policy Perspective, 1996
- The National Agricultural Policy 2000
- The National Population Policy 2000
- The National Policy and Macro-level Strategy and Action Plan on Biodiversity 2000
- The National Environmental Policy 2006 etc.

Principles for sustainable development

1. **Human beings are at the centre –**

    Human beings are at the centre of concerns for sustainable development. They are entitled to a healthy and productive life in harmony with nature.

2. **Inclusive growth, poverty eradication and gender equality –**

    Equal opportunities Inclusive growth, poverty eradication and gender equality are indispensable requirements for sustainable development of India, and must be addressed in all policies, plans and programmes.

3. **Balanced development and intergenerational justice -**

    Developmental sectors and activities must be planned in a balanced manner to meet economic, social and environmental needs of present and future generations, and should aim to minimize any large-scale displacement of population.

4. **Efficient utilisation of resources and mitigation of impacts –**

    Long term planning for optimum utilisation of land and saving scarce land resource is essential and as far as possible, projects should be set up on recycled lands, wastelands, degraded lands or un-irrigated lands provided these are not performing any other important function like bio-diversity, water resources etc.

5. **Integrated and comprehensive development planning –**

   Development planning must be comprehensive, sustainable, and integrated vertically (national, state, regional and local levels) taking into consideration the interest of all other sectors and stakeholders.

6. **The States are custodian of land –**

   States are custodian of land and state governments must eliminate unsustainable patterns of land utilization / land management and provide necessary legal and institutional support to facilitate capacity building and participatory, transparent and comprehensive land use planning.

7. **Harmonization with existing policy, legislative and regulatory framework –**

   The existing constitutional provisions and rights, the existing laws, rules, standards, procedures, guidelines and stipulations brought out by various ministries, departments and institutions of the Government of India, as applicable to land utilization policy shall continue to be in force for taking decisions on land matters and land use changes.

Goal of National Land Utilization Policy

The goal of the National Land Utilization Policy is to achieve improvement of livelihood, food and water security, and best possible realization of various developmental targets so as to ensure sustainable development of India. To ensure optimal utilization of the limited land resources in India for achieving sustainable development, addressing social, economic and environmental considerations and to provide a framework for the States to formulate their respective land utilisation policies incorporating state-specific concerns and priorities.

The specific objectives are given below:

Objectives related to social concerns:

1. Protection of agricultural lands from land use conversions so as to ensure food security

and to meet consumption needs of a growing population and to meet livelihood needs of the dependent population.

2. To identify and protect lands that are required to promote and support social development, particularly of tribal communities and poor section of society for their livelihood.

3. To preserve historic and cultural heritage by protecting, places/sites of religious, archaeological, scenic and tourist importance.

Objectives related to environmental concerns:

4. To preserve and conserve lands under important environmental functions such as those declared as National Parks, Wild Life Sanctuaries, Reserved Forests, Eco Sensitive Zones, etc. and guide land uses around such preserved and conserved areas so as not to have land use conflicts or negative environmental impacts.

5. To preserve the areas of natural environment and its resources that provide ecosystem services.

Objectives related to developmental/ economic concerns:

6. To promote properly guided and coordinated development in a sustainable manner of all developmental sectors including agriculture, urban, industrial, infrastructure and mining so as to minimise land use conflicts or negative environmental impacts.

Objectives related to enforcement and implementation of the policy:

7. To suggest a general implementation framework for implementing land utilization policy by all concerned at different levels, viz. national, state, regional and local, and undertaking capacity building.

Weather forecasting

Weather and climatic information plays a major role before and during the cropping season and if provided in advance can be helpful in inspiring the farmer to organize and activate their own resources in order to reap the benefits.

The Agro-meteorological Advisory Service (AAS) rendered by India Meteorological Department (IMD), Ministry of Earth Sciences (MoES) is a mechanism to apply relevant meteorological information to help the farmer make the most efficient use of natural resources, with the aim of improving agricultural production; both in quantity and quality. It becomes more and more important to supply climatological information blended with seasonal climate forecasts before the start of the cropping season in order to adapt the agricultural system to increased weather variability. Subsequent to this, short and medium range weather forecast based Agro-meteorological advisories become vital to stabilize their yields through management of agro-climatic resources as well as other inputs such as irrigation, fertilizer and pesticides. Agro- meteorological service rendered by IMD, MoES is a step to contribute to weather information based crop/livestock management strategies and operations dedicated to enhancing crop production and food security.

The main emphasis of the existing AAS system is to collect and organize climate/weather, soil and crop information, and to amalgamate them with weather forecast to assist farmers in taking management decisions. This has helped to develop and apply operational tools to manage weather related uncertainties through agro-meteorological applications for efficient agriculture in rapidly changing environments.

The information support systems under AAS

The information support systems under AAS include provision of weather, climate, crop/soil and pest disease data to identify biotic and abiotic stress for on-farm strategic and tactical decisions:

- » Provide district specific (Pan India) weather forecast (Rainfall, cloudiness, maximum/minimum temperature, wind speed, wind direction, maximum/minimum relative humidity) up to 5 days with outlook for rainfall for remaining two days of a week.

- » Translate weather and climate information into farm advisories using existing research knowledge on making more efficient use of climate and soil resources through applications of medium range weather forecast to maximize benefits of benevolent weather conditions and alleviate the adverse impacts of malevolent weather events .

A broad spectrum of advisories includes:

- » Weather sensitive farm operations such as sowing/ transplanting of crops, fertilizer application based on wind condition & intensity of rain, pest and disease control, intercultural operations, quantum and timing of irrigation using meteorological threshold and advisories for timely harvest of crops.

- » Introduction of technologies such as crop simulation model based decision support system for agro-meteorologists to adapt agricultural production systems to changing weather & climate variability and to the increasing scarcity of input such as water, seed, fertilizer, pesticide etc.

- » Develop effective mechanism to on time dissemination of agro-Met advisories to farmers.

- » Effective training, education and extension on all aspects of agricultural meteorology.

The District level Agromet Advisory Service (DAAS) run by IMD is a multidisciplinary and multi- institutional project. It involves all stake holders such as Indian Council for Agriculture Research (ICAR), State agricultural Universities (SAUs), Krishi Vigyan Kendras (KVKs), Department of Agriculture & Cooperation, State Departments of Agriculture/ Horticulture/ Animal Husbandry/ Forestry (Up to District level offices), NGOs, Media Agencies, etc. This project is being implemented through five tier structure to set up different components of the service spectrum. It includes meteorological (weather observing & forecasting), agricultural (identifying weather sensitive stress & preparing suitable advisory using weather forecast), extension (two way communication with user) and information dissemination (Media, Information Technology, Telecom) agencies.

Flow chart of district level Agro-Met Advisory Service System (DAAS)

IMD has started issuing quantitative district level (612 districts) weather forecast upto 5 days from 1st June, 2008.

- » The products comprise quantitative forecasts for 7 weather parameters viz., rainfall, maximum and minimum temperatures, wind speed and direction, relative humidity and cloudiness, besides weekly cumulative rainfall forecast.

- » IMD, New Delhi generates these products based on a Multi Model Ensemble (MME) technique (35*35 km grid) using forecast products available from a number models of India and other countries.

- » Ensemble products have skill better than individual members of the ensemble group.

- » The average skill for rainfall during monsoon season has been found to be 70 to 80 %.

- » Individual members include: T-254 model of NCMRWF, T-799 model of European Centre for Medium Range Weather Forecasting (ECMWF); United Kingdom Met Office (UKMO), National Centre for Environmental Prediction (NCEP), USA and

Japan Meteorological Agency (JMA). The products are disseminated to Regional Meteorological Centres and Meteorological Centres of IMD located in different states.

These offices undertake value addition to these products twice a week on Tuesday and Friday and communicate to 130 AgroMet Field Units (AMFUs) located with State Agriculture Universities (SAUs), ICAR etc.

- » The Regional Meteorological Centers and Meteorological Centers of IMD located in different states undertake value addition to these products.

- » The value addition is based on the inputs from very high resolution meso-scale model (WRF) model, synoptic knowledge, bias correction of district forecast etc.

- » These Centres run the WRF model using initial conditions generated from global model for detailed analysis of rain-bearing systems at higher resolution (9*9 Km grid).

- » Data used in these numerical weather modeling are upper air soundings, land surface (including network of automatic weather stations and automatic rain gauge at sub- district scale), marine surface buoys, aircraft observations, wind profilers, and satellite-data (wind, radiance, rain-rate, etc.).

Performance evaluation of MME generated district level rainfall forecasts for day-1 and day- 5 are given for some selected states like Orissa, Rajasthan, Maharashtra, Gujarat and Kerala, which represent east central India the domain of monsoon low; northwest India region of less rainfall; west India; region of mid-tropospheric circulation and extreme south east Peninsula. The results show that performance skill of day-1 district level rainfall forecast for the rainfall amount of moderate range (>10 mm and <65 mm) is reasonably good for all these states, where Probability of Detection (POD) is more than 0.4. For rain/no rain case, POD for these states is more than 0.6. For the heavy rainfall event (>65 mm) POD significantly deteriorates. Among these states, Kerala has the best performance, followed by Orissa, Gujarat, Maharastra and Rajasthan. POD of Kerala for rain/no rain case has been above 1.0 at all the districts, for light rain it is around 0.3, for moderate rainfall it is around 0.9 and for heavy rainfall close to zero. For Orissa, POD for rain/no rain case has been above 0.8 at all the districts, for light rain it is around 0.6, for moderate rainfall it is between 0.3 to 0.4 and for heavy rainfall it is close to zero.

For Rajasthan , POD for rain/no rain case has been between 0.4 to 0.7 at most of the districts, for light rain it ranges between 0.3 to 0.5 for moderate rainfall it is between 0.1 to 0.3 and for heavy rainfall it is zero. At the day-5 forecast, performance deteriorates but pattern remains the same.

Based on the above forecast products and the crop information available from districts, the AMFU prepares district-wise agro-advisories. Ministry of Earth Sciences has set up a network of 130 AMFUs covering the agro-climatic zones of the country. Each agro-climatic zones cover on average, 4 to 6 districts. These units are operated at State Agriculture Universities (SAUs), Indian Council of Agricultural Research institutions (ICAR), Indian Institute of Technology (IIT) by providing grant-in-aid from IMD. These units are responsible for recording agro meteorological observations, preparing forecast based Agromet advisories for the districts falling under precinct of concerned agroclimatic zone and dissemination of the same. Concerned university/institute has appointed Nodal Officer and Technical Officers, who prepare the advisory bulletins in consultation with the panel of experts already created at these units. The Agromet bulletins include specific advice on field crops, horticultural crops and livestock, etc. which farmers need to act upon. Its frequency is twice a week i.e. Tuesday and Friday.

The agromet district advisories, generated by 130 AMFUs

The advisories are being disseminated to the farmers through mass media (Radio, Print and TV), Internet, etc.

- » A mechanism has also been developed to obtain feedback from the farmers on quality of weather forecast, relevance and content of agromet advisory and effectiveness of information dissemination system.

- » A multi- media system for dissemination of agro-meteorological advisories to the farming community has been put in place in which beside the conventional modes e.g. radio, television & print media, concerted efforts are made to reach farmers through emerging modes of communication such as mobile phones and the internet.

- » Short message service (SMS) and voice messages are being send to subscribing farmers by Govt. and private companies such as Reuters Market Light, IFFCO Kisan Sanchar Limited, MahaAgri, Vritti Solutions, NOKIA, eFresh and State Govt. agencies.

- » Beside these agencies there are many companies such as TATA Consultancy Service, NABARD, ICT, Infosys; Infronics etc. are like to start the service in near future.

- » SMS service as on date covers 3.0 million users spread across 16 states while IVRS covers around 30000 farmers spread across 5 states.

- » The agro-meteorological advisory service has also developed a mechanism to assess the users' needs and strive to meet them in order to play an efficient role for the improvement of the agricultural production.

» A study aiming to assess Economic Impact of AAS carried out during 2003 to 2007 at 15 AAS units covering 3 *kharif* and 3 *rabi* seasons, concluded that the farmers could save significant losses of farm input like seeds, water, pesticides and fertilizers and reaped better harvest and made their farming more profitable by using the AAS.

» In general there is net gain ranging from 8 to 10 percent by those farmers who used the information provided by the AAS system.

» Reliable and timely forecast provide important and useful input for proper, foresighted and informed planning.

» Early prediction of crop yield is important for planning and for taking various policy decisions.

» Final production estimates based on the sampling method become available long after the crops are actually harvested.

» In recent past Department of Agriculture and Cooperation (DAC), Min. of Agriculture in collaboration with Department of Space, IMD and Institute for Economic Growth launched FASAL (Forecasting Agricultural output using Space, Agrometeorology and Land based observations) scheme for crop acreage estimation and crop yield forecast by integrating technological advancement and adoption of emerging methodologies, in particular, those of remote sensing and geographic information.

» Objective is to develop, validate and issue multiple crop yield forecast for major crops at mid season (F2) and pre-harvest stage (F3) of crops.

The crop yield forecasting models are developed at different stages of the crops combining statistical, crop growth simulation and other methods/techniques, which take into account the influence of weather and technological advances on crop yield.

» The model requires the meteorological data (rainfall, temperature, humidity, bright hours of sunshine with wind speed, wet spell etc.) as well as technological data as the inputs to estimate the yield. Using agromet models and database developed, IMD is providing the yield forecast at district/ state/ national scale to DAC since monsoon 2011. Although, a good progress has been made in this regard but much is required to be done.

- » There is need to develop methodologies for Remote Sensing and Conventional Data Merging.
- » Concerted efforts are needed for ground based data collection, satellite data collection, GIS software applications, operational applications of meteorological satellite data, weather radars and the monitoring of cropping season by meteorological and remote sensing data to equip AAS units to generate better advisories.
- » The system is working to improve these aspects of the service.

14 Principles, Significance And Future Of Conservation Agriculture

Conservation agriculture (CA) can be defined by a statement given by the Food and Agricultural Organization of the United Nations as "a concept for resource-saving agricultural crop production that strives to achieve acceptable profits together with high and sustained production levels while concurrently conserving the environment".

Agriculture according to the New Standard Encyclopedia is "one of the most important sectors in the economies of most nations". At the same time conservation is the use of resources in a manner that safely maintains a resource that can be used by humans. Conservation has become critical on the fact that the world population has increased over the years and more food needs to be produced every year.

Principles of CA

The Food and Agricultural Organization of the United Nations (FAO) has determined that CA has three key principles that producers (farmers) can proceed through in the process of CA. These three principles outline what conservationists and producers believe can be done to conserve what we use for a longer period of time.

1. The first key principle in CA is practicing minimum mechanical soil disturbance which is essential to maintaining minerals within the soil, stopping erosion, and preventing water loss from occurring within the soil. In the past agriculture has looked at soil tillage as a main process in the introduction of new crops to an area. It was believed that tilling the soil would increase fertility within the soil through mineralization that takes place in the soil. Also tilling of soil can cause severe erosion and crusting which will lead to a

decrease in soil fertility. Today tillage is seen as a way as destroying organic matter that can be provided within the soil cover.

No-till farming has caught on as processes that can save soils organic levels for a longer period and still allow the soil to be productive for longer periods as given by FAO, 2007. Also with the process of tilling causes the time and labor for producing that crop. Producers can save 30 to 40% of time and labor by practicing the no-till process.

When no-till practices are followed, the producer sees a reduction in production cost for a certain crop. Tillage of the ground required more money due got fuel for tractors or feed for the animals pulling the plough. The producer sees a reduction in labor because he or she does not have to be in the fields as long as a conventional farmer.

2. The second key principle in CA is much like the first in dealing with protecting the soil. The principle of managing the top soil to create a permanent organic soil cover can allow for growth of organisms within the soil structure. This growth will break down the mulch that is left on the soil surface. The breaking down of this mulch will produce a high organic matter level which will act as a fertilizer for the soil surface. If the practices of CA were being done for many years and enough organic matter was being built up at the surface, then a layer of mulch would start to form. This layer helps prevent soil erosion from taking place and ruining the soils profile or layout.

In the article "The role of conservation agriculture and sustainable agriculture" the layer of mulch that is built up over time will start to become like a buffer zone between soil and mulch that will help reduce wind and water erosion. With this, comes the protection of a soils surface when rain is falling to the ground. Rainfall on land that is not protected by a layer of mulch is left open to the elements. But when soils are covered under a layer of mulch, the ground is protected so that the ground is not directly impacted by rainfall. This type of ground cover also helps keep the temperature and moisture levels of the soil at a higher level rather than if it was tilled every year.

3. The third principle is the practice of crop rotation with more than two species. The crop rotation can be used best as a disease control against other preferred crops. This process will not allow pests such as insects and weeds to be set into a rotation with specific crops. Rotational crops will act as a natural insecticide and herbicide against specific crops. Not allowing insects or weeds to establish a pattern will help to eliminate problems with yield reduction and infestations within fields. Crop rotation can also help build up a soils infrastructure. Establishing crops in a rotation allows for an extensive build up of

rooting zones which will allow for better water infiltration.

The breakdown of organic molecules in the soil into phosphates, nitrates and all the other "ates" are then in a form which plants can use. Plowing increases the amount of oxygen in the soil and increases the aerobic processes, hastening the breakdown of organic material. Thus more nutrients are available for the next crop but, at the same time, the soil is depleted more quickly of its nutrient reserves.

Benefits

In the field of conservation agriculture (CA) there are many benefits that can be obtained by both the producer and conservationist.

- » On the side of the conservationist, CA can be seen as beneficial because there is an effort to conserve what people use every day. Since agriculture is one of the most destructive forces against biodiversity, CA can change the way humans produce food and energy. With conservation come environmental benefits of CA. These benefits include less erosion possibilities, better water conservation, improvement in air quality due to less emission being produced, and a chance for larger biodiversity in a given area.

- » On the side of the producer and/or farmer, CA can eventually do all that is done in conventional agriculture, and it can conserve better than conventional agriculture. CA according to Theodor Friedrich, who is a specialist in CA, believes "Farmers like it because it gives them a means of conserving, improving and making more efficient use of their natural resources". Producers will find that the benefits of CA will come later rather than sooner. Since CA takes time to build up enough organic matter and have soils become their own fertilizer, the process does not start to work over night. But if producers make it through the first few years of production, results will start to become more satisfactory.

- » CA is shown to have even higher yields and higher outputs than conventional agriculture once CA has been establish over long periods. Also, a producer has the benefit of knowing that the soil in which his crops are grown is a renewable resource. According to New Standard Encyclopedia, soils are a renewable resource, which means that whatever is taken out of the soil can be put back over time. As long as good soil upkeep is done, the soil will continue to renew itself. This could be very beneficial to a producer who is practicing CA and is looking to keep soils at a productive level for an extended time.

> The farmer and/or producer can use this same land in another way when crops have been harvested. The introduction of grazing livestock to a field that once held crops can be beneficial for the producer and also the field itself. Livestock can be used as a natural fertilizer for a producer's field which will then be beneficial for the producer the next year when crops are planted once again. The practice of grazing livestock in a CA helps the farmer who raises crops on that field and the farmer who raises the livestock that graze off that field. Livestock produce compost or manures which are a great help in producing soil fertility. With the practices of CA and grazing livestock on a field for many years can allow for better yields in the following years as long a practices are continued to be followed.

The FAO believes that there are three major benefits from CA in brief.

> Within fields that are controlled by CA the producer will see an increase in organic matter.

> The second benefit is an increase in water conservation due layer of organic matter and ground cover to help eliminate transportation and access runoff.

> The third benefit is an improvement of soil structure and rooting zone.

Future developments

As in any other businesses, producers and conservationist are always looking towards the future. In this case CA is a very important process to be looked at for future generations to have a chance to produce. There are many organizations that have been created to help educate and inform producers and conservationist in the world of CA. These organizations can help to inform, conduct research and buy land in order to preserve animals and plants.

Another way in which CA is looking to the future is through prevention. The producers are looking for ways to reduce leaching problems within their fields. These producers are using the same principles within CA, in that they are leaving cover over their fields in order to save fields from erosion and leaching of chemicals out of fields.

The producers and conservationist are looking towards the future as the circulation of plant nutrients can be a vital part to conserving for the future. An example of this would be the use of animal manure. This process has been done for quite some time now but the future is looking towards how to handle and conserve the nutrients within manure for a longer time. But besides just animal waste also food and urbanized waste are being looked towards as a way to use growth within CA. Turning these products from waste to being used to grow crops and improve yields is something that would be beneficial for conservationists and producers.

Problems

As much as conservation agriculture can benefit the world, there are some problems that come with CA. There are many reasons why conservation agriculture cannot always be a win-win situation.

- » There are not enough people who can financially turn from a conventional farmer to a conservationist.

- » Within the process of CA comes time; when a producer first starts to process as a conservationist the results can be a financial loss to that certain producer.

- » Since CA is based upon establishing an organic layer and producing its own fertilizer, then this may take time to produce that layer. It can be many years before a producer will start to see better yields than he/she has had previously before.

- » Another financial undertaking is purchasing of new equipment. When starting CA a producer may have to buy new planters or drills in order to produce effectively, also comes the responsibility of harvesting a crop. These financial tasks are ones that may impact whether or not a producer would want to conserve or not.

- » With the struggle to adapt comes the struggle to make CA grow across the globe. CA has not spread as quickly as most conservationists would like. The reasons for this is because there is not enough pressure for producers in places such as North America to change their way of living to a more conservative outlook. But in the tropics there is more of a pressure to change to conservation areas because of the limited resources that are available. Places like Europe have also started to catch onto the ideas and principles of CA but still nothing much is being done to change due to their being a minimal amount of pressure for people to change their ways of living.

- » With CA comes the idea of producing enough food. With cutting back in fertilizer, not tilling of ground and among other processes come the responsibility to feed the world. According to the Population Reference Bureau, at the 2000 census count of the world population there were around 6.08 billion people on earth. By 2050 there will be an estimated 9.1 billion people. With this increase comes the responsibility for producers to increase food supply with the same or even less amounts of land to do it on. With CA problems arise in the fact that if farms do not produce as much as conventional ways then this leaves the world with less food for more people.

Resource conservation technologies

Rice and wheat are the staple food crops occupying nearly 13.5 million hectares of the Indo-Gangetic plains (IGP) of South Asia covering Pakistan, India, Bangladesh and Nepal. These crops contribute more than 80% of the total cereal production and are critically important to employment and food security for several millions of rural families. The demand for these two cereals is expected to grow between 2 and 2.5 per cent per year until 2020, requiring continued efforts to increase productivity while ensuring sustainability. Starting from the 1960s, expansion of area and intensification of rice–wheat productions system based on the adoption of Green Revolution (GR) technologies, incorporating the use of high-yielding varieties, fertilizers and irrigation, led to increased production and productivity of both these crops. However, continued intensive use of GR technologies in recent years has resulted in lower marginal returns and in some locations to salinization, overexploitation of groundwater, physical and chemical deterioration of the soil and pest problems. Field studies results show that the resource conserving technologies, an exponent of conservation agriculture, improve yields, reduce water consumption and reduce negative impacts on the environmental quality.

Emerging challenges

Over the years, the rice–wheat systems in the north-western part of Indo-Gangetic Plain (IGP) have become largely mechanized, input-intensive, and dependent on the conjunctive use of surface and groundwater. In contrast, the rice–wheat systems of the eastern IGP have remained largely labour-intensive and less mechanized. Farmers use low inputs because of socio-economic constraints and serious problems of drainage congestion and rainwater management. In all parts of the IGP farmers rely on tube-well irrigation to supplement rain water or to meet full water requirements for crop production.

Evidence is now emerging that continued intensification of input use since the adoption of Green Revolution Technologies is providing lower marginal returns. At the same time, it is known that inappropriate use of applied inputs and overexploitation of natural resource base, principally land and water, is in many situations leading to secondary salinization in low-quality aquifer zones, groundwater table recession in fresh water aquifer zones, physical and chemical deterioration of the soil and water quality due to nutrient mining and pollution of ground water in some locations due to over application of nitrogenous fertilizers and of environment through crop residue burning and pesticide use. Consequently, there are now serious concerns about the future potential for productivity growth and long-term sustainability of the irrigated rice–wheat systems of the IGP. A study has revealed that rice–wheat systems suffer from stagnation in productivity in spite of large production potential yet to be tapped in large areas of middle and lower Gangetic plains. The total factor productivity (TPF) index of the crops was 1.4, 0.9, 0.43 and 3.1 per cent in trans-, upper- middle- and lower-Gangetic plains, respectively. A significant increase in the number of districts where input, output and TFP growth has turned negative or stagnated. Thus, the major challenge for South Asian countries is to continue to look for technological innovations coupled with socio-economic adjustments and policy reforms to sustain increases in productivity and production of the rice–wheat systems.

Technologies for Sustainable Management of Natural Resources

1. No-tillage and crop establishment

The conventional system for establishment of wheat crops includes repeated ploughing (6–8 ploughing), cultivating, planking, and pulverizing of topsoil. This has been substituted with direct drilling of wheat using zero-till seed drills fitted with inverted T-openers to place seed and fertilizers into a narrow slot with only minimal of soil disturbance and without land preparation. Substitution of conventional tillage with zero or minimum tillage for wheat planting in rice–wheat system in IGP is a development of regional significance and contributes to the global application of resource conservation technologies (RCTs) in to a new eco-system. Rice crop is conventionally established as a puddled transplanted crop. Joint efforts of the public institutions and the small-scale private entrepreneurs are giving promising results for development of 'double no-till' system where both rice and wheat crops are drilled with minimum cultivation. This required development of new seed drill fitted with either a double disk openers or mechanical dibbler- 'punch planter', shredders-spreader (Happy Seeder) or roto-coulter type disk-drills. Experiments have been undertaken with direct-drilling of rice and wheat crops in both flat and raised bed planting systems.

In the IGP, new resource conserving technologies and development of appropriate machinery is being combined with novel land and water management approaches for greater efficiency and sustainability of the rice–wheat systems. At the same time, these technologies are generating alternative sources of productivity growth through diversification and intensification of production systems. For example many farmers are now practicing intercropping in raised bed system. In this system wheat is planted on the raised beds and mint or sugarcane in the furrows. Inter-cropping systems such as maize + potato/onion/red-beets or sugarcane + chickpea/Indian-mustard are also becoming popular with farmers in India.

2. Water management

The total annual irrigation water requirement of the rice–wheat system ranges from 1100 to 1600 mm/yr. Work initiated in close collaboration with the private sector, and later supported by (Rice-wheat crop establishment method) RWC, has successfully adapted the technique of laser land levelling for use in the rice–wheat system. Laser assisted precision land levelling facilitates application of less water more uniformly under flood irrigation, reduces leaching losses and improves crop-stand and yields. In rice–wheat system, precision land levelling saves irrigation water in wheat season by up to 25%; reduces labor requirements by up to 35 per cent; leads to about 2% increase in the area irrigated due to removal and/or reduction in size of bunds made to impound water for rice cultivation; and increases crop yields by up to 20 per cent. Further work is now in progress in all the RWC countries to integrate other land-preparation and crop-establishment methods with laser levelling to reduce water use at the field/farm/basin levels.

3. Nutrient management

In the case of nitrogen, findings from research on matching site-specific capacities of the soil to supply nutrients and to the demand of crop(s) in the system have been reflected in the development of a leaf color chart (LCC) to help farmers select the right dose and time of application for optimum response in rice. Efforts have also been made to extend the LCC technology to wheat crop by synchronising N application with irrigation practices. LCC technology has the potential to save about 15–20 per cent of N fertiliser application in rice. The work on other nutrients is less advanced at the farm level although a careful examination of long-term experiments undertaken in the consortium countries by the RWC is identifying nutrient mining (such as of K) and imbalances, along with the loss of C in some situations, as contributing factors to reduced yields. These nutrient management strategies are now being adapted to new crop and tillage systems in presence of residues retained on the soil surface.

4. Crop improvement and management

The research on the rice–wheat system is providing useful information as a result rice breeders have given greater attention to such traits as early maturity to allow earlier wheat planting to open opportunities for introduction of short-season crops, e.g. pulses, potatoes. More recent commodity research programs in wheat and rice are examining the genotype × tillage interactions of cultivars under zero-till, raised-bed and surface seeding situations for their ability to compete with weeds. These developments are also contributing to a broader debate about the need for modification of selection criterion in the breeding programs to accommodate new crop establishment and management practices.

As more farmers use the new resource conservation technologies (RCTs,) there will be a need to adapt crop, variety, fertilizer, water and pest management practices to new systems in relation to local needs. This is already beginning to happen in the management of the herbicide resistant weed *Phalaris minor*. In India *P. minor* is an important weed in the wheat crop. Continuous use of isoproturon for the control of this weed has led to development of severe resistance to this herbicide. To overcome this problem integrated approaches involving rotations of crops and other herbicides (e.g. clodinafop, fenoxaprop, sulfosulfuron, tralkoxydim) have been recommended. The use of zero-tillage for wheat planting is emerging as a new tool in integrated weed management. In the short-term it reduces weed population due to elimination of tillage and in conjunction with new herbicides provides effective weed control at lower rates, especially when closer row spacing is adopted.

Impact of Resource Conservation Technologies (RCTs)

1. Crop yield

Researchers from both Pakistan and India are reporting higher wheat yields following adoption of zero-tillage in rice–wheat rotations. On-farm trials in the rice-growing belt higher yields observed with zero-tillage. This is largely due to the time saved in land preparation that enabled a more timely planting of wheat crop. It has been reported from the simulation study that planting time of wheat regulates yield, governed by the climatic parameters mainly through temperature and delayed planting results in significant losses in yield.

From the several studies the zero-tillage gave higher yields compared to conventional tillage averaged at around 270 kg/ha (wheat yield of 5380 and 5110 kg/ha for zero till and conventional tillage, respectively). In eastern Gangetic Plains, where late planting of wheat is quite common, yield losses can be much higher due to shorter winter season window for wheat growth. In this situation productivity gains due to advancement in wheat planting through adoption of zero-tillage, as widely observed in rice-wheat crop establishment (RWC managed trials), can be in the range of 400–1000 kg/ha. Research has shown that the zero-till system advances crop planting by at least 1 week thereby reducing yield losses by 1–1.5 per cent /day after optimum wheat-sowing time. In an on-station trials considerable variations in the performance between cultivars of wheat was observed under flat and raised bed systems and planting densities, highlighting the need for additional research in breeding and development of cultivars appropriate for the raised bed planting system. It should be noted, however, that under field situation raised-bed and furrow dimension is usually governed by the width of the tractor. It has been observed that two beds formed per tractor pass (134 cm wide, Indian tractors), restricts soil compaction to tractor lanes and facilitates formation of a permanent raised bed planting system. The timing of the first irrigation for wheat on beds may need to be earlier than the recommended practice for wheat on flat layouts on the coarse textured soils of north-west India.

2. Cost comparison under reduced tillage systems with conventional practices

Net benefits were observed on the basis of several trials average around 10500/ha. Contributory factors to cost savings included: higher yields and reduced cost of cultivation (about half of that for the conventional tillage system). More information on cost comparison of zero-tillage was recorded over conventional cultivation based on a survey of farmers' perceptions and research findings.

3. Impact on the environment

Straw retained on the soil surface reduces weed seed germination and growth, moderates soil temperature and reduces loss of water through evaporation. In addition, crop residue is also an important source of fodder for animals in the Indo-Gangetic Plain (IGP) countries. Despite these potential benefits, however, large quantities of straw (left over after rice and wheat harvesting) are burnt each year by farmers to facilitate land preparation for crop planting. It is estimated that the burning of one ton of straw releases 3 kg particulate matters, 60 kg CO, 1460 kg CO_2, 199 kg ash and 2 kg SO_2. With the development of new drills which are able to cut through crop residue, for zero-tillage crop planting, burning of straw can be avoided which amounts to as much as 10 tons per hectare, potentially reducing release of some 13–14 tons of carbon dioxide. Elimination of burning on just 5 million hectares would reduce the huge flux of yearly CO_2 emissions by 43.3 million tons (including 0.8 million ton CO_2 produced upon burning of fossil fuel in tillage). Zero-tillage on an average saves about 60 l of fuel per hectare thus reducing emission of CO_2 by 156 kg/ha/year.

Adoption of resource conservation technologies (RCTs) which allow alterations in water, tillage and surface residue management practices can have a direct effect on emissions of greenhouse gases (GHGs) and enhance the carbon stocks of the soil. Soil submergence in rice cultivation leads to unique processes that influence ecosystem sustainability and environmental services such as carbon storage, nutrient cycling and water quality. For example the submergence of soils promotes the production of methane by anaerobic decomposition of organic matter. However, worries that such rice systems are a major contributor to global warming were allayed through a wide-scale study in the region. It has been noticed that methane emissions from rice fields range from 16.2 to 45.4 kg/ha during the entire season, whereas nitrous oxide emission under rice and wheat crops amounts to 0.8 and 0.7 kg/ha, respectively. Incorporation of straw increases methane emissions under flooded conditions but surface management of the straw under aerated conditions and temporary aeration of the soils can mitigate these effects. Thus, adoption of aerobic mulch management with reduced tillage is likely to reduce methane emissions from the system.

The water regime can strongly affect the emission of nitrous oxide, another green house gas (GHG), which increases under submergence and is negligible under aeration. Any agronomic activity that increased nitrous oxide emission by 1 kg/ha needs to be offset by sequestering 275 kg/ha of carbon, or reducing methane production by 62 kg/ha. Adoption of RCTs would favour the decrease of this GHG.

In order to minimize nitrate pollution of ground water, volatilization losses of fertilizer N in rice/wheat and address issues of crop residue burning, receding water table and emission of GHG, measures such as introduction of a legume crop (Mungbean) between wheat and rice, deep placement of nitrogenous fertilisers and raised bed planting and laser land levelling have been developed. With further refinement of double disk planters, punch planter and roto-disk-drill it has become easier to plant crops with through retained residues. These implements are now being evaluated in farmer participatory trials along with modified fertilizer and irrigation practices.

Given the potential of RCTs to influence all the major GHGs and underground water reserves and its quality, in planning future research it is important that due consideration is given to potential positive and negative impacts of agronomic and crop management practices on the environmental quality.

4. Extent of uptake of resource conservation technologies (RCTs)

The farmers have rapidly adopted zero-tillage for planting of wheat and other crops after rice. It has been estimated that close to 2 m ha have been covered with zero-till planting of winter season crops (wheat, maize, lentil, chickpea, peas, etc.) in year 2004, mainly in India and Pakistan. The acreage of zero-till/reduced-till winter season crops was nearly 1.7 m ha during the same period in Indian part of the IGP. The area of wheat planted to zero-till in 2002 and 2003 were estimated to be 0.37 and 1.1 m ha respectively. The emerging trends of zero/reduced tillage are observed in South Asia (India, Pakistan, Nepal and Bangladesh). The spread of RCTs has also encouraged the growth of private sector input providers. It is estimated that in 2004 nearly 100 small private entrepreneurs were manufacturing direct drills in India and 50 in Pakistan. In the year 2000 this number was just 5 in the two countries. A survey shows that even resource-poor smallholders have started to benefit from this technology by using contractors to direct-drill their crops. Some more progressive farmers are now experimenting with the option of establishing rice either as a transplanted or a direct dry seeded crop in unpuddled no-till fields with a view to 'double no-till' rice–wheat-planting system.

Future Directions for Resource Conservation Technologies (RCTs)

The natural resource management issues impacting on sustainability of the rice–wheat systems are complex and differ significantly between different parts of IGP transects. Locally adapted RCTs appropriate to resource endowments of farmers and the biophysical environment hold potential to improve management of natural resources and provide sustainable increases in productivity.

The increasing demand for basic cereals in the future would need to be met largely through increased productivity, allowing some land (and other resources) for diversification for greater income generation. Clearly, market forces and national and state policies will drive the pace and form of the diversification. An additional factor influencing the diversification of RWSs would be the new 'platform' made possible by the RCTs.

Socio-economic and biological research to determine the feasibility of changing the culture of rice to enhance productivity, diversity and sustainability, particularly regarding water use, are needed to determine under what circumstances such changes are appropriate.

Changes in tillage as well as land and water management practice, and a better understanding of drivers of the diversification process may require adjustments in all of the component technology for the new systems. This will involve examination of such issues as to which rice-based ecology to diversify, in which season and how best to address the multidimensional nature of poverty, including consideration of issues related to risk management, improved livelihoods, food security and nutrition.

The long-term trials, set up at the beginning of the green revolution era to understand nutrient mining in the system and to develop nutrient management strategies have provided valuable information to develop future strategies. Appropriate long-term monitoring must continue, and be relevant to future changes in tillage and water management practices. In addition, benefits of changes in the tillage system and stubble management to the soil ecosystem need to be better understood.

The main contribution in integrated pest management (IPM) research so far has been in the control of *P. minor*. However, the new tillage systems with reliance on herbicide inputs are likely to change the weed species and expose the system to more herbicide resistance. Gaps remain in the IPM agenda for the systems of today and there is a need for anticipatory IPM research (e.g., integrated weed, insect and disease management, the emerging role of nematodes in a more diversified and aerated system) in the context of the new RCT systems.

The changes in the RWSs have the potential to change the balance in global warming gases. Reduced tillage increases carbon accumulation in the soil and reduces fuel-based emissions. Soil submergence is the dominant feature of present rice cultivation in the IGP and leads to unique biogeochemical processes that influence methane and nitrogen gas emissions and nutrient availability. Changes in rice culture to a more aerated system could change the balance of these gases for the better.

Education in Farmer's Community

In the country that has the most farmers in the world, educating and training farmers has got to be the most essential way of supporting agriculture. It is the prerequisite for other kinds of support for this sector. With the country's entry into the WTO, Indian farmers are facing more psychological pressure, because they, by themselves, are not able to cope with the influx of cheap foreign farm produce. The government must step in to help them improve their educational level and ease the pressure of outside competition.

The government should consider establishing special schools for farmers under the agricultural, economic, and science colleges, where farmers can acquire farming and marketing knowledge. By equipping farmers with knowledge about technology, marketing and law, the government will raise their competitiveness and boost the long-term growth of agriculture and the rural economy. Farmer education should be recognized as crucial to the support of agriculture. And this way of supporting agriculture is more in line with the spirit of the WTO rules.

Policy Challenges And Potential Solutions Of Residue Burning In India

The agricultural industry plays a major role in the overall economic growth of the world. However, there is limited discussion on the management of agricultural waste in the published literature. It could be related to the fact that agriculture industry is not regulated as the municipal solid waste (MSW). The MSW is mainly governed by public entities such as municipalities and hence the generation and management data are collected, recorded, and analyzed in the public domain. Agricultural waste is predominantly handled by the owners of the agricultural land which is predominantly in the private sector, with little public sector involvement.

The growing demand for food in developing countries has led to tremendous increase in food production around the world. Hence, agro-based activities represent profitable businesses, both in developing as well as developed countries. The multitude of agricultural activities increases the amount of agro-products produced and this has led to an overall increase in environmental pollution and waste generation. The nature of the activities deployed, and the waste generated depends on the geographical and cultural factors of a country. Large stretches of wasteland have been converted to arable lands due to developments in water management systems, modern agro-technologies and large-scale agrochemical deployment. These measures have resulted in global environmental pollution and increased complexity in the disposal of agricultural waste. However, the national agencies are continuously developing policies and possible options to manage these wastes, which include their conversion to reusable resources.

Waste materials derived from various agricultural operations are defined as agricultural wastes. As per the United Nations, agricultural waste usually includes manure and other wastes from farms, poultry houses and slaughterhouses; harvest waste; fertilizer run-off from fields; pesticides that enter water, air or soils; salt and silt drained from fields. According to the world energy council, in addition to all above, agricultural waste can also comprise of spoiled food waste. The harvest waste, which is more popularly termed as crop residue can contain both the field residues that are left in an agricultural field or orchard after the crop has been harvested and the process residues that are left after the crop is processed into a usable resource. Stalks and stubble (stems), leaves, and seed pods are some common examples for field residues. Sugarcane bagasse and molasses are some good examples for process residue.

According to the Indian Ministry of New and Renewable Energy (MNRE), India generates on an average 500 Million tons (Mt here after) of crop residue per year. The same report shows that a majority of this crop residue is in fact used as fodder, fuel for other domestic and industrial purposes. However, there is still a surplus of 140 Mt out of which 92 Mt is burned each year. Table 1 compares the agricultural waste generated by selected Asian countries in Mt/year. It is also interesting to note that the portion burnt as agricultural waste in India, in volume is much larger than the entire production of agricultural waste in other countries in the region.

Table 1. Generation agricultural waste in India compared to other countries

Country agricultural waste generated (million tons/year)

Country	Waste (million tons/year)
India	500
Bangladesh	72
Indonesia	55
Myanmar	19

Waste from the agricultural industry can be beneficially utilized in various agro-based applications and other industrial processing. However, the cost of collection, processing and transportation can be much higher than the revenue from the beneficial use of such waste. The classic example of how economic reasons have prevented attaining the sustainable use of agricultural waste and lead to environmental chaos in India. This topic is important to the wider audience beyond India for two reasons: first, crop residues are an important constituent of agricultural waste that can actually be used for the benefit of the society due to its organic composition. The other important reason is that the volume of crop residue, with unsustainable management practices creates high adverse environmental impacts, which go far beyond India. Specifically, India is the second largest producer of rice and wheat in the world, two crops that usually produce large volume of residue.

Paddy straw is a major field-based residue that is produced in large amounts in Asia. In fact the total amount equaling 668 Mt could produce theoretically 187 gallons of bioethanol if the technology were available. However, an increasing proportion of this paddy straw undergoes field burning. This waste of energy seems inapt, given the high fuel prices and the great demand for reducing greenhouse gas emissions as well as air pollution. As climate change is extensively recognized as a threat to development, there is a growing interest in alternative uses of field-based residues for energy applications.

Punjab produces around 23 Mt of paddy straw and 17 Mt of wheat straw annually. More than 80 % of paddy straw (18.4 Mt) and almost 50 per cent wheat straw (8.5 Mt) produced in the state is being burnt in fields. Almost whole of paddy straw, except Basmati rice is burnt in the field to enable early sowing of next crop. Lately, the farmers have extended this practice to wheat crop also. Though part of the wheat straw is used as dry fodder for the milch cattle, the remaining straw is usually burnt for quick disposal.

There are primarily two types of residues from rice cultivation that have potential in terms of energy-straw and husk. Although the technology of using rice husk is well established in many Asian countries, paddy straw as of now is rarely used as a source of renewable energy. One of the principal reasons for the preferred use of husk is its easy procurement, i.e., it is available at the rice mills. In the case of paddy straw, however, its collection is a tedious task and its availability is limited to harvest time. The logistics of collection could be improved through baling but the necessary equipment is expensive and buying it is uneconomical for most rice farmers. Thus, technologies for energy use of straw must be efficient to compensate for the high costs involved in straw collection.

The presents disposal pattern of paddy straw giving details of alternate uses of agriculture waste, viz., rice residue as fodder for animals, its use in bio-thermal power plants, and its use for bedding material for animals, mushroom cultivation and so on.

Crop Residue: Composition and Decomposing Mechanisms

General types of crop residues produced by the main cereal crops and sugar cane are summarized in the table 2. These crop residues, specifically as a field residue is a natural resource that traditionally contributed to the soil stability and fertility through ploughing directly into the soil, or by composting.

Good management of field residues can also increase irrigation efficiency and erosion control. However, the mass scale and rapid pace of crop production have imposed economic and practical limitations to such traditional sustainable practices. It is a common practice in many of the developing countries, especially in Asia to burn the surplus crop residue. While burning creates environmental issues, ploughing field residue into the ground for millions of hectares within a short time requires new and expensive technical assistance.

Table 2. Crop residues produced by major crops

Source	Composition
Rice	Husk, bran
Wheat	Bran, straw
Maize	Stover, husk, skins
Millet	Stover
Sugarcane	Sugarcane tops, bagasse, molasses

Plant biomass is mainly comprised of cellulose, hemicellulose and lignin with smaller amounts of pectin, protein extractives, sugars, and nitrogenous material, chlorophyll and inorganic waste. Compared to cellulose and hemicellulose, lignin provides the structural support and it is almost impermeable. Lignin resist fermentation as it is very resistant to chemical and biological degradation. The non-food-based portion of crops such as the stalks, straw and husk are categorized under lignocellulosic biomass. The major agricultural crops grown in the world are maize, wheat, rice and sugarcane, respectively, account for most of the lignocellulosic biomass. Lignocellulosic biomass composed of cellulose, hemicellulose, and lignin, are increasingly recognized as a valuable commodity, due to its abundant availability as a raw material for the production of biofuels.

The crop residues generated due to agricultural activities are exploited by several countries in different ways. They are utilized in processed or unprocessed form, depending on the end use. The possible options include its use as animal feed, composting, production of bio-energy and deployment in other extended agricultural activities such as mushroom cultivation]. Many countries such as China, Indonesia, Nepal, Thailand, Malaysia, Japan, Nigeria and Philippines utilize their crop residues to generate bio energy and compost.

Numerous researchers have worked on lignocellulosic biomass pretreatment techniques for bio-fuel conversion. Because of its resistance to chemical and biological degradation by fungi, bacteria and enzymes, the lignin layer is usually pretreated or acted upon by the lignin degrading microorganisms to break down the lignin layer and degrade cellulose and hemicellulose matter to the corresponding monomers and sugars for effective biomass to fuel conversion. The pretreatment could be mechanical, chemical, physico-chemical and biological. These methods result in increase of the accessible surface area, porosity and decrease in crystallinity of cellulose and hemicellulose and degree of polymerization.

The management of agricultural waste using microbes could also be an excellent option for the detoxification of the soil and mitigation of environmental pollution. Microbial populations degrade the complex substances present in the biomass to simpler ones that can be reused or recycled through environmental processes. The techniques adopted can either be aerobic or anaerobic, depending on the nature of bacteria, fungi or algae involved in the degradation. The microbial degradation techniques reduce the soil toxicity, promote plant growth through provision of growth accelerating metabolites and provide plant nutrients through sequestration from soil. Thus, the bioremediation of the agricultural waste could be effectively carried out by anaerobic and aerobic processes, through some of the associated techniques like composting, vermicomposting, biogas production, bio-methanation and bio pile farming.

Anaerobic digesters can turn biomass into biogas, a renewable energy source, containing approximately 50 per cent methane, and a final solid residue usable as a fertilizer rich in nutrients. Anaerobic digestion is a promising valorization technology due to its ability to convert almost all sources of biomass, including different types of organic waste, slurry and manure into highly energetic bioga. It is an effective and environmentally attractive pathway and promising option for recycling agricultural by-products because these contain high percentage of biodegradable materials. Anaerobic digestion involves microbial conversion in aqueous environment and could be processed without any pretreatment. Further it encompasses a complex biological process, involving anaerobic degradation of the biomass. The degradation and conversion continue in individual phases carried out by different groups of specific symbiotic micro-organisms. It involves controlled substrate and methanogenic bacteria for methane fermentation. The anaerobic digestion proceeds through three phases, with the hydrolytic bacteria degrading polymeric organic matter into monomers (sugars, amino acids) in the first phase. Followed by monomer degradation to fatty acids, (acetate, formate) as the second stage and in the third phase, the acids are reduced to carbon dioxide andmethane by acetotrophs, methylotrophs and hydrogenotrophs bacteria.

The past governmental interventions mainly focus on the use of crop residue as a source of energy in the form of biogas as well as a supplement for thermal power plants. Biogas generated through anaerobic biodegradation of municipal solid waste and agricultural waste, contains around 40–70 per cent methane, this is usually augmented to natural gas quality with methane content of 70–99 per cent. Further it can be injected into the natural gas grid or used as fuel for transportation. The methane production potential of wheat straw ranges from 0.145to 0.390 m^3/kg for dry organic mass fed to the digester. Rice straw has a methane production potential ranging from 0.241 to 0.367 m^3/kg. It has been reported a biogas production potential of around 0.550 to 0.620 m^3/kg for rice straw biomass with around 50 per cent methane content. Similarly, the reported biochemical methane production from sugarcane biomass varies from 0.266–0.314 m^3/kg.

Burning of Crop Residue in India

India accounts for about 2.4 per cent of the world's geographical area and 4.2 per cent of its water resources, but supports about 18 per cent of its population which highlights the fact that our natural resources are under considerable strain. The need for providing food grains for a growing population, while sustaining the natural resource base, has emerged as one of our main challenges. Food grains are a major source of energy and are thus vital for food and nutritional security. As such food grains would continue to be the main pillar of food security and out of various crops grown, rice, wheat, and pulses are still part of the staple diet of most of the rural population. Directorate of Economics & Statistics, MOA, DAC, New Delhi estimated during 2012-13, India produced about 94 million tons (Mt) of wheat, 105Mt of rice, 22 Mt of maize, 16Mt of millets (jowar, bajra, ragi and small millet), 341Mt of sugarcane, 8Mt of fiber crops (jute, mesta, cotton), 18 Mt of pulses and 31Mt of oil seed crops (Table 3). Out of various crops grown, rice, wheat and sugarcane are prone to crop residue burning. These crops are preferred by farmers since they provide higher economic return as compared to other crops.

Table 3. Production estimate of major crops in India during 2012-13

Crop	Estimate of Production (Mt)
Rice	105
Wheat	94
Sugarcane	341
Cotton	35
Oil seeds	31
Maize	22
Pulses	18
Millets (jowar, bajra, ragi and small millet)	16
Fiber crops (jute, mesta, cotton, others)	08

National policy for management of crop residues (NPMCR) provides the details of the state-wise statistics of crop residue generated and excess residue burned. Based on NPMCR, it is evident that the generation of crop residues is highest in the state of Uttar Pradesh (60 Mt) followed by the other states Punjab (51 Mt) and Maharashtra (46 Mt) with a grand total of 500 Mt per year out of which 92 Mt is burned. Among different crops, cereals generate maximum residues (352 Mt), followed by fibres (66 Mt), oilseeds (29 Mt), pulses (13 Mt) and sugarcane (12 Mt). Rice and wheat contribute nearly 70 per cent of the crop residues while rice crop alone contributes 34 per cent to the crop residues. Sugarcane residues consisting of top and leaves generate 12 Mt, i.e., 2 per cent of the crop residues in India. Out of the total waste generated, the surplus residue refers to the waste that remains after utilizing the residue for various other purposes. A part of the surplus waste is burned, and the remains are left in the field. The Intergovernmental Panel on Climate Change (IPCC), the highest contribution to the amount of residue burned on the farm is from the states of Uttar Pradesh, followed by Punjab and Haryana. According to IPCC, over 25 per cent of the total crop residues were burnt on the farm. It has also been reported that the fraction of crop residue burned ranged from 8–80 per cent for paddy waste across all states. Among different crop residue, major contribution was 43 per cent of rice, followed by wheat to around 21 per cent, sugarcane to 19 per cent and oilseed crops around 5 per cent.

The Ministry of Agriculture attributes the increase in the on-farm crop residue burning to the shortage of human labor. It has been reported in the literature that 80 per cent of the crop residue burning took place during the post-harvest period of April-May and November-December. The reason behind this is attributed to the crop patterns used to ensure higher economic returns, which leave limited time between two consecutive crop cultivations. Some farmers even resort to a cycle of three crops a year with a short gap between harvesting and sowing.

The disposal pattern of paddy straw by the farmers depends on the market value of the by-product. The methods adopted for end-use of paddy straw as mentioned in various studies (Table 4). From the table, it is clear that on an average, three fourth of the paddy straw is burnt openly in the fields. The above ratio implies that in the year 2007–08 around 11,930–15,858 thousand tonnes of paddy straw was burnt in the open field. Burning in Punjab involves partial and full burning. Partial burning entails running of combine harvester followed by burning of small stalks while complete burning entails setting the entire field on fire. The latter practice is mostly followed by the farmers in Punjab. Both the practices cause pollution but the impact is more severe in the case of complete burning. The farmers in the region are resorting to burning of straw, because they don't have other equal or more remunerative alternatives available to them.

Table 4. Different report of end use of paddy straw

S. No.	Disposal pattern	Author(s)
1.	80 % of the crop residue burning took place	Jitendra et al., 2017
2.	75–80 % area is machine harvested, 75 % of straw is burnt	Badarinath and Chand Kiran, 2006
3.	30–40 % straw burnt (IGP)	Venkataraman et al., 2006
4.	81 % of paddy burnt and 48 % of wheat burnt, fodder (7 % of rice and 45 % of wheat), rope making (4 % of rice and 0 % of wheat), incorporated in soil (1 % of rice and less than 1 % of wheat), miscellaneous (7 % each of rice and wheat)	Sidhu and Beri, 2005
5.	75 % combine harvested and 100 % burnt	Sarkar et al. (1999)
Average 75 % of paddy is burnt		

There are many environmental risks associated with stubble burning. If followed continuously burning can reduce soil quality and make land more susceptible to erosion. Moreover, continuous burning is not a sustainable agricultural practice. Smoke from burning straw also contributes to increased carbon dioxide levels in the atmosphere which may affect greenhouse gas build-up.

The Department of Science, Technology and Environment and Non-Conventional Sources of Energy, Government of Punjab, constituted a task force in September, 2006 for formulation of policy to mitigate the problem due to the severity of burning of agricultural waste in the open fields after harvest and its consequent effects on soil, ambient air and health effects on living organism. The task force has suggested promotion of agronomic practices and technological measures for better utilization of agricultural wastes. These include use of happy seeder, developed by PAU in collaboration with Australian Centre for International Agriculture Research (ACIAR) and use of paddy straw for power generation.

Impact of Crop Residue Burning on the Environment

The burning of crop residues generates numerous environmental problems. The main adverse effects of crop residue burning include the emission of greenhouse gases (GHGs) that contributes to the global warming, increased levels of particulate matter (PM) and smog that cause health hazards, loss of biodiversity of agricultural lands, and the deterioration of soil fertility. Crop residue burning significantly increases the quantity of air pollutants such as CO_2, CO, N_2O, NH_3, NOX, SOX, Non-methane hydrocarbon (NMHC), volatile organic compounds (VOCs), semi volatile organic compounds (SVOCs) and PM. This basically accounts for the loss of organic carbon, nitrogen, and other nutrients, which would otherwise have retained in soil. It was reported in literature that burning of 98.4 Mt of crop residue has resulted in emission of nearly 8.57 Mt of CO, 141.15 Mt of CO_2, 0.037 Mt of SOx, 0.23 Mt of NOx, 0.12 Mt of NH_3 and 1.46 Mt NMVOC, 0.65 Mt of NMHC, 1.21 Mt of PM during 2008–2009, where CO_2 is 91.6 per cent of the total emissions. Remaining 8.43 per cent consisted of 66 per cent CO, 2.2 per cent NO, 5 per cent NMHC and 11 per cent NMVOC. The PM emitted from burning of crop residues in Delhi is 17 times that from all other sources such as vehicle emissions, garbage burning and industries. As such the residue burning in the northwest part of India contributes to about 20 per cent of organic carbon and elemental carbon towards the overall national budget of emission from agricultural waste burning.

This has estimated that approximately 730 Mt of biomass was burned annually from both anthropogenic and natural resources in Asia and 18% of that is from India. Crop burning increases the PM in the atmosphere and contributes significantly to climate change. One contributor to global climate change is the release of fine black and also brown carbon (primary and secondary) that contributes to the change in light absorption. Usually PM in the air is categorized as PM2.5 and PM10 based on the aerodynamic diameter and chemical composition (PM2.5 or fine, particulate matter with aerodynamic diameter <2.5_m and PM10 or coarse, particulate matter with aerodynamic diameter <10_m). Lightweight particulate matter can stay suspended in the air for a longer time and can travel a longer distance with the wind. The effect of particulate matter gets worsened by the weather conditions, as the particles are lightweight, stay in air for a longer time and causes smog. The annual contribution of PM2.5 due to burning of paddy residue in the Patiala district of Punjab was estimated to be around 60 to 390 mg/m^3. During the period of October 2017, smoke from crop residue burning in Punjab and Haryana blows across northern India and Pakistan. With the onset of cooler weather in November, the smoke, mixed with fog, dust, and industrial pollution, forms a thick haze. Wind usually helps disperse air pollution, and the lack of it, worsens the problem for several days as was the case during November 2017. Several major cities including Lahore, New Delhi, Lucknow, and Kanpur faced elevated levels of pollution. On 7 November 2017, the Moderate Resolution Imaging Spectro-Radiometer (MODIS) of NASA's Aqua satellite captured a natural-color image of haze and fog blanketing the northern states region of India.

The WHO standard for permissible levels of PM2.5 in the air is 10 µg/m^3, and according to the India's National Ambient Air Quality Standard, the permissible level for PM2.5 is set at 40 µg/m^3. However, the National Capital territory of Delhi recorded a mean value of 98 µg/m^3, which is at least twice more than the Indian standard and ten times higher than the WHO standard. In addition to the emission of gases and aerosols, there is continuous deterioration of soil fertility due to burning.

Heat from burning of residues raises the soil temperature and causes depletion of the bacterial and fungal population. The residue burning increases the subsoil temperatures to nearly 33.8-42.2 °C at 10 mm depth and long-term effects can even reach up to 15 cm of the top soil. Frequent burning reduces nitrogen and carbon potential of the soil and kills the microflora and fauna beneficial to the soil and further removes the large portion of the organic matter. With crop burning the carbon-nitrogen equilibrium of the soil is completely lost.

It is estimated that upon burning, carbon (C) present in rice straw is emitted as CO_2 (70% of Carbon present), CO (7%) and CH4 (0.66%) while 2.09% of Nitrogen (N) in straw is emitted as N_2O. Some scientists predicted that cumulative CO, CO_2, N_2O and NOx emissions from rice and heat straw burning are 0.11, 2.306, 0.002 and 0.084 Mt, respectively. According to NPMCR, it is reported that burning of one ton of straw accounts for the loss of entire amount of organic carbon, 5.5 kg of nitrogen, 2.3 kg of phosphorous, 25 kg of potassium and 1.2 kg of sulphur. On an average crop residue of different crops contain approximately 80 per cent of nitrogen (N), 25 per cent of phosphorus (P), 50 per cent of sulphur (S) and 20 per cent of potassium (K). If the crop residue is retained in the soil itself, it can enrich the soil with C, N, P and K as well.

Government Intervention

Stringent measures to mitigate crop burning and further to regulate crop waste management require involvement of the appropriate Government agencies. Several attempts were made by the Government of India to introduce and educate the agricultural community about the best practices of agricultural waste management through Government-initiated projects. Numerous forums and proposals were also formulated by environmentalists and Government officials to curb crop residue burning and to promote the usage of alternative sustainable management methods. Some of the laws that are in operation pertaining to crop residue burning are: The Section 144 of the Civil Procedure Code (CPC) to ban burning of paddy; The Air Prevention and Control of Pollution Act, 1981; The Environment Protection Act, 1986; The National Tribunal Act, 1995; and The National Environment Appellate Authority Act, 1997. Particularly, in the states of Rajasthan, Uttar Pradesh, Haryana and Punjab stringent measures have been taken by the National Green Tribunal (NGT) to limit the crop residue burning standard for permissible levels of PM2.5 in the air is 10 g/m^3, and according to the India's National Ambient Air Quality Standard, the permissible level for PM2.5 is set at 40 g/m^3. However, the National Capital territory of Delhi recorded a mean value of 98g/m^3, which is at least twice more than the Indian standard and ten times higher than the WHO standard. In addition to the emission of gases and aerosols, there is continuous deterioration of soil fertility due to burning.

Initiative towards biogas plants

Biogas plants are a progressive step taken by the Government of India to curb crop burning and to prevent pollution. The biogas technologies have been in vogue since the 1970s and several programs are run by the National Biogas and Manure Management Program-off grid biogas power generation program to provide renewable energy for electricity generation, cooking and lighting purpose. These programs were implemented by the Government under the "waste to energy mission". This is also a part of India's action plan on climate change.

Large scale industrial biogas plants generate 5000m^3 of bio gas per day. Nearly 400 off-grid biogas power plants have been set up with a power generation capacity of 5.5 MW. Currently there are 56 biogas-based power plants operational in India; the majority of them are in the states of Maharashtra, Kerala and Karnataka. Small family type biogas plants have also been introduced in the rural areas, which can generate 1 to 10 m^3 biogas per day. Nearly five million family biogas plants have been installed by MNRE under the biogas development program. Recent developments in technology have opened the possibility of using paddy straw and other crop residue other than dung and vegetable waste for biogas generation in an integrated approach. It has been reported the setting up of a biogas plant combined with commercial farms and processing units that was set up in Fazilka, Punjab as a novel initiative towards green energy. This plant generates biogas using rice straw through bio-methanation technology. The biogas plant having been certified by the premier academic institutes like the Indian Institute of Technology, Delhi and Punjab Agricultural University, generates around 4000 m3 of biogas from 10 tons of agricultural residue. In another biogas enterprise, a 12 MW rice-straw power plant can consume 120,000 tons of stubble collected from nearly 15,000 farmers. These private enterprises generated around 700,000 jobs for the farming population. The secondary users such as bio-gas plants offered farmers Rs. 600 to Rs. 1600 (8 to 22 USD) per ton of straw. Through some of these measures implemented by the Government agencies and private sectors, crop burning has been reduced but not completely stopped.

National schemes and policies

The Government of India recently directed the National Thermal Power Corporation (NTPC) to mix crop residue pellets (nearly 10%) with coal for power generation. This helped the farmers with a monetary return of approximately Rs. 5500 (77 USD) per ton of crop residue. These lucrative measures are yet to be in action and it can be profitably exploited by the farmers. Few measures, associated with bio-composting are run by the Indian government. The Rashtriya Krishi Vikas Yogna (RKVY), State Plan Scheme of Additional Central Assistance launched in August 2007 is a government initiative, as a part of the 11th Five Year Plan by the Government of India. Under this scheme eight demonstration and training projects were established in different villages of Azamgarh and Marinath Bhanjam districts of eastern Uttar Pradesh. Around 456 farmers were trained for agro-waste bio-conversion and bio-compost production. These large-scale efforts supported farmers in gaining economic advantages.

In addition to above, the Ministry of Agriculture of India recently developed a National Policy for Management of Crop Residue (NPMCR). The following are the main objectives of the NPMCR:

» Promote the technologies for optimum utilization and in-situ management of crop residue, to prevent loss of valuable soil nutrients, and diversify uses of crop residue in industrial applications.

» Develop and promote appropriate crop machinery in farming practices such as modification of the grain recovery machines (harvesters with twin cutters to cut the straw). Provide discounts and incentives for purchase of mechanized sowing machinery such as the happy seeder, turbo seeder, shredder and baling machines.

» Use satellite-based remote sensing technologies to monitor crop residue management with the National Remote Sensing Agency (NRSA) and Central Pollution Control Board (CPCB).

» Provide financial support through multidisciplinary approach and fund mobilization in various ministries for innovative ideas and project proposals to accomplish above.

No significant information is reported in the literature yet on any new interventions by the government to achieve objectives 1, 2, or 4 above, however, the new policy did help with the objective 3 on monitoring and enforcing the measures taken by the Central Government in collaboration with the State Governments. One such example comes from Punjab. In an effort to identify and locate the exact crop burning locations, the Punjab Pollution control

Board (PPCB) and the Environmental Prevention and Control Authority (EPCA) (National Agency) used remote sensing techniques and aerial surveillance.

The burning areas were identified as red dots in the imagery. A typical case is shown in the aerial photograph taken on November of 2015, which depicts the farming lands in Punjab and Haryana after the rice harvesting period. Localized red spots seen indicate the areas of crop burning. Also, during the same year the crop burning problem became dominant and gained national and international attention after the NASA alert and subsequent alarming rise of air pollution levels in the city of Delhi. As a consequence, states like Rajasthan, Punjab and Haryana imposed fine between Rs. 2500 to 15,000 (35 to 210 USD) on farmers who are indulging in crop-burning. The National Green Tribunal, a government enterprise, established under the National Green Tribunal Act laid down stringent directives to the states to curb crop burning through recycling initiatives and spread proper awareness among the people.

Crop burning areas in Punjab and Haryana, as captured by NASA agencies, the states of Punjab and Haryana have witnessed a reduction of 38 and 25 per cent in crop the satellite imageries from the Punjab Remote Sensing Centre, were also able to locate the crop burning areas and levy fine on the farmers. The total recorded current cases for the year of 2018 was 1816 compared to 4710 for the year of 2017 with nearly 38 per cent reduction. Similar actions were implemented by the Haryana Government, which witnessed a 25 per cent reduction. Some of the farmers in these states were awarded incentives, rewards and subsidies for practicing the control measures.

Sustainable Management Practices for Crop Residue

Vigilance of government agencies: Most of the government interventions thus far have mainly with the vigilance of government agencies, the states of Punjab and Haryana has witnessed a reduction of 38 and 25 per cent in crop stubble burning, respectively. Punjab Pollution Control Board through the satellite imageries from the Punjab Remote Sensing Centre, were also able to locate the crop burning areas and levy fine on the farmers. The total recorded current cases for the year of 2018 was 1816 compared to 4710 for the year of 2017 with nearly 38 per cent reduction. Similar actions were implemented by the Haryana Government, which witnessed a 25 per cent reduction. Some of the farmers in these states were awarded incentives, rewards and subsidies for practicing the control measures.

Energy production: Most of the government interventions thus far have mainly focused on the energy production out of crop residue, particularly biogas production. Specifically, in the states of Tamil Nadu, Bihar, Assam, West Bengal and Jammu and Kashmir, were crop residues are being used as a source for animal feed. Some of the residues are processed to be used in construction applications, such as the use of rice husk ash in cement mixes. Banana peels and sugarcane waste are being utilized in the paper industry, while husk and bagasse ash are utilized for mushroom cultivation. Alternative measures have long been suggested by scientists and agriculturalists over the past decade to counter crop residue burning, but due to a lack of awareness and social consciousness among the farmers these measures have not been fully implemented. This could be one of the reasons why biogas production has prospered while other alternatives such as using crop residue as raw material for animal feed, paper industry, construction industry have not become very popular. If a solution involves making another product out of crop residue, such a product should have a secured market for this solution to succeed. In certain cases, logistic issues in transportation of the materials to larger distances also add to the cost. In this context, it is believed that the best alternatives could be the ones that make end-products to be used by the agricultural industry itself, and on-site if possible. In this section information on three such agricultural applications that have either been overlooked or skipped due to various reasons are presented. They are: composting, biochar, and in-situ management through mechanical intensification.

Use of rice residue in bio thermal power plants in Punjab

Use of rice residue that is being encouraged by various institutions and departments is the use of rice residue for generation of electricity in Punjab. A 10 MW biomass based power plant at village Jalkheri, Fatehgarh Sahib with paddy straw as fuel was set up in the year 1992 (Box 4.1). The plant is operational since 2001, after the PSEB entered into a lease-cum-power purchase agreement with Jalkheri Power Private Limited (JPPL). The original system installed by BHEL i.e. firing the boiler with paddy straw in baled form, used to create innumerable problems like ash melting, snagging, super heater choking, clinkerisation, drop in boiler temperature due to moisture in the bales, etc. Hence, the fuel was changed from paddy straw to rice husk, wood chips, cotton waste, etc., in mixed form or rice husk alone to achieve the desired parameters. The total requirement of biomass is estimated to be 82,500 Mt/annum at 100 per cent capacity utilization for optimum plant activity. Crop residues are bought from the farmers at Rs. 35 per quintal (which would otherwise have remained unutilized or burnt in the field). The farmers are being made aware of this offer through newspapers and other awareness activities. Apart from the generation of electricity for supply to state grid to meet the ever-increasing demand for energy in the state, the plant also reduces the Green House Gases (GHGs) emissions. As per Cleaner Development Mechanism (CDM) estimates, the plant would supply energy equivalent of approximately 417.9 million kWh to the grid in a period of 10 years (2002–2012), thereby resulting in total CO_2 emission reduction of 0.3 million tonnes.

Box 4.1 Case study of generation of electricity from agri-waste in Punjab

The thermal plant at Jalkheri, District Fatehgarh Sahib is the first plant in India, which is based on use of Biomass i.e. renewable energy source. This plant can utilize rice husk, waste wood chips, straw of various plants e.g. paddy, wheat, etc. This plant was commissioned in June, 1992 on turn-key basis by M/s BHEL for PSEB to utilize rice straw at a cost of Rs. 47.2 crores.

Some teething problems were experienced initially being an experimental project but with modifications, full 10 MW capacity has been achieved. As harvesting pattern in Punjab has changed and farmers found it convenient to harvest crop with mechanical means and non-availability of adequate quantity of hand cut rice, the plant was further modified to accept any bio-mass e.g. any straw, rice husk, wood chips etc. The plant has been given on lease and is being operated at 10 MW i.e. full capacities on sustainable basis.

One 10–15 MW agri-waste based power project has been set up jointly by Punjab Biomass Power, Bermaco Energy, Archean Granites and Gammon Infrastructure projects Limited in Punjab. The project uses locally available agricultural waste such as rice straw and sugar cane trash for fuel. The total annual fuel requirement is around 120,000 t of biomass, all of which will be sourced locally. Punjab produces around 20–25 million tonnes of rice straw annually. As rice straw is a poor fodder and fuel, farmers burn it in the fields and make way for the Rabi wheat crop. With the development of technology now there is an option to use this waste for generating electricity. The project is expected to provide additional income to 15,000 farmers from the sale of agricultural waste. The project will be a major milestone in environment protection—converting agricultural waste to energy. Secondly, it will reduce the release of smoke and other pollutants caused by burning of wastes which could now be used for earning carbon credits.

Another biomass based power project of 7.5 MW was initiated by Malwa Power Pvt. Ltd. at village Gulabewalla in district Mukatsar in 2002. The project was commissioned in May 2005 and is operating satisfactorily. The plant is selling electricity to PSEB through power purchase agreement. The plant is using crop residues available in the area like cotton stalks, mustard stalks, lops and tops of Eucalyptus, Poplar and *Prosopis juliflora* and some quantity of agro waste such as rice husk and saw dust. The total requirement of biomass is estimated to be 65,043 MT per annum at 90 per cent capacity utilization and 72,270 MT per annum at 100 % capacity utilization.

As per estimates for Clean Development Mechanism (CDM), this plant would supply energy equivalent of approximately 465.10 million kWh to the grid in a period of 10 years (2005–2015) and would result in reduction of 0.43 million tonnes total of CO_2 emission. Both these power plants are obtaining Carbon Credits under CDM. Further, in August, 2006, PSEB has signed two agreements with M/S Punjab Biomass Power Limited for setting up 12 MW paddy straw based power plants at village Baghaura near Rajpura and Village Sawai Singh near Patiala. The company intends to collect paddy straw from command area of 25 km² around each village and would use 1 lakh Mt per annum paddy straw for generation of 12 MW of electricity. The company has entered into an agreement with farmers on barter system and farmers will be provided electricity in lieu of supplying paddy straw. The plants were expected to start operations in 2009. Land at Baghaura village has already been purchased.

The State of Punjab has been a victim of acute power famines, load shedding and power cuts, year after year. Agricultural requirement for power is highest during June to September for the purpose of paddy cultivation. Biomass, such as agricultural residue, bagasse, cotton stalks, rice husk, etc., is emerging as a viable source of power for rural electrification in India. Direct burning of such waste is inefficient and leads to pollution. When combusted is in a gasifier at low oxygen and high temperature, biomass can be converted into a gaseous fuel known as producer gas. This gas has a lower calorific value compared to natural gas or liquefied petroleum gas, but can be burnt with high efficiency and without emitting smoke.

The advantages of utilizing crop residue over and above the conventional resources are that such residue is renewable, readily available and can be used successfully by burning in boilers with the efficiency of 99 per cent. Further, they are available at low cost as compared to that of coal while ash contents is much less (as compared to 36 % ash content of coal) and at the same time the calorific value of both, coal and paddy straw are comparable, i.e., 4,200 and 3,590 kcal/kg, respectively. Additional income to the farmers from the sale of straw is an added advantage. At the same time, the agencies involved/state could also take advantage of carbon credit policy set up under the UNFCCC (United Nation Framework Convention on Climate Change) from developed countries. The policy involves emission credit for programmes which help in curbing global warming. The government should encourage private parties/agencies to take advantage of this carbon credit policy of UNFCCC.

According to Dr. A.K. Rajvanshi, who runs the non-profit Nimbkar Agriculture Research Institute, Phaltan, Maharashtra, it is feasible to set up a bio-mass-based power plant of 10–20 MW capacity in every Taluka (a block of about 100 villages). This can meet energy needs of villages and employ thousands of people. Similarly, in Punjab the developers of biomass energy can sell their power to PSEB, which will be purchased as per, 'New &

Renewable Sources of Energy Policy' notified by the government from time to time and distributed as per usual norms.

Kirangatevalu village in Karnataka has set an example in this regard. Electrification of the village earlier meant supply of power to a few homes and farms for 4–5 hr a day. The transformation of the village is the result of an initiative taken by a private firm that has set up a power plant using agricultural waste such as sugarcane refuse and coconut fronds that are plentiful in the area. Villagers sell their agro waste to the plant and get access to quality power at commercial rate. A supply chain to procure agricultural waste from villages in a radius of 10 km has been established to ensure the supply of agricultural waste throughout the year. The waste that was burnt in open fields has now become a source of income and jobs. The 4.5 MW power plants set up by Malaballi Power Plant Private Limited supplies electricity to 48 villages inhabited by 120,000 people in Mandya district in Karnataka.

In Punjab in the 1980s PSEB had set up a 10 MW power plant based on paddy straw at Village Jalkheri, District Fatehgarh Sahib in which 250–3,000 TPD of fuel is burnt in a boiler furnace of steam generation capacity of 50 TPH. The plant earlier used paddy straw but due to clinkerisation of boiler, paddy straw was replaced with rice husk, cow dung and other agro waste. This plant has since been leased out by PSEB to M/S Jalkheri Power Private Limited. Now these plants will be using improved technology and M/S Punjab Biomass Power Limited has signed two agreements with PSEB for setting up 12 MW paddy straws based power plants at Baghaura in Rajpura Tehsil and Sawai Singh village in Patiala Tehsil. A total amount of 0.1 million ton paddy straw would be collected from a command area of 25 km^2 around each unit and a barter system of providing electricity will be worked out with the farmers. The units will be run on BOO basis. DPRs have been prepared and land is being purchased.

The bottlenecks apprehended by PSEB in generation of power from paddy straw are the availability of paddy straw for power generation in case the Happy Seeder technology succeeds in the State. Hence, it is recommended that areas around these power plants could be reserved to ensure enough availability of straw. Further, techniques to collect and store paddy straw may also be developed and incentives provided.

Energy technologies: The transportation of biomass is one of the key cost factors for its use as a source of renewable energy. Decentralized energy systems provide an opportunity to use biomass to meet local energy requirements that are, heat and electricity. In contrast to straw, the use of rice husk for energy has been realized faster. One important factor is that rice mills can use husk to serve their internal energy requirement. As an alternative, rice millers could sell the husk to a power-plant operator. The propagation of rice husk use for energy was accelerated by energy providers, who deal with a relatively small number of rice millers for supplying husk, which is an easier task than dealing with thousands of farmers supplying paddy straw.

As a new trend, electricity is now often produced by the millers themselves and then sold to a power grid. This setup has to be seen as the most promising option in terms of logistics and transportation for energy generation. Transportation costs of straw are a major constraint to its use as an energy source. As a rule of thumb, transportation distances beyond a 25–50 km radius (depending on local infrastructure) are uneconomical. For long distances, straw could be compressed as bales or briquettes in the field, rendering transport to the site of use a viable option. Nevertheless, the logistics of a supply chain is more complicated in the case of straw.

Although five different energy conversion technologies seem to be applicable for paddy straw in principle only combustion technology is currently commercialized and the other technologies are at different stages of development. As a general rule for energy use, each step in the chain consumes a certain amount of energy and thus reduces the net energy at the end product. The following sections describe the principal features of the possible energy conversion technologies, experiences and technical difficulties in the use of paddy straw.

Thermal combustion: Paddy straw can either be used alone or mixed with other biomass materials (the latter is called co-firing or co-combustion) in direct combustion. In this technology, combustion boilers are used in combination with steam turbines to produce electricity and heat. In thermal combustion, air is injected into the combustion chamber to ensure that the biomass is completely burned in the combustion chamber. Fluidized bed technology is one of the direct combustion techniques in which solid fuel is burned in suspension by forced air supply into the combustion chamber to achieve complete combustion. A proper air-to-fuel ratio is maintained and, in the absence of a sufficient air supply, boiler operation encounters various problems.

In straw combustion at high temperatures, potassium is transformed and combines with other alkali earth materials such as calcium. This in turn reacts with silicates, leading to the formation of tightly sintered structures on the grates and at the furnace wall. Alkali earths are also important in the formation of slags and deposits. This means that fuels with lower alkali content are less problematic when fired in a boiler. The byproducts are fly ash and bottom ash, which have an economic value and could be used in cement and/or brick manufacturing, construction of roads and embankments, etc.

National Biomass Assessment Project of Ministry of New and Renewable Energy, Government of India conducted a biomass study in which 29 Tehsil were surveyed which was started in the late eighties and continued till 1995–1996. Total 36 Talukas were included from different districts. The total estimated power generating potential was estimated AT 342 MW. Biomass Power project has the following inherent advantages over thermal power generation:

- It is environmentally friendly because of relatively lower CO_2 and particulate emissions
- It displaces fossil fuels such as coal
- It is a decentralised, load based means of generation, because it is produced and consumed locally, losses associated with transmission and distribution are reduced
- It offers employment opportunities to locals
- It has a low gestation period and low capital investment
- It helps in local revenue generation and upliftment of the rural population
- It is an established and commercially viable technology option.
- Punjab has substantial availability of Biomass/Agro-waste in the state sufficient to produce about 1,000 MW of electricity.

PEDA has planned to develop some of the available potential talukas/tehsils with the aim to promote and install biomass/agro waste based projects. PEDA has so far allocated 30 sites/tehsils for setting up of total 332.5 MW capacity Biomass/Agro waste based power projects under three phases. In different phases the biomass power project were allocated.

- In Phase I agreement is already done with two companies- M/s Turbo Atom TPS and M/S. Green Field Energen Pvt. Ltd., in New Delhi and Chandigarh, respectively for two Tehsils, Ferozepur and Patti with a total capacity of 56 MW.
- In Phase II three companies were there for Abohar, Sunam and Ajnala, respectively. The two companies of Sunam and Ajnala are cancelled having 41 MW. With the capacity of 8 MW the company M/s Dee Development in Abohar Tehsil is commissioned.
- In Phase III there are six companies which are based differently. The M/s Green

Planet of Chandigarh is based on paddy stubble which is planned in 14 Tehsils with total 146.5 MW of capacity. Out of which Garhshankar with 10 MW capacities is likely to begin. The M/s Univeral Biomass of Mukatsar which is mostly based on cotton stock with 14.5 MW in Malout Tehsil is commissioned. The Malwa Power Ltd., in the village Gulabevala in the district of Muketsar was started before PEDA took over with 6 MW. Other three companies had total capacity of 65 MW.

Thus, PEDA has so far allocated 30 sites/tehsils for setting up of total 332.5 MW capacity Biomass/Agro-waste based power projects during three phases (details in Appendix).

Composting

Composting is not a new concept to India. While small scale backyard composting and making compost from organic material in MSW is common, there is no information in the literature to prove that it is also the case for the agriculture industry in India. The common challenges faced by the organic waste composting, are mostly not technical but economical as the end-product does not always secure a steady market. This is one of the challenges the agricultural community does not have to worry about if they make compost on-site out of their own crop residue as it can be easily fed back to the same agricultural lands. The high organic content in crop residue makes it an ideal raw material for compost similar to animal manure and food waste.

Composting is the natural process of rotting or decomposition of organic matter by micro-organisms under controlled conditions. As a rich source of organic matter, compost plays an important role in sustaining soil fertility and thereby helping to achieves stainable agricultural productivity. Addition of compost to the soil improves physio-chemical and biological properties of the soil and can completely replace application of agricultural chemicals such as fertilizer and pesticides. Higher potential for increased yields and resistance to external factors such as drought, disease and toxicity are the beneficial effects of compost amended soil. These techniques also help in higher nutrient uptake, and active nutrient cycling due to enhanced microbial activity in the soil.

Composting is mediated by different micro-organisms actuating in aerobic environment. Bacteria, fungi, acitnomycetes, algae, and protozoa are naturally present in organic biomass or added artificially in order to facilitate decomposition. It is the biological maturity under aerobic condition, where organic matter of animal or plant origin is decomposed to materials with shorter molecular chains. More stable, hygienic, humus rich compost, beneficial for agricultural crops and for recycling of soil organic matter is ultimately formed.

Composting in complete in two phases

Degradation: The first phase of degradation starts with breakdown of easily degradable organics like sugars, amino and organic acids. The aerobic microorganisms consume oxygen and release carbon dioxide and energy. The first thermophilic phase is dominated by high temperature, high pH and humidity, essential for activating the microorganisms and proceeds for several weeks to months. During this phase, it is also ensured that the substrate is properly cooled with sufficient supply of oxygen.

Maturation: The second phase continues for few weeks, with breakdown of more complex organic molecules followed by decrease in microbial population. There is a change from thermophilic to mesophilic phase with a decrease in temperature to 40–45 ^{0}C. Further at the final stage, temperature declines to an ambient value and the system becomes biologically less active. Finally, a dark brown to black color soil-like material is produced. This soil-like material also exhibits an increased humus content and decreased carbon-nitrogen ratio with a neutralized pH. Aerobic composting is affected by many factors, such as the amount of oxygen, moisture content, nutrient supply, temperature, pH and lignin content. The nutrient supply or ratio of C: N should be optimum in the range of 20:1 to 40:1 for proper growth of microorganisms. The temperature plays a vital role during composting, higher temperatures in the thermophilic range contributes to the destruction of the pathogens and disinfects the organic matter. Eventually the biomass is transformed to a material rich in nutrients, which can improve the structural characteristics of the soil. Aerobic process also involves a large release of energy. It was reported an experimental and observational bio-composting study performed in Uttar Pradesh, India. Wheat straw, rice straw, vegetable crops, leaves of garden plants constituted the total weight of the biomass for this study. The final bio-compost contained 45 per cent of total solids, 26.7 per cent organic matter, 15.3 per cent carbon and 1.36 per cent total nitrogen reflecting a rich compost of carbon and organic matter. They found a significant increase in the agronomic properties of the rice and wheat crops they experimented. Nutrients like nitrogen (N) and phosphorous (P) provided to the crops by the bio-compost is of significant importance to the crop production strategy. This also increases the microbial population and native microflora and fauna necessary for the soil health. The same study reported that a one-inch thick bio compost layer added approximately 1.0 ton/ha of total Nitrogen, 13.3 ton/ha of carbon, 24 t/ha of organic carbon and 1.02 t/ha of organic nitrogen in the soil besides imparting nutrients such as P, K, Ca, Mg, S, Iron, Zn, etc.

Biochar as soil amendment

As a measure for controlling GHG emissions, the agricultural research community is constantly looking for ways to effectively enhance natural rates of carbon sequestration in the soil. This has made an increased interest in applying charcoal, black carbon and biochar as soil amendment to stabilize soil organic content. These techniques are viewed as a viable option to mitigate the GHG emissions while considerably reducing the volume of agricultural waste. The process of carbon sequestration essentially requires increased residence time and resistance to chemical oxidation of biomass to CO_2 or reduction to methane, which leads to reduction of CO_2 or methane release to the atmosphere. The partially burnt products are pyrogenic carbon/carbon black and become a long-term carbon sink with a very slow chemical transformation, ideal for soil amendment.

Biochar is a fine-grained carbon rich porous product obtained from the thermo-chemical conversion called the pyrolysis at low temperatures in an oxygen free environment. It is a mix of carbon (C), hydrogen (H), oxygen (O), nitrogen (N), sulphur (S) and ash in different proportions. When amended to soil, highly porous nature of the biochar helps in improved water retention and increased soil surface area. It mainly interacts with the soil matrix, soil microbes, and plant roots, helps in nutrient retention and sets off a wide range of biogeochemical processes. Many researchers have reported an increase in pH, increase in earthworm population and decreased fertilizer usage. Specifically, biochar is used in various applications such as the water treatment, construction industry, food industry, cosmetic industry, metallurgy, treatment of waste water and many other chemical applications. In India currently, the biochar application is limited and mainly seen in villages and small towns. Based on its wide applicability, it could be more valuable to promote biochar method in India.

In-situ management crop residue

In-situ application of the crop residue is adopted by many farmers as it is a natural process. This method also imparts certain benefits to the soil. There are two main way of conducting field applications but both methods involve leaving crop residue on the farmland after harvesting. How they differ is based on what happens with tillage in the next season. In the first method, planting in the next season is carried out without tillage or with less tillage and in the other method crop residue is incorporated into the soil by mechanical means during tillage. While in-situ management of crop residues can offer long-term cost savings on equipment and labor, both methods need special (new) equipment, e.g., machinery for crop residue incorporation into soils or no-till seeing equipment.

Crop residue retention with no-tillage is mostly practiced in the North America and about 40% of the cropland across the United States alone is cultivated with no-till practice. This method has many advantages for the soil such as cooling effect, increased moisture, source of carbon, and erosion protection. However, this method also finds some negative implications for example, microbial infestation, formation of phytotoxins and nutrient immobilization. This results in a reduced yield, which may warrant additional use of agricultural chemicals. For improving the soil organic matter, crop residue is incorporated into the soil by plowing. Adding nitrogen fertilizers while plowing at a depth of 20–30 cm can enrich the soil with humus and prevent nitrogen depression.

The National Policy for Management of Crop Residue specifically mentions in-situ management through methods such as direct incorporation into soils and mulching as methods that should be promoted in India not only to control crop residue burning but also to prevent environmental degradation in the croplands. Any specific follow-ups or government-supported interventions since the establishment of this national policy, has not yet been reported in the literature. However, it is worth noting that the National conference on Agriculture for *Kharif* Campaign that took place in 2017 re-emphasized on the same facts and listed mechanization practices to avoid crop residue burning among the recommendations made by the focus groups.

Paddy straw

The technical measures are 'straw incorporation' and 'straw mulching'. In both these measures, the residue is incorporated in the field itself and is thus used to increase the nutrient value or fertility of the soil. In the first measure, the residue is allowed to decompose in the field itself through a chemically developed process (available at PAU), and in the second measure, incorporation is done with the help of a properly designed machine along with seeding (know-how developed at PAU). The second measure is more useful as there is no weeding in this process and it is less expensive.

Another study reveals that, incorporation of paddy straw in soil immobilized native as well as added fertilizer N and about half of the immobilized N was mineralized after 90 days of straw incorporation. Straw and N application alone or in combination increased biomass carbon, phosphates and respiratory activities of the soil. Microbial biomass carbon and phosphate activities were observed maximum at 30 days of straw decomposition. In field trials, incorporation of paddy straw 3 weeks before sowing of wheat significantly increased the wheat yield at Sonepat district in a clay loam soil while no such beneficial effect was observed in a sandy loam soil at Hissar.

The incorporation of the straw in the soil has a favorable effect on the soil's physical, chemical and biological properties such as pH, Organic carbon, water holding capacity and bulk density of the soil. On a long-term basis it has been seen to increase the availability of zinc, copper, iron and manganese content in the soil and it also prevents the leaching of nitrates. By increasing organic carbon it increases bacteria and fungi in the soil. In a rice-wheat rotation, it has been observed that soil treated with crop residues held 5–10 times more aerobic bacteria and 1.5–11 times more fungi than soil from which residues were either burnt or removed. Due to increase in microbial population, the activity of soil enzymes responsible for conversion of unavailable to available form of nutrients also increases. Mulching with paddy straw has been shown to have a favorable effect on the yield of maize, soybean and sugarcane crops. It also results in substantial savings in irrigation and fertilizers. It is reported to add 36 kg per hectare of nitrogen and 4.8 kg per hectare of phosphorous (6 g of Nitrogen and 0.8 g of phosphorous per kg of paddy straw) leading to savings of 15–20 % of total fertilizer use.

Use of rice residue for Mushroom cultivation in Punjab

Paddy straw can be used for the cultivation of Agaricus bisporus, *Volvariella Volvacea* and Pleurotus spp. one kg of paddy straw yields 300, 120–150 and 600 g of these mushrooms, respectively. At present, about 20,000 metric tonnes of straw is being used for cultivation of mushrooms in the state.

Paddy Straw Mushrooms (*Volvariella Volvacea*) also known as grass mushrooms are so named for their cultivation on paddy straw used in South Asia. Paddy straw is high temperature mushroom grown largely in tropical and subtropical regions of Asia, e.g. China, Taiwan, Thailand, Indonesia, India, and Madagascar. In Indonesia and Malaysia, mushroom growers just leave thoroughly moistened paddy straw under trees and wait for harvest. This mushroom can be grown on a variety of agricultural wastes (the cultivation method of this mushroom is similar to that of *Agaricus bisporus*) for preparation of the substrate such as water hyacinth, oil palm bunch waste, dried banana leaves, cotton or wood waste, though with lower yield than with paddy straw, which is most successful. Paddy straw mushroom accounts for 16 % of total production of cultivated mushroom in the world.

Use of rice residue as fodder for animals

The rice residue as fodder for animals is not a very popular practice among farmers in Punjab. This is mainly because of the high silica content in the rice residue. It is believed that more than 90 % of the wheat straw produced in the state is used as wheat *bhusa* (dry fodder) for animals. However, to encourage the use of rice residue as fodder for animals, under the pilot project which was taken up by PSCST at PAU trials were conducted on natural fermentation of paddy straw for use as protein enriched livestock feed. The cattle fed with this feed showed improvement in health and milk production. The department of Animal Husbandry, Punjab had propagated the technology in the state.

Highest imbalance of livestock and consumption of paddy residue is noted in Rajasthan with zero consumption per animal. Other such low ranked state with least consumption rate is Madhya Pradesh, Himachal Pradesh, Maharashtra and Sikkim. In north, Punjab has got highest ratio of consumption, followed by Kerala and North Eastern state Tripura and Manipur. Uttar Pradesh has highest concentration of livestock, which is followed by Rajasthan, Madhya Pradesh and Maharashtra. The residue is found highest in West Bengal and Arunachal Pradesh.

The availability of crop residue in India is shortfall of about 40 per cent against the requirement. On the other hand, the availability of green fodder is about 36 per cent short fall against the requirement in India. It can be noted that only in Punjab and Mizoram there is surplus in case of crop residues. The availability of crop residue is highest in Uttar Pradesh followed by Maharashtra, Bihar, Rajasthan and Andhra Pradesh. Excepting Assam almost all the north Eastern States and Kerala have least availability of crop residue. As in the case of availability, the highest requirement of crop residue is in Uttar Pradesh. States like Punjab, Haryana and Bihar has higher per animal availability as compared to other states of India.

Use of rice residue as bedding material for cattle in Punjab

The farmers of the state have been advised to use paddy straw as bedding material for cross bred cows during winters as per results of a study conducted by the Department of Livestock Production and Management, College of Veterinary Sciences, Punjab Agricultural University. It has been found that the use of paddy straw bedding during winter helped in improving the quality and quantity of milk as it contributed to animals' comfort, udder health and leg health. Paddy straw bedding helped the animals keep themselves warm and maintain reasonable rates of heat loss from the body. It also provides clean, hygienic, dry, comfortable and non-slippery environment, which prevents the chances of injury and lameness. Healthy legs and hooves ensure enhancement of milk production and reproductive efficiency of animals. The paddy straw used for bedding could be subsequently used in biogas plants. The use of paddy straw was also found to result in increased net profit of Rs. 200-1000 per animal per month from the sale of additional amount of milk produced by cows provided with bedding. The PAU has been demonstrating this technology to farmers through training courses, radio/TV talks and by distributing leaflets.

Use of rice residue in paper production

The paddy straw is also being used in conjunction with wheat straw in 40:60 ratios for paper production. The sludge can be subjected to bio-methanization for energy production. The technology is already operational in some paper mills, which are meeting 60 per cent of their energy requirement through this method. Paddy straw is also used as an ideal raw material for paper and pulp board manufacturing. As per information provided by PAU, more than 50 % pulp board mills are using paddy straw as their raw material.

Use of rice residue for making Bio gas in Punjab

The PSFC has been coordinating a project for processing of farm residue into biogas based on the technology developed by Sardar Patel Renewable Energy Research Institute (SPRERI). A power plant of 1 MW is proposed to be set up at Ladhowal on pilot basis on land provided by PAU. The new technology will generate 300 m^3 of biogas from 1 t of paddy straw.

Production of Bio-oil from paddy straw and other agricultural wastes

Bio-oil is a high density liquid obtained from biomass through rapid pyrolysis technology. It has a heating value of approximately 55 per cent % as compared to diesel. It can be stored, pumped and transported like petroleum based product and can be combusted directly in boilers, gas turbines and slow and medium speed diesels for heat and power applications, including transportation. Further, bio-oil is free from SO_2 emissions and produces low NO_2. Certain Canadian companies (like Dyna Motive Canada Inc.) have patented technologies to produce bio-oil from biomass including agricultural waste. Though their major experience is with bagasse, wheat straw and rice hulls, feasibility of this technology with paddy straw needs to be assessed. The state government, through PSCST and PEDA, could promote further studies in this direction.

Policy issues

Crop residue burning has become an environmental catastrophe, not only for India but for the Asian region as well. Then the policy and implementation steps taken by the Indian government were briefly presented. Using crop residue for making compost/biochar or incorporating into soils were also briefly introduced as they are three key technical solutions, that have not yet been widely considered by the policy sector in India. The sequence of the process of understanding the crop burning issues, looking for potential solutions and implementation the solutions, seem very logical. However, the smog experienced by millions of people in the country each year clearly suggests that the crop burning issue has not been sufficiently addressed by any of the previous interventions. There is need to identify the better alternatives, which will help for implementation.

The Policy Issues

Proper mechanism development

When MSW is generated in households it is properly managed. This is in line with the popular term "Polluter Pays" used in the environmental law, which simply explains who should be responsible for producing pollution. Fortunately, community living has already formed a mechanism to manage MSW. The MSW generated by the households in a community is collected, treated and disposed (or at least supposed to be) by the municipality. This implies that there exists a known process to handle the MSW, irrespective of how efficient or sustainable it is. The responsibility of keeping the community clean and safe rests with the municipality and the residents pay taxes and/or other fees as their contribution. It is also worth noting that before intervention by local and municipal governments, MSW was also burned in urban and peri-urban communities. When crop residue is produced by the farmers, who should take the responsibility of managing it? As per the polluter pays principle, it should be the farmer. This often works well with large-scale agricultural businesses, and especially in the developed countries where environmental laws are strictly enforced. However, when it comes to small-scale farming in developing countries, the individual farmers do not have the capacity, or the means to handle their own waste. Going by the MSW example, what is lacking here is a mechanism of an organized effort, such as the municipality, to manage the crop residue. However, government intervention could provide the necessary support for the farmers in establishing an organized network. For example, the local government/municipality can establish a service to manage the crop residue for a reasonable fee charged to the farmer. Until farmers get used to the concept, it is even worth considering providing such service at a government subsidized price.

Based on the needs of the community or the region, the Government agencies can offer different options such as to collect and transport the crop residue from the fields to where it is needed/utilized as a raw material such as a composting, biogas, biochar manufacturing plant. Alternatively, the same entity can establish a service to rent the machinery for those who need equipment to incorporate the crop residue into the soil before the next season. The local government does not necessarily have to take the responsibility. Instead it can be delegated to a community organization such as a farmers' association, based on the educational and organizational skills of the farmer community. This method has been in use for many years in some water-stressed countries where water distribution must be overseen and controlled. Some of the examples of community involvement in managing recycled wastewater as irrigation water.

Empowering stakeholders

The Indian government has initiated some pilot projects to raise awareness about crop residue burning and to promote its sustainable utilization as a resource. While these efforts should be praised, one should also question why the efforts have not made any significant impact yet. One reason could be due to the difference between how much work is done by the government versus how much of it was felt or understood by the farming communities. Educating and empowering the farming stakeholders are crucially important steps to make a significant impact. While there is information on implementation of pilot projects, the literature does not provide details on how these projects were communicated to the stakeholders. As of now the thinking of the farmers has been shaped up by what they have seen for generations: they are only responsible for producing crops but the crop residue is not their responsibility and it is ok to get rid of it with the least cost option. This thinking needs to change, and the farmers should feel responsible for the residue they produce, which is only possible through proper awareness raising campaigns. However, raising the technical knowledge does not mean much until it is packaged with a practical solution to answer their questions on how to handle the crop residue without costing a fortune, or even better, how to make money out of it by using it as a resource. For example, technical knowledge on how to incorporate residue into soils and how much nutrients they can receive by that, will not make a significant impact, when they find out the equipment that they must rent for such operations cost thousands of Rupees (14 USD). The farmers also should be educated about the advantage of reduced agrochemical cost due to the utilization of crop residue in agricultural land. Therefore, awareness raising campaigns should always run parallel to implementation of a practical solution that empowers them not only technically but also economically.

Avoid sectarian thinking: Focus on nexus thinking

It is true that the culprit is in the agricultural sector. However, is the issue completely an agricultural issue? Based on what was discussed in the previous paragraphs, crop residue burning is an issue that goes way beyond agriculture. Some of the issues such as the environmental impact is clearly visible, thanks to smog. But for the farmers who cannot afford to spend more money on proper management of crop residue, it has always been an economic issue which is relatively invisible. When the price for equipment rental is thousands of rupees (14 USD) versus the price of a box of matches is a just a few rupees (0.01 USD), from the economic standpoint of the farmer, it is an easy decision to make even if they are knowledgeable about the environmental damage it can cause. However, what really happens in the big picture is that the farmers are burning a resourceful biomass due to a combination or economic and social issues such as lack of education or awareness. Even though crop residue burning touches upon many sectors, such as environment, agriculture, economy, social aspects, education, and energy, the governmental efforts are mainly revolving only around the agriculture and energy. This sectarian thinking does not help much as exhibited by the slow progress with the previous governmental interventions. Waste material that can be resourceful to the agriculture (soil/food) and the energy sectors gets wasted simply because of sectarian thinking. This is where the government of India can benefit on the emerging concept of nexus thinking in managing environmental resources. What nexus thinking promotes is a higher-level integration that goes beyond the disciplinary boundaries. One excellent example is the wastewater recycling. Wastewater that originates in the waste sector, is used by many water stressed countries for irrigation purpose after treatment (and in some unfortunate cases, without treatment). This way waste is helping to alleviate the water supply issues faced by the water sector and water demand issues in the agricultural sector. In addition, crops also benefit from the nutrients recycled though wastewater when used wisely. Compost, biochar, or biogas can be best examples to explain how the nexus thinking can be put to good use while combatting crop residue burning.

16 Type, Importance, History, Emerging Technology of Renewable Energy

RENEWABLE AND NON-RENEWABLE ENERGY

Energy exists freely in nature. Some of them exist infinitely (never run out, called renewable, the rest have finite amounts (they took millions of years to form and will run out one day called non-renewable is generally defined as energy that comes from resources which are naturally replenished on a human timescale such as sunlight, wind, rain, tides, waves and geothermal heat. Renewable energy replaces conventional fuels in four distinct areas: electricity generation, hot water/space heating, motor fuels, and rural (off-grid) energy services.

About 16 per cent of global final energy consumption presently comes from renewable resources, with 10 per cent of all energy from traditional biomass, mainly used for heating, and 3.4 per cent from hydroelectricity. New renewables (small hydro, modern biomass, wind, solar, geothermal, and biofuels) account for another 3 per cent and are growing rapidly. At the national level, at least 30 nations around the world already have renewable energy contributing more than 20 per cent of energy supply. National renewable energy markets are projected to continue to grow strongly in the coming decade and beyond. Wind power, for example, is growing at the rate of 30 per cent annually, with a worldwide installed capacity of 282,482 megawatts (MW) at the end of 2012.

Renewable energy resources exist over wide geographical areas, in contrast to other energy sources, which are concentrated in a limited number of countries. Rapid deployment of renewable energy and energy efficiency is resulting in significant energy security, climate change mitigation, and economic benefits. In international public opinion surveys there is strong support for promoting renewable sources such as solar power and wind power.

While many renewable energy projects are large-scale, renewable technologies are also suited to rural and remote areas and developing countries, where energy is often crucial in human development. United Nations› Secretary-General Ban Ki-moon has said that renewable energy has the ability to lift the poorest nations to new levels of prosperity.

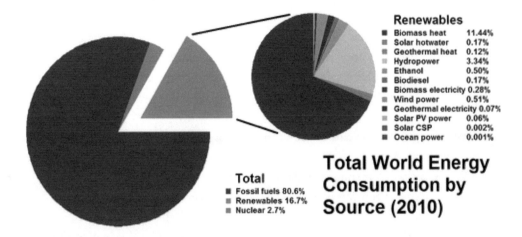

Total energy consumption in 2010 in the world is depicted in the above figure. Renewable energy flows involve natural phenomena such as sunlight, wind, tides, plant growth, and geothermal heat, as the International Energy Agency explains:

Renewable energy is derived from natural processes that are replenished constantly. In its various forms, it derives directly from the sun, or from heat generated deep within the earth. Included in the definition is electricity and heat generated from solar, wind, ocean, hydropower, biomass, geothermal resources, and biofuels and hydrogen derived from renewable resources.

Types of Renewable Energy

The United States currently relies heavily on coal, oil, and natural gas for its energy. Fossil fuels are non-renewable, that is, they draw on finite resources that will eventually dwindle, becoming too expensive or too environmentally damaging to retrieve. In contrast, the many types of renewable energy resources-such as wind and solar energy-are constantly replenished and will never run out.

Sunlight or solar energy: Most renewable energy comes either directly or indirectly from the sun. Sunlight, or solar energy, can be used directly for heating and lighting homes and other buildings, for generating electricity, and for hot water heating, solar cooling, and a variety of commercial and industrial uses.

Wind energy: The sun's heat also drives the winds, whose energy is captured with wind turbines. Then, the winds and the sun's heat cause water to evaporate. When this water vapor turns into rain or snow and flows downhill into rivers or streams, its energy can be captured using hydroelectric power.

Along with the rain and snow, sunlight causes plants to grow. The organic matter that makes up those plants is known as biomass. Biomass can be used to produce electricity, transportation fuels, or chemicals. The use of biomass for any of these purposes is called bioenergy.

Hydrogen energy: This also can be found in many organic compounds, as well as water. It's the most abundant element on the Earth. But it doesn't occur naturally as a gas. It's always combined with other elements, such as with oxygen to make water. Once separated from another element, hydrogen can be burned as a fuel or converted into electricity.

Not all renewable energy resources come from the sun. Geothermal energy taps the Earth's internal heat for a variety of uses, including electric power production, and the heating and cooling of buildings. And the energy of the ocean's tides come from the gravitational pull of the moon and the sun upon the Earth.

In fact, **ocean energy** comes from a number of sources. In addition to tidal energy, there's the energy of the ocean's waves, which are driven by both the tides and the winds. The sun also warms the surface of the ocean more than the ocean depths, creating a temperature difference that can be used as an energy source. All these forms of ocean energy can be used to produce electricity.

Importance of Renewable Energy

Renewable energy is important because of the benefits it provides. The benefits are:

a. **Environmental benefits:** Renewable energy technologies are clean sources of energy that have a much lower environmental impact than conventional energy technologies.

b. **Energy for our children's children:** Renewable energy will not run out ever. Other sources of energy are finite and will someday be depleted.

c. **Jobs and the economy:** Most renewable energy investments are spent on materials and workmanship to build and maintain the facilities rather than on costly energy imports. Renewable energy investments are usually spent within the United States, frequently in the same state and often in the same town. This means your energy dollars stay home to create jobs and fuel local economies, rather than going overseas. Meanwhile, renewable energy technologies developed and built in the United States are being sold overseas, providing a boost to the U.S. trade deficit.

d. **Energy security:** After the oil supply disruptions of the early 1970s, our nation has increased its dependence on foreign oil supplies instead of decreasing it. This increased dependence impacts more than just our national energy policy.

Wind power is growing at the rate of 30 per cent annually, with a worldwide installed capacity of 282,482 megawatts (MW) at the end of 2012, and is widely used in Europe, Asia, and the United States. At the end of 2012 the photovoltaic (PV) capacity worldwide was 100,000 MW, and PV power stations are popular in Germany and Italy. Solar thermal power stations operate in the USA and Spain, and the largest of these is the 354 MW SEGS power plant in the Mojave Desert. The world›s largest geothermal power installation is The Geysers in California, with a rated capacity of 750 MW. Brazil has one of the largest renewable energy programs in the world, involving production of ethanol fuel from sugar cane, and ethanol now provides 18% of the country's automotive fuel. Ethanol fuel is also widely available in the USA.

As of 2011, small solar PV systems provide electricity to a few million households, and micro-hydro configured into mini-grids serves many more. Over 44 million households use biogas made in household-scale, digesters for lighting and/or cooking, and more than 166 million households rely on a new generation of more-efficient biomass cook stoves. United Nations› Secretary-General Ban Ki-moon has said that renewable energy has the ability to lift the poorest nations to new levels of prosperity.

Renewable energy resources and significant opportunities for energy efficiency exist over wide geographical areas, in contrast to other energy sources, which are concentrated in a limited number of countries. Rapid deployment of renewable energy and energy efficiency, and technological diversification of energy sources, would result in significant energy security and economic benefits.

Renewable energy replaces conventional fuels in four distinct areas: electricity generation, hot water/space heating, motor fuels, and rural (off-grid) energy services:

Power generation: Renewable energy provides 19 per cent of electricity generation worldwide. Renewable power generators are spread across many countries, and wind power alone already provides a significant share of electricity in some areas: for example, 14 per cent in the U.S. state of Iowa, 40 per cent in the northern German state of Schleswig-Holstein, and 49 per cent in Denmark. Some countries get most of their power from renewables, including Iceland (100%), Norway (98%), Brazil (86%), Austria (62%), New Zealand (65%), and Sweden (54%).

Heating: Solar hot water makes an important contribution to renewable heat in many countries, most notably in China, which now has 70 per cent of the global total (180 GWth). Most of these systems are installed on multi-family apartment buildings and meet a portion of the hot water needs of an estimated 50–60 million households in China. Worldwide, total installed solar water heating systems meet a portion of the water heating needs of over 70 million households. The use of biomass for heating continues to grow as well. In Sweden, national use of biomass energy has surpassed that of oil. Direct geothermal for heating is also growing rapidly.

Transport fuels: Renewable biofuels have contributed to a significant decline in oil consumption in the United States since 2006. The 93 billion liters of biofuels produced worldwide in 2009 displaced the equivalent of an estimated 68 billion liters of gasoline, equal to about 5 per cent of world gasoline production.

At the national level, at least 30 nations around the world already have renewable energy contributing more than 20 per cent of energy supply. National renewable energy markets are projected to continue to grow strongly in the coming decade and beyond, and some 120 countries have various policy targets for longer-term shares of renewable energy, including a 20 per cent target of all electricity generated for the European Union by 2020. Some countries have much higher long-term policy targets of up to 100 per cent renewables. Outside Europe, a diverse group of 20 or more other countries target renewable energy shares in the 2020–2030 time frame that range from 10 to 50 per cent.

In international public opinion surveys there is strong support for promoting renewable sources such as solar power and wind power, requiring utilities to use more renewable energy (even if this increases the cost), and providing tax incentives to encourage the development and use of such technologies. There is substantial optimism that renewable energy investments will pay off economically in the long term.

Climate change and global warming concerns, coupled with high oil prices, peak oil, and increasing government support, are driving increasing renewable energy legislation, incentives and commercialization. New government spending, regulation and policies helped the industry weather the global financial crisis better than many other sectors. According to a 2011 projection by the International Energy Agency, solar power generators may produce most of the world's electricity within 50 years, dramatically reducing the emissions of greenhouse gases that harm the environment.

Renewable energy sources, that derive their energy from the sun, either directly or indirectly, such as Hydro and wind, are expected to be capable of supplying humanity energy for almost another 1 billion years, at which point the predicted increase in heat from the sun is expected to make the surface of the Earth too hot for liquid water to exist.

History of Renewable Energy

Prior to the development of coal in the mid 19th century, nearly all energy used was renewable. Almost without a doubt the oldest known use of renewable energy, in the form of traditional biomass to fuel fires, dates from 790,000 years ago. Use of biomass for fire did not become commonplace until many hundreds of thousands of years later, sometime between 200,000 and 400,000 years ago.

Probably the second oldest usage of renewable energy is harnessing the wind in order to drive ships over water. This practice can be traced back some 7000 years, to ships on the Nile.

Moving into the time of recorded history, the primary sources of traditional renewable energy were human labor, animal power, water power, wind, in grain crushing windmills, and firewood, a traditional biomass. A graph of energy use in the United States up until 1900 shows oil and natural gas with about the same importance in 1900 as wind and solar played in 2010.

By 1873, concerns of running out of coal prompted experiments with using solar energy. Development of solar engines continued until the outbreak of World War I. The importance of solar energy was recognized in a 1911 *Scientific American* article: "in the far distant future, natural fuels having been exhausted [solar power] will remain as the only means of existence of the human race".

The theory of peak oil was published in 1956. In the 1970s environmentalists promoted the development of renewable energy both as a replacement for the eventual depletion of oil, as well as for an escape from dependence on oil, and the first electricity generating wind turbines appeared. Solar had long been used for heating and cooling, but solar panels were too costly to build solar farms until 1980.

Wind power

Wind power: Airflows can be used to run wind turbines. Modern utility-scale wind turbines range from around 600 kW to 5 MW of rated power, although turbines with rated output of 1.5–3 MW have become the most common for commercial use; the power available from the wind is a function of the cube of the wind speed, so as wind speed increases, power output increases dramatically up to the maximum output for the particular turbine. Areas where winds are stronger and more constant, such as offshore and high altitude sites are preferred locations for wind farms. Typical capacity factors are 20-40 per cent, with values at the upper end of the range in particularly favourable sites.

Globally, the long-term technical potential of wind energy is believed to be five times total current global energy production, or 40 times current electricity demand, assuming all practical barriers needed were overcome. This would require wind turbines to be installed over large areas, particularly in areas of higher wind resources, such as offshore. As offshore wind speeds average ~90% greater than that of land, so offshore resources can contribute substantially more energy than land stationed turbines.

Hydropower: Energy in water can be harnessed and used. Since water is about 800 times denser than air, even a slow flowing stream of water, or moderate sea swell, can yield considerable amounts of energy. There are many forms of water energy:

» Hydroelectric energy is a term usually reserved for large-scale hydroelectric dams. The largest of which is the Three Gorges Dam in China and a smaller example is the Akosombo Dam in Ghana.

» Micro hydro systems are hydroelectric power installations that typically produce up to 100 kW of power. They are often used in water rich areas as a remote-area power supply (RAPS).

» Run-of-the-river hydroelectricity systems derive kinetic energy from rivers and oceans without the creation of a large reservoir.

Hydropower is produced in 150 countries, with the Asia-Pacific region generating 32 percent of global hydropower in 2010. China is the largest hydroelectricity producer, with 721 terawatt-hours of production in 2010, representing around 17 percent of domestic electricity use. There are now three hydroelectricity plants larger than 10 GW: the Three Gorges Dam in China, Itaipu Dam across the Brazil/Paraguay border, and Guri Dam in Venezuela.

Solar energy

Solar energy, radiant light and heat from the sun, is harnessed using a range of ever-evolving technologies such as solar heating, solar photovoltaics, solar thermal electricity, solar architecture and artificial photosynthesis.

Solar technologies are broadly characterized as either passive solar or active solar depending on the way they capture, convert and distribute solar energy. Active solar techniques include the use of photovoltaic panels and solar thermal collectors to harness the energy. Passive solar techniques include orienting a building to the Sun, selecting materials with favorable thermal mass or light dispersing properties, and designing spaces that naturally circulate air.

Solar power is the conversion of sunlight into electricity, either directly using photovoltaics (PV), or indirectly using concentrated solar power (CSP). Concentrated solar power systems use lenses or mirrors and tracking systems to focus a large area of sunlight into a small beam. Commercial concentrated solar power plants were first developed in the 1980s. Photovoltaics convert light into electric current using the photoelectric effect. Photovoltaics are an important and relatively inexpensive source of electrical energy where grid power is inconvenient, unreasonably expensive to connect, or simply unavailable. However, as the cost of solar electricity is falling, solar power is also increasingly being used even in grid-connected situations as a way to feed low-carbon energy into the grid.

In 2011, the International Energy Agency said that "the development of affordable, inexhaustible and clean solar energy technologies will have huge longer-term benefits. It will increase countries' energy security through reliance on an indigenous, inexhaustible and mostly import-independent resource, enhance sustainability, reduce pollution, lower the costs of mitigating climate change, and keep fossil fuel prices lower than otherwise. These advantages are global. Hence the additional costs of the incentives for early deployment should be considered learning investments; they must be wisely spent and need to be widely shared".

Biomass

A cogeneration plant in Metz, France, this station uses waste wood biomass as energy source, and provides electricity and heat for 30,000 dwellings.

Biomass is biological material derived from living, or recently living organisms. It most often refers to plants or plant-derived materials, which are specifically called lignocellulosic biomass. As an energy source, biomass can either be used directly via combustion to produce heat, or indirectly after converting it to various forms of biofuel. Conversion of biomass to biofuel can be achieved by different methods, which are broadly classified into: thermal, chemical, and biochemical methods.

Wood remains the largest biomass energy source today; examples include forest residues (such as dead trees, branches and tree stumps), yard clippings, wood chips and even municipal solid waste. In the second sense, biomass includes plant or animal matter that can be converted into fibers or other industrial chemicals, including biofuels. Industrial biomass can be grown from numerous types of plants, including miscanthus, switchgrass, hemp, corn, poplar, willow, sorghum, sugarcane, bamboo, and a variety of tree species, ranging from eucalyptus to oil palm (palm oil).

Plant energy is produced by crops specifically grown for use as fuel that offer high biomass output per hectare with low input energy. Some examples of these plants are wheat, which typically yield 7.5–8 tons of grain per hectare, and straw, which typically yield 3.5–5 tons per hectare in the UK. The grain can be used for liquid transportation fuels while the straw can be burned to produce heat or electricity. Plant biomass can also be degraded from cellulose to glucose through a series of chemical treatments, and the resulting sugar can then be used as a first generation biofuel.

Biomass can be converted to other usable forms of energy like methane gas or transportation fuels like ethanol and biodiesel. Rotting garbage, and agricultural and human waste, all release methane gas—also called «landfill gas» or «biogas.» Crops, such as corn and sugar cane, can be fermented to produce the transportation fuel, ethanol. Biodiesel, another transportation fuel, can be produced from left-over food products like vegetable oils and animal fats. Also, biomass to liquids (BTLs) and cellulosic ethanol are still under research.

There is a great deal of research involving algal, or algae-derived, biomass due to the fact that it's a non-food resource and can be produced at rates 5 to 10 times those of other types of land-based agriculture, such as corn and soy. Once harvested, it can be fermented to produce biofuels such as ethanol, butanol, and methane, as well as biodiesel and hydrogen.

The biomass used for electricity generation varies by region. Forest by-products, such as wood residues, are common in the United States. Agricultural waste is common in Mauritius (sugar cane residue) and Southeast Asia (rice husks). Animal husbandry residues, such as poultry litter, are common in the UK.

Biofuel

Biofuels include a wide range of fuels which are derived from biomass. The term covers solid biofuels, liquid biofuels, and gaseous biofuels. Liquid biofuels include bioalcohols, such as bioethanol, and oils, such as biodiesel. Gaseous biofuels include biogas, landfill gas and synthetic gas.

Bioethanol is an alcohol made by fermenting the sugar components of plant materials and it is made mostly from sugar and starch crops. These include maize, sugar cane and, more recently, sweet sorghum. The latter crop is particularly suitable for growing in dryland conditions, and is being investigated by ICRISAT for its potential to provide fuel, along with food and animal feed, in arid parts of Asia and Africa.

With advanced technology being developed, cellulosic biomass, such as trees and grasses, are also used as feedstocks for ethanol production. Ethanol can be used as a fuel for vehicles in its pure form, but it is usually used as agasoline additive to increase octane and improve vehicle emissions. Bioethanol is widely used in the USA and in Brazil. The energy costs for producing bio-ethanol are almost equal to, the energy yields from bio-ethanol. However, according to the European Environment Agency, biofuels do not address global warming concerns.

Biodiesel is made from vegetable oils, animal fats or recycled greases. Biodiesel can be used as a fuel for vehicles in its pure form, but it is usually used as a diesel additive to reduce levels of particulates, carbon monoxide, and hydrocarbons from diesel-powered vehicles. Biodiesel is produced from oils or fats using transesterification and is the most common biofuel in Europe. Biofuels provided 2.7% of the world's transport fuel in 2010.

Geothermal energy

Geothermal energy is from thermal energy generated and stored in the Earth. Thermal energy is the energy that determines the temperature of matter. Earth's geothermal energy originates from the original formation of the planet (20%) and from radioactive decay of minerals (80%). The geothermal gradient, which is the difference in temperature between the core of the planet and its surface, drives a continuous conduction of thermal energy in the form of heat from the core to the surface. The adjective *geothermal* originates from the Greek roots *geo*, meaning earth, and *thermos*, meaning heat.

The heat that is used for geothermal energy can be from deep within the Earth, all the way down to Earth's core – 4,000 miles (6,400 km) down. At the core, temperatures may reach over 9,000 °F (5,000 °C). Heat conducts from the core to surrounding rock. Extremely high temperature and pressure cause some rock to melt, which is commonly known as magma. Magma convects upward since it is lighter than the solid rock. This magma then heats rock and water in the crust, sometimes up to 700 °F (371 °C).

From hot springs, geothermal energy has been used for bathing since Paleolithic times and for space heating since ancient Roman times, but it is now better known for electricity generation.

Hydroelectricity

The Three Gorges Dam in Hubei, China, has the world›s largest instantaneous generating capacity (22,500 MW), with the Itaipu Dam in Brazil/Paraguay in second place (14,000 MW). The Three Gorges Dam is operated jointly with the much smaller Gezhouba Dam (3,115 MW). As of 2012, the total generating capacity of this two-dam complex is 25,615 MW. In 2008, this complex generated 98 TWh of electricity (81 TWh from the Three Gorges Dam and 17 TWh from the Gezhouba Dam), which is 3% more power in one year than the 95 TWh generated by Itaipu in 2008.

Wind power development

Wind power is growing at over 20 per cent annually, with a worldwide installed capacityof 238,000 MW at the end of 2011, and is widely used in Europe, Asia, and the United States. Several countries have achieved relatively high levels of wind power penetration, such as 21per cent of stationary electricity production in Denmark, 18per cent in Portugal, 16 per cent in Spain, 14 per cent in Ireland and 9 per cent in Germany in 2010. As of 2011, 83 countries around the world are using wind power on a commercial basis.

Wind power capacity in different countries at the of 2012

Country	Wind power capacity (MW)	Total (%) of world
China	75,564	26.8
United States	60,007	21.2
Germany	31,332	11.1
Spain	22,796	8.1
India	18,421	6.5

United Kingdom	8,845	3.0
Italy	8,144	2.9
France	7,196	2.5
Canada	6,200	2.2
Portugal	4,525	1.6
(Rest of world)	39,853	14.1
World Total	2,82,482	100

As of 2012, the Alta Wind Energy Center (California, 1,020 MW) is the world's largest wind farm. The London Array (630 MW) is the largest offshore wind farm in the world. Phase 1 is complete; it is intended to introduce more turbines for Phase 2. The United Kingdom is the world's leading generator of offshore wind power, followed by Denmark.

There are many large wind farms under construction and these include Anholt Offshore Wind Farm (400 MW), BARD Offshore 1 (400 MW), Clyde Wind Farm (548 MW), Fântânele-Cogealac Wind Farm (600 MW), Greater Gabbard wind farm (500 MW), Lincs Wind Farm (270 MW), London Array (1000 MW), Lower Snake River Wind Project (343 MW),Macarthur Wind Farm (420 MW), Shepherds Flat Wind Farm (845 MW), and the Sheringham Shoal (317 MW).

Solar thermal

Ivanpah Solar Electric Generating System with all three towers under load in San Bernardino County, California. The Clark Mountain Range can be seen in the distance.

The United States conducted much early research in photovoltaics and concentrated solar power. The U.S. is among the top countries in the world in electricity generated by the Sun and several of the world›s largest utility-scale installations are located in the desert Southwest. The oldest solar power plantin the world is the 354 MW SEGS thermal power plant, in California. The Ivanpah Solar Electric Generating System is a solar thermal power project in the California Mojave Desert, 40 miles (64 km) southwest of Las Vegas, with a gross capacity of 392 megawatts (MW). The 280 MW Solana Generating Station is a solar power plant near Gila Bend, Arizona, about 70 miles (110 km) southwest of Phoenix, completed in 2013. When commissioned it were the largest parabolic trough plant in the world and the first U.S. solar plant with molten salt thermal energy storage.

The solar thermal power industry is growing rapidly with 1.3 GW under construction in 2012 and more planned. Spain is the epicenter of solar thermal power development with 873 MW under construction, and a further 271 MW under development. In the United States, 5,600 MW of solar thermal power projects have been announced. In developing countries, three World Bank projects for integrated solar thermal/combined-cycle gas-turbine power plants in Egypt, Mexico, and Morocco have been approved.

Photovoltaic power stations

Solar photovoltaic cells (PV) convert sunlight into electricity and photovoltaic production has been increasing by an average of more than 20 per cent each year since 2002, making it a fast-growing energy technology. While wind is often cited as the fastest growing energy source photovoltaics since 2007, has been increasing at twice the rate of wind an average of 63.6 per cent/year, due to the reduction in cost. At the end of 2011 the photovoltaic (PV) capacity world-wide was 67.4 GW, a 69.8 per cent annual increase. Top capacity countries were, in GW: Germany 24.7, Italy 12.8, Japan 4.7, Spain 4.4, the USA 4.4, and China 3.1.

Many solar photovoltaic power stations have been built, mainly in Europe. As of May 2012, the largest photovoltaic (PV) power plants in the world are the Agua Caliente Solar Project (USA, 247 MW), Charanka Solar Park (India, 214 MW), Golmud Solar Park (China, 200 MW), Perovo Solar Park (Ukraine, 100 MW), Sarnia Photovoltaic Power Plant (Canada, 97 MW), Brandenburg-Briest Solarpark (Germany, 91 MW), Solarpark Finow Tower (Germany, 84.7 MW), Montalto di Castro Photovoltaic Power Station (Italy, 84.2 MW), and the Eggebek Solar Park (Germany, 83.6 MW).

There are also many large plants under construction. The Desert Sunlight Solar Farm is a 550 MW solar power plant under construction in Riverside County, California, that will use thin-film solar photovoltaic modules made by First Solar. The Topaz Solar Farm is a 550 MW photovoltaic power plant, being built in San Luis Obispo County, California. The Blythe Solar Power Project is a 500 MW photovoltaic station under construction in Riverside County, California. The California Valley Solar Ranch (CVSR) is a 250 MW solar photovoltaic power plant, which is being built by Sun Power in the Carrizo Plain, northeast of California Valley. The 230 MW Antelope Valley Solar Ranch is a First Solar photovoltaic project which is under construction in the Antelope Valley area of the Western Mojave Desert, and due to be completed in 2013.

Many of these plants are integrated with agriculture and some use tracking systems that follow the sun's daily path across the sky to generate more electricity than fixed-mounted systems. There are no fuel costs or emissions during operation of the power stations.

However, when it comes to renewable energy systems and PV, it is not just large systems that matter. Building-integrated photovoltaics or "onsite" PV systems use existing land and structures and generate power close to where it is consumed.

Biofuel development

Brazil has bioethanol made from sugarcane available throughout the country. Shown a typical Petrobras gas station at São Paulo with dual fuel service, marked A for alcohol (ethanol)and G for gasoline.

Biofuels provided 3 per cent of the world›s transport fuel in 2010. Mandates for blending biofuels exist in 31 countries at the national level and in 29 states/provinces. According to the International Energy Agency, biofuels have the potential to meet more than a quarter of world demand for transportation fuels by 2050.

Since the 1970s, Brazil has had an ethanol fuel program which has allowed the country to become the world's second largest producer of ethanol (after the United States) and the world's largest exporter. Brazil's ethanol fuel program uses modern equipment and cheap sugarcane as feedstock, and the residual cane-waste (bagasse) is used to produce heat and power. There are no longer light vehicles in Brazil running on pure gasoline. By the end of 2008 there were 35,000 filling stations throughout Brazil with at least one ethanol pump.

Nearly all the gasoline sold in the United States today is mixed with 10% ethanol, a mix known as E10, and motor vehicle manufacturers already produce vehicles designed to run on much higher ethanol blends. Ford, Daimler AG, and GM are among the automobile companies that sell "flexible-fuel" cars, trucks, and minivans that can use gasoline and ethanol blends ranging from pure gasoline up to 85% ethanol (E85). By mid-2006, there were approximately 6 million E85-compatible vehicles on U.S. roads. The challenge is to expand the market for biofuels beyond the farm states where they have been most popular to date. Flex-fuel vehicles are assisting in this transition because they allow drivers to choose different fuels based on price and availability. The *Energy Policy Act of 2005*, which calls for 7.5 billion US gallons (28,000,000 m^3) of biofuels to be used annually by 2012, will also help to expand the market.

Renewable energy

The incentive to use 100 per cent renewable energy has been created by global warming and other ecological as well as economic concerns. Renewable energy use has grown much faster than anyone anticipated. The Intergovernmental Panel on Climate Change has said that there are few fundamental technological limits to integrating a portfolio of renewable energy technologies to meet most of total global energy demand. Mark Z. Jacobson says producing all new energy with wind power, solar power, and hydropower by 2030 is feasible and existing energy supply arrangements could be replaced by 2050. Barriers to implementing the renewable energy plan are seen to be "primarily social and political, not technological or economic". Jacobson says that energy costs with a wind, solar, water system should be similar to today's energy costs. Critics of the "100 per cent renewable energy» approach include Vaclav Smil and James Hansen.

Emerging technologies

Other renewable energy technologies are still under development, and include cellulosic ethanol, hot-dry-rock geothermal power, and ocean energy. These technologies are not yet widely demonstrated or have limited commercialization. Many are on the horizon and may have potential comparable to other renewable energy technologies, but still depend on attracting sufficient attention and research, development and demonstration (RD&D) funding.

There are numerous organizations within the academic, federal, and commercial sectors conducting large scale advanced research in the field of renewable energy. This research spans several areas of focus across the renewable energy spectrum. Most of the research is targeted at improving efficiency and increasing overall energy yields. Multiple federally supported research organizations have focused on renewable energy in recent years. Two of the most prominent of these labs are Sandia National Laboratories and the National Renewable Energy Laboratory(NREL), both of which are funded by the United States Department of Energy and supported by various corporate partners. Sandia has a total budget of $2.4 billion while NREL has a budget of $375 million.

Cellulosic ethanol

Companies such as Iogen, POET, and Abengoa are building refineries that can process biomass and turn it into ethanol, while companies such as the Verenium Corporation, Novozymes, and Dyadic International are producing enzymes which could enable a cellulosic ethanol future. The shift from food crop feed stocks to waste residues and native grasses offers significant opportunities for a range of players, from farmers to biotechnology firms, and from project developers to investors.

Selected commercial cellulosic ethanol plants in the US

Company	Location	Feedstock
Abengoa Bioenergy	Hugoton, KS	Wheat straw
Blue Fire Renewables	Irvine, CA	Multiple sources
Gulf Coast Energy	Mossy Head, FL	Wood waste
Mascoma	Lansing, MI	Wood
POET	Emmetsburg, IA	Corn cobs
SunOpta	Little Falls, MN	Wood chips
Xethanol	Auburndale, FL	Citrus peels

Carbon-neutral and negative fuels

Carbon-neutral fuels are synthetic fuels (including methane, gasoline, diesel fuel, jet fuel or ammonia) produced by hydrogenating waste carbon dioxide recycled from power plant flue-gas emissions, recovered from automotive exhaust gas, or derived from carbonic acid in seawater. Such fuels are considered carbon-neutral because they do not result in a net increase in atmospheric greenhouse gases. To the extent that synthetic fuels displace fossil fuels, or if they are produced from waste carbon or seawater carbonic acid, and their combustion is subject to carbon capture at the flue or exhaust pipe, they result in negative carbon dioxide emission and net carbon dioxide removal from the atmosphere, and thus constitute a form of greenhouse gas remediation.

Such renewable fuels alleviate the costs and dependency issues of imported fossil fuels without requiring either electrification of the vehicle fleet or conversion to hydrogen or other fuels, enabling continued compatible and affordable vehicles. Carbon-neutral fuels offer relatively low cost energy storage, alleviating the problems of wind and solar intermittency, and they enable distribution of wind, water, and solar power through existing natural gas pipelines. Night time wind power is considered the most economical form of electrical power with which to synthesize fuel, because the load curve for electricity peaks sharply during the warmest hours of the day, but wind tends to blow slightly more at night than during the day, so, the price of night time wind power is often much less expensive than any alternative. Germany has built a 250 kilowatt synthetic methane plant which they are scaling up to 10 megawatts.

The George Olah carbon dioxide recycling plant in Grindavík, Iceland has been producing

2 million liters of methanol transportation fuel per year from flue exhaust of the Svartsengi Power Station since 2011. It has the capacity to produce 5 million liters per year.

Estimated potential and installed capacity of major renewable energy technologies in India

Technology	Potential	Cumulative installation upto April 2000
Family size biogas plants	12 million	3.02 million
Improved cookstoves	120 million	32 million
Solar	20MWkm^2	-
Solar thermal	-	
Solar hot water systems (m^2)	-	0.5 million
Solar cookers	-	0.485 million
Solar photovoltaic	-	57 MW
Water pumps	-	3371
Water units	-	KWP
Community lights street lights	-	40,000
Home systems	-	100,000
Lanterns	-	250,000
Biomass	17,000 MW	-
Biomass Gasifier	-	34.36 MW
Bagasse Cogeneration	3500 MW	222 MW
Windfarms	20,000 MW	1167 MW
Minimicro hydel	10,000 MW	217 MW
Waste to energy	1700 MW	15.21 MW

Source: Annual Report 1999-2000, Ministry of Non-Conventional Energy Sources, Government of India

Elements, Technologies, Approaches Of Precision Farming

Agriculture is the backbone of our country and currently accounts for about 14.2 per cent of GDP. More than 60 per cent of India's population earns livelihood directly or indirectly from agriculture sector. In India, green revolution contributed to increase the farmer's income, yield of major crops and made India self-reliant in food production with the introduction of high-yielding varieties and by the use of synthetic fertilizers and pesticides. During the post-green revolution period, the agricultural production has become stagnant and expansion of cultivable lands has become limited due to burgeoning population and industrialization. The loss of subsidies, increased international competition and fundamental shifts in the farm and global economics will bring big changes to the business of farming. As participants in probably the second-oldest human industry (after tool making), farmers are often miscast as conservatives hopelessly wedded to traditional practices and technology. But it just isn't so. Farmers are constantly looking for ways to boost production, reduce inputs of labour and chemicals and build margins while producing a cornucopia of crops.

Looking the present situation of agriculture Dr. M S Swaminathan has also called it as exploitative agriculture. He has also warned that it may prove harmful for environment if it is not followed according to scientific principles of soil, water and crop management. According to him green revolution should be transformed to evergreen revolution in which required targets can be achieved without much harm to environment. Modern agriculture requires state-of-the-art technology for its practitioners to survive. Greatest challenge of the new millennium is maintaining the sustainability, environmental safety and economic profitability. As the availability of land has decreased, application of fertilizers and pesticides has become necessary to increase production. But simultaneously, the dependency of agriculture on chemicals has resulted in deterioration of the environment. In this situation, it is essential to develop eco-friendly technologies for maintaining crop productivity. An information and technology based farm management system to identify, analyze and manage variability within fields for optimum profitability, sustainability and protection of the land resources is need of the hour.

Precision farming is a current buzz word among agricultural circles. The term precision farming means carefully tailoring soil and crop management to fit the different conditions found in each field. In lay man language it can be defined as a need based or variable rate application system. According to Dr. Raj Khosla, PF can be defined in the form of 5 R's i.e. application of Right input at Right time on Right place in Right amount with a Right method.

Statistically it can be defined as

$$P = 1 - SD$$

Where,
SD is standard deviation
$P = 1$, indicates highly homogeneous field
$P = 0$ describes maximum variability of field

Precision farming is also known as -

» Prescription farming

» Farming by inch

» Variable rate technology (VRT)

» Site-specific farming (SSF)

Need of precision farming

During the 1950s, the food grain production of India was about 50 MT, which was not sufficient to make our nation self-sufficient. So main emphasize was to hike the production and productivity of crop by collaboration of traditional and scientific ways. Consequently Green Revolution stepped in India during 1965-66 due to which brought the India to self-sufficient front with respect to food grain availability. Although there was doubtless success of Green Revolution but simultaneously there was non-judicious, non-regulated use of agricultural inputs which created many flaws in the system. It can also say that there was exploitation of natural resources by keeping the health of environment, humans on the anvil. So to meet the food grain requirements of the increasing population and hunger of higher crop yields have created many issues, which are self-explanatory for the need of precision farming in India:

- » Decline/stagnation in crop productivity
- » Degradation of soil health/fertility
- » Environmental and social concerns like ground water pollution, human health problems etc.
- » Increasing problem of soil salinity and alkalinity
- » Depleting water table
- » Increasing soil erosion
- » Degradation of quality of agricultural produce
- » Increased Incidence of diseases and pests.

Precision farming cycle

A precision farming cycle represents the yield monitoring system. In which yield maps are prepared from the data taken by yield monitor while harvesting of the crop. According to these yield maps soil analysis is done for a factor, which can be seeding, fertilizer application, plant protection. Soil analysis data is used to prepare prescription maps in the computer for the variable rate application of these factors. Prescription maps are used by making data cards, which are fitted in the machinery.

Elements of precision farming (PF)

There are mainly three elements of precision farming, which play a pivotal role for the successful accomplishment of the whole system. These are as follows:

- » Information
- » Technology
- » Management

Information: In any area where someone wants to do precision crop management, first thing to be required is information of that particular area. It may be related to crop characteristics, soil properties, nutrient availability, topography etc. This is the most important element of PF because whole of the precision farming system depends on the availability as well as accuracy of the information.

Technology: This is the technology part, processing of available information/data is done. For example, from the data various types of maps are prepared like nutrient availability map, soil pH map, topographical map etc. A soil pH map of an area prepared and variation in soil pH displayed in the form of different colors. Highest pH value may be shown as purple color while lowest is shown as deep red colour.

Management: This element consists of Decision Support System in which step by step movement of whole cycle is decided. After collection and processing pf the data, appropriate management strategies are followed.

Steps in precision farming: The three basic steps in precision farming are Assessing Variability, Managing Variability and Evaluation, which are described as:

a. **Assessing variability:** In precision farming, inputs are to be applied precisely in accordance with the existing variability. Therefore, assessing the infield variability is very crucial and first step of precision agriculture. Spatial variability of all the determinants of crop yield should be well recognized adequately quantified and properly located. Construction of condition mapping on the basis of the variability is a critical component of precision farming. Condition maps can be generated through (i) Surveys, (ii) Point sampling & interpolation, (iii) Remote sensing (high resolution), and (iv) Modeling.

b. **Managing variability:** After assessing the spatial variability, the same can be taken care of through precision land leveling, variable rate technology, site specific planting, site specific nutrient management, precision water management, site specific weed management and precision pests and disease management etc. These precision farming practice aims at managing the variability by applying and making farm inputs available only in required quantities, at particular time and at specific locations.

c. **Evaluation:** There are three important issues regarding precision agriculture evaluation.

- » Economics
- » Environment

» Technology transfer

The most important fact regarding the analysis of profitability of precision agriculture is that the value comes from the application of the data and not from the use of the technology. Potential improvements in environmental quality are often cited as a reason for using precision agriculture. Reduced agrochemical use, higher nutrient use efficiencies, increased efficiency of managed inputs and increased production of soils from degradation are frequently cited as potential benefits to the environment. Enabling technologies can make precision agriculture feasible, agronomic principles and decision rules can make it applicable and enhanced production efficiency or other forms of value can make it profitable.

The term technology transfer could imply that precision agriculture occurs when individuals or firms simply acquire and use the enabling technologies. While precision agriculture does involve the application of enabling technologies and agronomic principles to manage spatial and temporal variability, the key term is manage. Much of the attention in what is called technology transfer has focused on how to communicate with the farmer. These issues associated with the managerial capability of the operator, the spatial distribution of infrastructure and the compatibility of technology to individual farms will change radically as precision agriculture continues to develop.

Technologies of precision farming

Precision farming consists of following technologies which work in an integration with each other:-

» Geographic information system (GIS)

» Global positioning system (GPS)

» Remote sensing

» Variable rate application

Geographic information system: A geographic information system (GIS) is a computer-based tool for mapping and analyzing spatial data. GIS technology integrates common database operations such as query and statistical analysis with the unique visualization and geographic analysis benefits offered by maps. These abilities distinguish GIS from other information systems and make it valuable to a wide range of public and private enterprises for explaining events, predicting outcomes, and planning strategies. GIS is considered to be one of the most important new technologies, with the potential to revolutionize many aspects of society through increased ability to make decisions and solve problems. The major challenges that we face in the world today overpopulation, pollution, deforestation, natural disasters all have a critical geographic dimension. Local problems also have a geographic component that can be visualized using GIS technology, whether finding the best soil for growing crops, determining the home range for an endangered species, or discovering the best way to dispose of hazardous waste. Careful analysis of spatial data using GIS can give insight into these problems and suggest ways in which they can be addressed.

Map making and geographic analysis are not new, but a GIS performs these tasks better and faster than do the old manual methods. And, before GIS technology, only a few people had the skills necessary to use geographic information to help with decision making and problem solving. Today, GIS is a multi-billion-dollar industry employing hundreds of thousands of people worldwide. GIS is taught in high schools, colleges, and universities throughout the world. Professionals in every field are increasingly aware of the advantages of thinking and working geographically.

Components of a geographic information system

A working Geographic Information System seamlessly integrates five key components: hardware, software, data, people, and methods.

Hardware: Hardware includes the computer on which a GIS operates the monitor on which results are displayed, and a printer for making hard copies of the results. Today, GIS software runs on a wide range of hardware types, from centralized computer servers to desktop computers used in stand-alone or networked configurations. The data files used in GIS are relatively large, so the computer must have a fast processing speed and a large hard drive capable of saving many files. Because a GIS outputs visual results, a large, high-resolution monitor and a high-quality printer are recommended.

Software: GIS software provides the functions and tools needed to store, analyze, and display geographic information. Key software components include tools for the input and manipulation of geographic information, a database management system (DBMS), tools that support geographic query, analysis, and visualization, and a graphical user interface (GUI) for easy access to tools. The industry leader is ARC/INFO, produced by Environmental Systems Research, Inc. The same company produces a more accessible product, ArcView, which is similar to ARCINFO in many ways.

Data: Possibly the most important component of a GIS is the data. A GIS will integrate spatial data with other data resources and can even use a database management system, used by most organizations to organize and maintain their data, to manage spatial data. There are three ways to obtain the data to be used in a GIS. Geographic data and related tabular data can be collected in-house or produced by digitizing images from aerial photographs or published maps. Data can also be purchased from commercial data provider. Finally, data can be obtained from the federal government at no cost.

People: GIS users range from technical specialists who design and maintain the system to those who use it to help them performs their everyday work. The basic techniques of GIS are simple enough to master that even students in elementary schools are learning to use GIS. Because the technology is used in so many ways, experienced GIS users have a tremendous advantage in today's job market.

GIS working

GIS store information about the world as a collection of thematic layers that can be linked together by geography. This simple but extremely powerful and versatile concept has proven invaluable for solving many real-world problems from modeling global atmospheric circulation, to predicting rural land use, and monitoring changes in rainforest ecosystems.

Geographic references

Geographic information contains either an explicit geographic reference such as a latitude and longitude or national grid coordinate, or an implicit reference such as an address, postal code, census tract name, forest stand identifier, or road name. An automated process called geocoding is used to create explicit geographic references (multiple locations) from implicit references (descriptions such as addresses). These geographic references can then be used to locate features, such as a business or forest stand, and events, such as an earthquake, on the Earth's surface for analysis.

GIS tasks

General purpose GIS performs following seven tasks:

- » Input of data
- » Map making

- » Manipulation of data
- » File management
- » Query and analysis
- » Visualization of results

Input of Data: Before geographic data can be used in a GIS, the data must be converted into a suitable digital format. The process of converting data from paper maps or aerial photographs into computer files is called digitizing. Modern GIS technology can automate this process fully for large projects using scanning technology; smaller jobs may require some manual digitizing which requires the use of a digitizing table. Today many types of geographic data already exist in GIS-compatible formats. These data can be loaded directly into a GIS.

Data structures: Digital data are collected and stored in different ways; the two data sources may not be entirely compatible. Therefore, a GIS must be able to convert data from one structure to another. Satellite image data that have been interpreted by a computer to produce a land use map can be "read into" the GIS in raster format. Raster data files consist of rows of uniform cells coded according to data values. An example is land cover classification. Raster files can be manipulated quickly by the computer, but they are often less detailed and may be less visually appealing than vector data files, which can approximate the appearance of more traditional hand-drafted maps. Vector digital data have been captured as points, lines (a series of point coordinates), or areas (shapes bounded by lines) . An example of data typically held in a vector file would be the property boundaries for a particular housing subdivision.

Data restructuring can be performed by a GIS to convert data between different formats. For example, a GIS can be used to convert a satellite image map to a vector structure by generating lines around all cells with the same classification, while determining the spatial relationships of the cell, such as adjacency or inclusion.

Vector based GIS

a. General definitions

Vector is a data structure, used to store spatial data. Vector data is comprised of lines or arcs, defined by beginning and end points, which meet at nodes. The locations of these nodes and the topological structure are usually stored explicitly. Features are defined by their boundaries only and curved lines are represented as a series of connecting arcs. Vector storage involves the storage of explicit topology, which raises overheads, however it only stores those points which define a feature and all space outside these features is 'non-existent'.

A vector based GIS is defined by the vectorial representation of its geographic data. According with the characteristics of this data model, geographic objects are explicitly represented and, within the spatial characteristics, the thematic aspects are associated. There are different ways of organizing this double data base (spatial and thematic). Usually, vectorial systems are composed of two components: the one that manages spatial data and the one that manages thematic data. This is the named hybrid organisation system, as it links a relational data base for the attributes with a topological one for the spatial data. A key element in these kinds of systems is the identifier of every object. This identifier is unique and different for each object and allows the system to connect both data bases.

b. Vector representation of data

In the vector based model, geospatial data is represented in the form of co-ordinates. In vector data, the basic units of spatial information are points, lines (arcs) and polygons. Each of these units is composed simply as a series of one or more co-ordinate points, for example, a line is a collection of related points, and a polygon is a collection of related lines.

i. **Co-ordinate:** Pairs of numbers are expressing horizontal distances along orthogonal axes or triplets of numbers measuring horizontal and vertical distances or n-numbers along n-axes expressing a precise location in n-dimensional space. Co-ordinates generally represent locations on the earth's surface relative to other locations.

ii. **Point**: A zero-dimensional abstraction of an object is represented by a single X, Y co-ordinate. A point normally represents a geographic feature too small to be displayed as a line or area; for example, the location of a building location on a small-scale map, or the location of a service covers on a medium scale map.

iii. **Line:** A set of ordered co-ordinates that represents the shape of geographic features too narrow to be displayed as an area at the given scale (contours, street centre lines, or streams), or linear features with no area (county boundary lines). A line is synonymous with an arc.

iv. **Arc:** An ARC/INFO term that is used synonymously expressed with line.

v. **Polygon:** A feature used to represent areas. A polygon is defined by the lines that make up its boundary and a point inside its boundary for identification. Polygons have attributes that describe the geographic feature they represent.

c. **Raster based GIS**

i. **Raster representation of data:** Raster is a method for the storage, processing and display of spatial data. Each area is divided into rows and columns, which form a regular grid structure. Each cell must be rectangular in shape, but not necessarily square. Each cell within this matrix contains location co-ordinates as well as an attribute value. The spatial location of each cell is implicitly contained within the ordering of the matrix, unlike a vector structure which stores topology explicitly. Areas containing the same attribute value are recognized as such, however, raster structures cannot identify the boundaries of such areas as polygons.

Raster data is an abstraction of the real world where spatial data is expressed as a matrix of cells or pixels, with spatial position implicit in the ordering of the pixels. With the raster data model, spatial data is not continuous but divided into discrete units. This makes raster data particularly suitable for certain types of spatial operation, for example overlays or area calculations. Raster structures may lead to increased storage in certain situations, since they store each cell in the matrix regardless of whether it is a feature or simply 'empty' space.

ii. **Grid size and resolution**: A pixel is the contraction of the words picture element commonly used in remote sensing to describe each unit in an image. In raster GIS the pixel equivalent is usually referred to as a cell element or grid cell. Pixel/cell refers to the smallest unit of information available in an image or raster map. This is the smallest element of a display device that can be independently assigned attributes such as colour.

iii. **Pixel size and number of rows and columns:** "The size of the pixel must be half of the smallest distance to be represented".

Map making: Maps have a special place in GIS. The process of making maps with GIS is much more flexible than are traditional manual or automated cartography approaches. It begins with database creation. Existing paper maps can be digitized and computer-compatible information can be translated into the GIS. The GIS-based cartographic database can be both continuous and scale free. Map products can then be created centered on any location, at any scale, and showing selected information symbolized effectively to highlight specific characteristics.

The characteristics of atlases and map series can be encoded in computer programs and compared with the database at final production time. Digital products for use in other GIS's can also be derived by simply copying data from the database. In a large organization, topographic databases can be used as reference frameworks by other departments.

Manipulation of data: It is likely that data types required for a particular GIS project will need to be transformed or manipulated in some way to make them compatible with your system. For example, geographic information is available at different scales (street centerline files might be available at a scale of 1:100,000; census boundaries at 1:50,000; and postal codes at 1:10,000). Before this information can be integrated, it must be transformed to the same scale. This could be a temporary transformation for display purposes or a permanent one required for analysis. GIS technology offers many tools for manipulating spatial data and for weeding out unnecessary data.

File management: For small GIS projects it may be sufficient to store geographic information as simple files. There comes a point, however, when data volumes become large and the number of data users becomes more than a few, that it is best to use a database management system (DBMS) to help store, organize, and manage data. A DBMS is nothing more than computer software for managing a database an integrated collection of data. There are many different designs of DBMS's, but in GIS the relational design has been the most useful. In the relational design, data are stored conceptually as a collection of tables. Common fields in different tables are used to link them together. This simple design has been widely used, primarily because of its flexibility and very wide deployment in applications both within and without GIS.

Analysis

Applications: Function of an Information system is to improve one's ability to make decisions. An Information system is a chain of operations starting from planning the observation and collection of data, to store and analysis of the data, to the use of the derived information in some decision making process. A GIS is an information system that is designed to work with data referenced to spatial or geographic coordinates. GIS is both a database system with specific capabilities for spatially referenced data, as well as a set of operation for working with data. There are three basic types of GIS applications which might also represent stages of development of a single GIS application.

Inventory Aapplication: Many times the first step in developing a GIS application is making an inventory of the features for a given geographic area. These features are represented in GIS as layers or themes of data. The emphasis at this stage of application development consists of updating and simple data retrieval.

Analysis application: Upon completion of the inventory stage, complex queries on multiple layers can be performed using spatial and a spatial analysis technique.

Management application: More advanced spatial and modelling techniques are required to support the decisions of managers and policy makers. This involves shifting of emphasis from basic geographic data handling to manipulation, analysis and modelling in order to solve real world problems.

Global positioning system: GPS is a satellite based navigation system consisting of a network of 24 satellites being launched by US Dept of Defense. By using the GPS system 2-D as well as 3-D location of any area in the field is identified-located. For 2-D location of any object, GPS receiver must be locked in to the signal of minimum three satellites. GPS works on the basis of triangulation method. In this system angle suspended by three known point on a particular object is measured. On the basis of which co-ordinates of that object are determined. Feature of satellites: They complete two rounds around earth within 24 hrs. They revolve at a height of 12000 mile from earth surface with a speed of 7000 mile/hr.

Remote sensing: Remote sensing is technique of getting information about an object without getting physical contact with it. For getting the information aerial pictures are taken of that particular area and objects are studied on the basis of their spectral signatures.

For taking the pictures various platforms are used which may be:

- » Satellite
- » Aircraft cameras
- » Hand held spectro-radiometers

Variable rate application system: This technology consists of that part which responds to the commands given by Decision Support System (DSS) for the need based and variable rate application of inputs. A variable rate application system which consists of applicators and control unit. This control unit responds to the signal given by information collecting and processing units and further controls the need based application through emitters.

Suitable Approaches of Precision Farming in India

Adoption of any advanced agricultural technology is not a single day work especially in developing countries like India due to many reasons like:

- » Small land holding
- » Low social and economic status of farmers
- » Traditional management practices have become bone marrow of the farmers.

According to the definition of PF in the form of 5 R's, it can be practiced manually. For the adoption of precision farming, use of small technologies can be taken as starting point. At present management of the two agricultural inputs are going to be the reason of major concerns i.e. nutrient and water. So following are some of the approaches that can be practiced in crop production without having much harmful impact on environment.

Precision nutrient management

In India, fertilizers are generally broadcast on the basis of so-called recommended doses. As these recommendations are made for either national or state basis, they are always erroneous for a particular field. Site-specific nutrient management (SSNM) should be followed to apply required amount of fertilizers. SSNM, a general concept for optimizing the supply and demand of nutrients according to their variation in time and space, is being tried in India for achieving the high yield targets. The within-field variability in relatively large fields can be managed using geo-referenced variable rate technology of nutrient management i.e. SSNM. A single field of usually 0.2-1.0 ha size is probably the smallest feasible management unit because managing with-in field variation is difficult without geo-referencing. Hence, smaller field units in India offer opportunity for site-specific nutrient management. Not only the location of nutrient placement, but also the amounts and the depth of placement has greater role in improving efficiency of applied fertilizer nutrients. In this direction, the low cost precision instruments or fertilizer drill must substitute the manual broadcasting which is aggravating the already existing variability. Hence there is greater scope of improving nutrient use efficiency through precision nutrient management or site-specific nutrient management in India.

Variable Rate Technology is the technique of applying the farm inputs such as seeds fertilizer, pesticides etc. in varying rates at the places and amounts where they are required to produce uniform yields throughout the entire field. The variable rate seed drills and planters, fertilizer spreaders and sprayer are commercially available. Farm machineries equipped with VRT generally possess a Differential Global Positioning System (DGPS) receiver to locate the spatial variability in the field and automatically regulate the rate of application. Due to high cost of integrated control systems, the farmers refer to rely on custom hiring for using VRT.

The most important and widely used form of VRT is variable rate fertilizer application. It is well-known that spatial variability in soil properties exists across the landscape. Therefore, uniform application of fertilizers results in under-fertilization of certain parts of a field and over fertilization in other areas. Under-fertilization results in yield loss and over-fertilization can be harmful to the crops and environment as well. With the invention of VRT, it has been possible to manage the soil nutrient variations through site-specific nutrient application. There are significant correlations between yield and soil properties viz, Slope, CEC, water-holding capacity etc., Spatial variability of P&K contents' in the soils affects yield and tuber quality in potato. They also showed that variable rate P& K application in potato produced similar yield as the conventional method (uniform rate). However, since lower amounts of fertilizer were applied in VRT, it reduced the input cost and enhanced the input use efficiency. In a study in Australia variable rate gypsum application on high sodic soils showed a benefit of $563 per hectare over five years as compared to uniform application.

Nitrogen fertilizer being a basic and widely applied nutrient in Indian agriculture need special attention for precision management practices. For improving N use efficiency, reducing N losses and environment pollutions, the use of simple precision tools viz, leaf colour chart (LCC) and chlorophyll meter (SPAD) has offered wide scope for farm level adoption of these technologies. The leaf colour chart (LCC) based real time N management can be used to optimize/synchronies N application with crop demand or to improve existing fixed spilt N recommendations. It was also reported that the net returns in rice- wheat system were increased by 19-31 per cent in LCC based N management than in fixed time N application.

Fertigation (application of fertilizer solution with drip irrigation) has the potential to ensure that the right combination of water and nutrients is available at the root zone, satisfying the plants total and temporal requirement of these two inputs. Fertigation saves fertilizer as it permits applying fertilizer in small quantities at a time matching with the plants nutrient need. Besides it is considered eco-friendly as it avoids leaching of fertilizers.

Precision water management

Water is a critical and most limiting input for crop production Crop yield generally shows linear correlation with amount of water transpired. When water is a limiting factor, the crop yield is much lower than the potential yield whereas excess watering may induce stresses in aeration and nutrient availability therefore the only option is precision water management which has four approaches viz,

    i. Judicious use of water

    ii. Variable rate irrigation

    iii. Soil-landscape water management

    iv. Drainage

i. Judicious use of water

Water loss reduction in field conveyance and distribution system: Percolation and seepage are two main losses in conveyance system. In the earthen channels (unlined channels) these losses may range from 25-40 per cent. The other losses in the distribution system are through evaporation, weed consumption and operational losses. The methods to reduce conveyance losses are by lining the channels and eradication of weeds. Lined channels minimize seepage losses, provides safety against breaks, provide control against damage by rodents and convey larger quantities of water in channels of a given size. The most common materials for channel lining include concrete, stone or brick masonry, clay tiles and polythene lining. To reduce the losses due to transit of irrigation water from the source to the irrigation water application, underground pipeline system can be used. This method practically eliminates conveyance losses. Pipelines are leak-proof and eliminate water loss by evaporation and seepage during conveyance.

Proper selection of an irrigation method

Table 3. The irrigation methods suitable for various crops are as under

Methods of Irrigation	Suitable Crops
Border strip method	Wheat, barseem and other fodders etc.
Furrow method	Cotton, sugarcane, potatoes, maize
Check Basin method	Paddy
Ring Basin method	Orchard crops

The conventional methods of water conveyance and irrigation being highly inefficient have led not only to wastage of water but also to several ecological problems like water logging, salinization and soil degradation. It has been recognized that use of modern irrigation methods viz. drip and sprinkler irrigation is the only alternative for efficient use of surface as well as ground water resources. The water use efficiency in these systems is much higher than the flood method of irrigation. The scheme on Micro irrigation which aims at increasing the area under efficient methods of irrigation viz. drip and sprinkler irrigation has been launched.

Paired row trench planting technique: Paired row trench planting technique in spring and autumn sugarcane saves irrigation water and gives higher cane yield and net returns as compared to conventional method.

Scheduling irrigation: The supply of irrigation water to a given crop at right time and in the right amount is called irrigation scheduling. By following optimal irrigation scheduling, not only optimal crop production can be obtained but also substantial amount of water, labour, time and energy can be saved under the given conditions.

Mulching: Application of straw mulch improves the water use efficiency. It reduces the evaporation losses from the soil surface. Mulching keeps the weed down and improves the soil structure and eventually increases the crop yield.

ii. **Variable rate irrigation:** It is a well managed site-specific irrigation system dependent on physical circumstances and socioeconomic condition it ensures optimum spatial and temporal distribution of water so as to promote the crop growth and yield. In India, irrigation is applied mainly on the basis of crop growth stages. But, it totally ignores the soil moisture variability. For precision water management, variable rate irrigation must be applied on the basis of soil moisture variability. Some simple instrument can be used to measure the soil moisture variability. Excess irrigation must be avoided particularly in canal-irrigated areas and provision of proper drainage has to be made there. The contribution of monsoon rain has to be taken into account while prescribing the level of irrigation.

iii. **Soil-landscape water management:** The variable rate irrigation alone cannot ensure precision water in due to differences in water availability which are governed by variation landspace different soil types and presence of soil degradation process such as erosion compaction salinisation etc it has been observe that the north facing slopes had 20 more available water than south facing slopices. Crop yields are often higher in lower slopes due to greater availability of nutrients in soil as well as compacted soil are reported to have lower infiltration rates. Redistribution of water within a landspace due to either runoff or subsurface horizontal flow is also responsible for causing spatial variability in water availability Laser land leveling soil amelioration etc may be employed for soil land space management. However the agronomic inputs including irrigation water have be applied in accordance with spatial variability in water availability within field.

The landscape variability in topography and physic-chemical properties trigger the disparity in water availability. Therefore, management of landscape variability should be the first step of precision water management. The general practice of managing landscape variability in India is leveling through animal/tractor drawn scrappers, which even after best efforts leads wide variability in the landscape. Precision land leveling through laser guided scrappers could manage the landscape variability and are in use in India.

Precision land levelling: The levelness of land is greatly influenced by the farming operations. Precision land leveling has been rated second only to development of high yielding varieties. As the leveling of land ensures higher resource use efficiency, it is in practice since time immemorial. However, the conventional leveling has not been upto the mark with many patches of dikes and ditches. Undulating topography hinders seedbed preparation, seed placement germination and proper distribution farm inputs. The soil moisture status of a field depends on levelness influences yield and input use efficiency, Significant improvement in crop productivity (5-15 %), water productivity (20-30 %), and nutrient use efficiency (15-20 %) under rice-wheat cropping system in Indo-Gangetic plains has been reported due to precision land leveling. Land grading/leveling in irrigated agriculture helps in uniform application of water, better water regulation and saving in irrigation time. Now a day it is possible to do precision land leveling on the fields, which seems to be leveled with naked eyes, with the help of Laser leveler (Fig. 4) which gives much better results than the earlier devices.

Resource conserving technologies

Zero-till seeding was noted to perform better on well leveled field as compared to unleveled or fairly leveled fields due to better seed placement, seed germination, and uniform distribution of plant nutrients as well as irrigation water. Laser leveling of the puddle field is the pre-requisite for mechanized rice transplanting

Tensiometer: The scientists of Punjab Agricultural University have been engaged in increasing the irrigation water use efficiency through many conservation irrigation techniques since a long time and have come out with many such technologies from time. The latest of these techniques is soil matric tension based irrigation scheduling to rice, which can help to save irrigation water.

Barriers to adoption of precision farming

One of the major problems going for precision farm management is that the majority of farmers in India Possess small size of farms (less than 1 hectare). There are many problems to adoption of precision farming in developing countries in general and India in particular. Some are common to those in other regions but the others are specific to Indian conditions are as follows.

Adoption patterns: As agriculture is an intrinsically conservative business, any major technological development takes much time and effort before it is adopted by a majority of farmers. It took 25 years for hybrid seeds to catch on and more than 30 years to see tractors fully utilized. A similar course must be expected for PF. Because diverse cropping systems, land quality, economic and geographic conditions characterize Indian farming, adoption patterns of PF are likely to be diverse. Even for selected crops, while some farmers may use the Internet to discover marketing opportunities, others may use PF solely for in-field decision making. Owing to relatively high cost and complexity, adoption process will not occur uniformly over time. Individual small farmers in India are likely to adopt them slowly, while estate owners and farmers' associations may adopt quickly. PF in India is likely to evolve as a combination of services and products mainly taken up by farm associations and the private sector, which can spread the capital costs over large area and over many years. Diffusion of PF is likely to be more rapid in areas surrounding cities and in areas with larger numbers of farm consultants and dealers.

Environmental implications: A more precise use of external inputs through PF may alleviate environmental pollution from Indian or Punjab agriculture in the long term. However, as in all countries, downstream effects of increased soil erosion, water use, nutrient and pesticide leaching are ordinarily of national concern and of little interest to individual farmers. Only those PF technologies that prove economically profitable are likely to be adopted first. The major challenge, therefore, is to design PF innovations and incentives that are financially rewarding to individual farmers as well as environmentally sound and beneficial to the society as a whole. Synergy between PF and biotechnology may also lead to environmental improvements. If there are yield penalties with genetically modified varieties for increased herbicide tolerance, those varieties can be sown in areas with high weed infestation. Likewise, varieties with insecticidal properties may be planted in potential problem areas with pests.

Economic implications: Although profits for large farms, which are likely to adopt PF technologies first, are expected to be high, small farmers too can get economic benefits over time through adopting only proven technologies. Adoption of profitable technologies also should lead to increases in the value of land.

Employment implications: PF requires significant supporting infrastructure in the form of skilled labour, software development and hardware availability. Therefore, new industries providing PF services, PF products and combination of PF services and products are likely to emerge, which in turn should provide additional employment. For example, agricultural consulting firms specializing in equipment, service, and technology delivery can develop to supply the needs of big farms adopting PF. In cooperation with local governments, they can assist in setting up farm GIS databases, process and analyze data to create maps, reports, statistics and recommendations for farmers in a given area. New opportunities may also arise in the form of VRT, equipment sales and recommendations, geoprocessing, etc. All these services will generate massive amounts of data, which obviously requires skilled personnel with knowledge and experience to interpret the data for guiding the decision process.

Future needs and research thrust

Precision farming is a direction of research rather than a destination. Extensive research on precision farming in India requires immediate attention of the researchers and policy makers. For practicing precision farming in India, farmer's participatory research has been taken up within the limits of available resources. It should not be misunderstood that precision farming because of its sophisticated nature is only suitable for large or resource rich farmers rather it could be more useful for a resource poor farmer. Suo motto research has to be carried out to suite the Indian farming situation and to meet the aspiration of Indian farmer. The imported version of precision agriculture must be Indianised through amalgamation of frontier technologies with low cost technologies even ITKs. Although few studies have been made in assessing the variability, primary research issue would be nationwide study of infield variability across the agro-eco-zones. The thrust area for research on precision farming should encompass site specific nutrient management for different crops and cropping system, study and management of landscape variability, site specific weed management through variable rate technology.

High accuracy sensing and data management tools must be developed and validated under our conditions. The limitation in data quality/availability has become a major obstacle in the demonstration and adoption of the precision technology. The research work on precision farming should be intensified in Punjab as well as in India. So the adoption of precision farming technology needs combined efforts of scientists, farmers and the government.

It is further suggested that the precision farming may help the Indian farmers to harvest the fruits of frontier technologies without comprising the quality of land and produce. The adoption of such a novel technique would trigger a techno green revolution in India which is the need of the hour. GIS based irrigation and fertigation scheduling should be done. About 20% of the farmers are having operational land holdings having size of farm more than 4 hectares. By using arial data, it has been noticed that in Patiala District of Punjab, more than 50% of contiguous field size are larger than 15 hectare. These contiguous fields can be considered a single field for the purpose of implementation of precision farming. There is a scope of implementing precision farming for major food grain crops such as rice, especially in the states of Punjab and Haryana.

Need to do for promotion of precision farming

In order to promote PF technologies to foster sustainable agricultural development in India or Punjab, many policies and technological initiatives have to be undertaken at regional, national and local levels. Some of those issues are discussed below:

- » Appropriate capabilities, policies and infrastructures must be created in the country to exploit PF technologies suitably and effectively. Many Asian countries started late in adopting the industrial revolution but they must not fall behind in adopting the information revolution in agriculture. Goal setting mechanisms must be evolved for identifying the most suitable PF approaches specific to various agro-ecological and socio-economic conditions. Partnership arrangements among research and extension agencies must be promoted to develop PF applications that are most economically beneficial and technically practicable, and to devise new methods of cost sharing and fine tuning of technology to make it affordable. Rapid developments in the Internet-based information delivery systems are expected to make some PF technologies affordable for farmers. Regional and inter-country cooperation pacts for conducting research on PF technologies of mutual interest should be reinforced.

- » Various regions within a country, research to identify the most suitable crops and farming systems for applying PF technologies must be intensified. The immediate needs of cropping systems must be considered while formulating PF research projects. Because PF technologies do not come with a guarantee of success in all situations, it is important to determine how to apply them especially at the level and scale of operation now possible in the Indian context. It is also necessary to localize and modify certain components of technology to meet specific requirements. While it is essential to study the latest innovative PF technologies all over the world, it is more important to determine what is relevant and what is not.

- » Inadequate research and institutional support is a major constraint for adoption of PF technologies. In order to make scientific developments in PF a commercial reality, due attention should be paid to the development of capabilities and institutions. It is must to make cost/benefit assessments and informed judgments on the outcome of introducing PF technologies in social, economic, environmental and institutional terms and in regard to sustainability.

- » Appropriate policies and measures must be established at various levels to enable PF adoption from a solely "technology-push" to an application-driven approach, and create the right environment for active partnerships among government, the private sector, R&D organizations and NGOs. Because no single establishment can take on the entire PF process -- which consists of data logging, point sampling, data analysis and spatial modeling. It is vital that various agencies join together and contribute their relative advantages in knowledge, skills and experience. Participatory learning is crucial to effective implementation of PF technologies. Therefore, local knowledge must be actively sought and incorporated during planning. This, however, requires close collaboration among farmers and farmers' associations, community groups, NGOs, spatial data providers, data analysis firms, machinery manufacturers, research and extension agencies, and local governments. Mechanisms to sustain such collaborative activities must be set up.

» As indigenous research capacity is vital for sustained progress in application of PF technologies, governments must allocate adequate budgets and ensure rational utilization of funds as per well-chosen priorities. As farmers and associations may not readily agree to share their datasets, government agencies and NGOs should promote models and templates for data sharing and provide examples of benefits of sharing and aggregating data.

» Governments should assist the private sector to create suitable PF applications tailored to the needs of specific agricultural conditions and design ways for effectively integrating technologies in current field operations in the country. For example, data collection hardware and software, information management software and GPS devices must be user-friendly to be operated by individuals who are novices to these technologies. There is also an urgent need to make fully functional farm GIS easier for non-specialists to learn and use.

» Farmers' associations and community-based organizations in the country can play a significant role in promoting sustainable agriculture through shared access to PF technologies. If PF is considered a series of discrete services such as map generation, crop scouting, etc., it is possible to fit these services within the structure of a progressive farm cooperative in the country.

» Remotely sensed high resolution imagery may become more important for assessing crop and soil variability than yield maps in Asia. Therefore, it is important to concentrate on developing robust image analysis methods for farming in the form of, for example, a good quality product of reflectance and an interpreted product based on which farmers can actually make decisions. New cost-effective methods for classifying fields and creating management zones within a village must also be developed.

- » As technology for gathering information is far ahead of our understanding of how to use it to help growers make decisions, knowledge-based systems must be developed quickly to realize full potential from PF. In this connection, development of the ability to use available field information and experience, to form relationships among the various GIS databases and to deduce reliable decision-making information is essential. By building farm databases and adding more information to create models, decision support systems that suit the needs and resources of farmers can be developed.

- » The long-range goal should be to assist farmers in the practical and cost-effective use of PF technologies so that they can maximize profits and make good crop production and management decisions.

Some methods to help farmers include:

- » Conducting pilot demonstration projects and offering training seminars for farmers to show correct and cost-effective use PF technologies

- » Providing institutional credit for adoption of PF technologies

- » Processing generated data and providing user-friendly summaries that will help farmers in making correct decisions and

- » Contributing the technical expertise to integrate various PF components for each farmer's specific application.

18. Challenges, Concept, Approaches, Management Of Site-Specific Nutrient

Site-specific nutrient management (SSNM) as developed in Asian rice producing countries provides an approach for 'feeding' rice with nutrients as and when needed. SSNM strives to enable farmers to dynamically adjust fertilizer use to optimally fill the deficit between the nutrient needs of a high yielding crop and the nutrient supply from naturally occurring indigenous sources, including soil, crop residues, manures, and irrigation water. The SSNM approach does not specifically aim to either reduce or increase fertilizer use. Instead, it aims to apply nutrients at optimal rates and times in order to achieve high rice yield and high efficiency of nutrient use by the rice, leading to high cash value of the harvest per unit of fertilizer invested.

Researchers developed the SSNM approach in the mid 1990s and evaluated it from 1997 to 2000 on about 200 irrigated rice farms at eight sites in Asia. Since 2001, the on-farm evaluation and promotion of SSNM have markedly increased. In 2003 to 2005, SSNM was evaluated and promoted with farmers at about 20 locations in tropical and subtropical Asia, each representing an area of intensive rice farming.

SSNM provides two complementary and equally effective options for improved N management using the LCC. In the 'real-time' N management option, farmers monitor the rice leaf color regularly (e.g. once a week) and apply fertilizer N whenever the leaves become more yellowish-green than the critical threshold value indicated on the LCC. In the 'fixed-time/adjustable dose' option, the time for N fertilization is pre-set at critical growth stages, and farmers adjust the dose of N upward or downward based on the leaf color.

In most irrigated rice-wheat areas of India, further yield increases are likely to occur in small, incremental steps that involve gradual buildup of soil fertility and fine-tuning of crop management. Future strategies for nutrient management in these systems must, therefore, become more site-specific and dynamic to manage spatially and temporally variable resources based on a quantitative understanding of the congruence between nutrient supply and crop demand. The SSNM concept has demonstrated promising agronomic and economic potential. It can be used for managing plant nutrients at any scale that is, ranging from a general recommendation for homogenous management of a larger domain to true management of between-field variability. Assessment of pest profiles in FFP and SSNM plots suggests that SSNM may also reduce pest incidence, particularly diseases that are often associated with excessive N use or unbalanced plant nutrition. Field-specific management of macronutrients increased nutrient accumulation, yields, and N use efficiency at majority of the 56 irrigated rice fields in Punjab, India. Grain yield increases averaged 0.9 Mg ha 21 across sites. On a global scale, such yield increases would be sufficient for matching about 7 to 12 yr of annual growth in rice demand in India. Yield increases were achieved with a decrease in average N rate, but were associated with increased N accumulation mainly due to improved N management. Compared with the current farmers' practices, N losses from fertilizer were typically reduced by 50 to 80 per cent and profitability increased by 14 per cent of the total average net return. Site-specific nutrient management requires little in the way of credit for financing or complex coordination among farmers. Significant performance differences among sites suggest further scope for improvement by alleviating other crop management constraints to nutrient use efficiency.

The major challenges for SSNM

» To retain the success of the approach

» To build on what has been already achieved using this approach

While reducing the complexity of the technology as it is disseminated to farmers. For example, despite an overall 50 to 80 per cent increase, N use efficiencies obtained with SSNM remained below an AE_N of > 20 kg kg^{-1} and RE_N of > 0.5 kg kg^{-1} typically achieved in irrigated rice with good crop management. Thus, there is a need to further refine location-specific N management strategies and test them against other forms of N management. The nature of the SSNM approach will need to be tailored to specific circumstances in different situations. In some areas, SSNM may be field- or farm-specific, but in many areas, it is likely to be just domain- and/or season specific. A simplified future SSNM approach should combine decisions that are made on a field-specific basis as well as decisions that are valid for somewhat larger recommendation domains with similar socio–economic and biophysical conditions. Estimates that allow placing a field into one of several broad categories of indigenous nutrient supply are probably sufficient for most SSNM applications and are easier to follow.

Blanket recommendations have served the purpose very well but these cannot help increase nutrient use efficiency beyond a limit as the assumption that need of a crop for nutrients is constant over time and over large areas is not true. The site-specific nutrient management (SSNM) approaches emphasizes on feeding rice and wheat with nutrients as and when needed to achieve high yields while optimizing use of nutrients from indigenous sources. The SSNM can be practiced using both soil-based and plant-based approaches. Soil based SSNM is not very effective, particularly in lowland rice based systems because soil-test analyses often do not effectively account for effects of soil submergence on soil nutrient supply and the needs of rice for supplemental nutrients. Soil-based approaches also do not provide with flexibility to adjust fertilization practices based on crop performance and climatic conditions during a given season. However, in the last five years, the initial SSNM concept was simplified to provide farmers and extension workers with easy-to-follow tools and guidelines for applying N, P, and K fertilizers. The plant based SSNM enables the pre-season determination of crop needs for fertilizer nitrogen, the within-season distribution of fertilizer N to meet crop needs, and the pre-season determination of fertilizer P and K rates to match crop needs and sustain soil fertility. The plant-based SSNM tools such as leaf colour chart, chlorophyll meter and optical sensors are becoming known to farm managers and the these fertilizer management strategies are being included in the package of practices given to farmers in different states in India.

Genetic improvements leading to development of highly fertilizer responsive rice and wheat varieties and improved management strategies resulted in a dramatic rise in productivity and production from rice-wheat system. By meeting the nutritional demand of high yielding varieties, fertilizers have helped sustain the production of rice and wheat and the burgeoning population and wealth growth in India since seventies. A rice-wheat sequence that yields 7 t ha^{-1} of rice and 4 t ha^{-1} of wheat removes more than 300 kg nitrogen, 30 kg phosphorus, and 300 kg ha^{-1} of potassium from the soil. The requirements of nitrogen, phosphorus, and potassium for high rice and wheat yields vary depending on soil and climatic conditions. In both low- and high-resource input conditions, these crops must use the nutrients contained in the soil as well as applied through fertilizers and manures efficiently to reach their yield potential.

Nutrient management is the art of managing the amount, form, placement, and timing of the application of nutrients (as fertilizer, manure, crop residues or any other form) to plants for optimum crop, fiber and forage yields with minimal adverse effects on water and air resources and to maintain and/or improve the condition of soil. In India, fertilizer recommendations for rice and wheat have generally been developed by state agricultural universities for their respective states. While soil testing is the principle science-based means of making nutrient recommendations for crop production, these are developed for large tracts having similar climate and landforms by conducting a number of field experiments to achieve optimum yields and high fertilizer use efficiency. These recommendations are generally worked out by developing response functions for a given nutrient when other nutrients are not limiting. As a result, a wide variation in N-P-K levels is observed between fertilizer recommendations made for rice and wheat by different states in the Indo-Gangetic plain, even when refinements in fertilizer recommendations are made based on soil tests and by working out nutrient needs based on crop removal and fertilizer use efficiency considerations. Depending upon the yield potentials of rice and wheat in a region, the blanket recommendations for N, P and K vary greatly among regions and states. For P the recommendation varies from 13 to 28 kg P ha^{-1} for rice in different states of Indo-Gangetic plain. The range for wheat is from 22 to 30.5 kg P ha^{-1} but in states like Punjab no P application to rice is recommended if it follows wheat adequately supplied (26 kg P ha^{-1}) with phosphate fertilizers.

Blanket recommendations have served the purpose very well but these cannot help increase nutrient use efficiency beyond a limit because these recommendations assume that need of a crop for nutrients is constant over time and over large areas. And many a times, to ensure high yields, ignorant farmers apply fertilizer (particularly N) doses even higher than the blanket recommendations. Due to large field-to-field variability of soil nutrient supply, efficient use of fertilizer is not possible when broad-based blanket fertilizer recommendations are used. Soil test based recommendations remain ignorant about the dynamics of nutrient release from crop residues, organic manures and irrigation water and are not very successful in rice and wheat. Thus farmers often apply fertilizers over and above the recommended doses to ensure higher yields by avoiding the risk of nutrient deficiency. The main reason of low N use efficiency is inefficient splitting of N applications, including the use of N in excess to the requirements. Need of the time is to synchronize nutrient applications with crop demand. In addition to field-to-field variability, fertilizer N management strategies must be responsive to temporal variations in nutrient demands by crops and soil nutrient supply in order to achieve supply-demand synchrony and to minimize N losses. When fertilizer N application is not synchronized with crop demand, N losses from the soil–plant system are large, leading to low N fertilizer use efficiency. The recovery efficiency of top dressed urea during panicle initiation stage could be as high as 78 per cent. Thus it seems that improving the synchrony between crop N demand and the N supply from soil and/or the applied N fertilizer is likely to be the most promising strategy to increase N use efficiency in rice and wheat. The site-specific nutrient management (SSNM) approach emphasizes on feeding rice and wheat with nutrients as and when needed. It aims to enable farmers to dynamically adjust fertilizer use to optimally fill the deficit between the nutrient needs of a crop and the nutrient supply from naturally occurring indigenous sources such as soil, organic amendments, crop residues, manures, and irrigation water. The SSNM approach does not strive to reduce or increase fertilizer use. Rather it helps to apply nutrients at optimal rates and times to achieve high yield and high efficiency of nutrient use by the crop.

CONCEPT AND APPROACHES OF SSNM

Growth and need of a crop for nutrients can vary greatly among fields, seasons and years because of changes in crop-growing conditions, crop and soil management and climate. Thus nutrients also need to be managed following an approach that enables adjustments in N, P and K applications as per field specific needs of the crop. The SSNM provides the principles and guidelines for tailoring nutrient management practices to specific field conditions. It can be practiced using both soil-based and plant-based approaches.

Soil-based SSNM: Soil-based site-specific nutrient management involves mapping field variability through assessment of soil chemical analysis based on large numbers of soil samples. More specifically, it consists of evaluating soil fertility and making fertilizer recommendations. In other words, soil-based SSNM approach attempts to tailor fertilizer recommendations to the soil nutrient supplying capacity of specific fields, as determined through soil-test analyses. The soil based SSNM approach involves two major steps. A series of fixation reaction studies on nutrient elements with soil are conducted in the first step. The data thus generated are used to determine the ratio of nutrient supply versus nutrient applied with a specific soil. This information provides a basis to know as to how much nutrient is required to be added to a soil to bring it to a level which is more than adequate for maximum growth, but less than that which would be toxic, or out of balance with other plant nutrients. In the second step a greenhouse study is carried out in which sorghum is grown as an indicator plant in a fertigation system. Complete nutrient treatment including all nutrients required for growth, and deletion treatments for each of the individual nutrients are tested in this study. The data pertaining to biomass of sorghum in this study provide an indication of when a potential positive, or negative, response is to be expected. This information makes a basis for field trials with the specific crop, or crops, providing the necessary insight as to what nutrients need to be considered in the evaluation. Fertilizer recommendations covering the full range of nutrients, including macro, secondary, and micronutrients are then formulated using yield goals and the soil analysis.

In India, the International Plant Nutrition Institute (IPNI) adopted a soil-based SSNM procedure and carried out a site-specific soil testing and assessment program, which can best be described as a 'systematic approach to soil fertility evaluation'. Fertilizer rates were established based on the concept of crop removal as well as adjustment for soil residual nutrients. The role that residual soil nutrients play in meeting crop nutrient requirements becomes a challenge when significant portions of crop biomass are not removed from the fields. For example, if a soil tests medium or low in most of the plant nutrients, then application of these nutrients based on target yield crop removal is going to address these nutrient demands. However, on soils where the soil nutrient analysis indicates a high level of nutrient supply the issue of whether to apply the nutrient at removal rates becomes a challenge. Another concern in the soil-based SSNM is that of balanced nutrition. Application of high rates of N, P, and K as guided by SSNM approach stimulates a deficiency of a secondary or micro-nutrient, which according to soil testing was considered adequate. For example, at a soil test K level once considered adequate, Recommendation for K application may be insufficient to balance the high rates of N and P being applied. Thus best option is to apply all macro and secondary nutrients which are required to meet crop yield removal, and those micronutrients which soil testing show to be marginal or deficient.

Plant-based SSNM

The plant based SSNM approach allows determining the amounts of N, P and K that best match the field specific needs of rice and wheat for supplemental application of these nutrients. Climatic factors (solar radiation and temperature) and indigenous supply of nutrients largely affect crop demand for nutrients. Direct relationship between crop yield and the need of the crop for a nutrient, as determined from the total amount of the nutrient in the crop at maturity forms the basis of the plant-based. An attainable yield target provides a measure of the total amount of nutrient requirement of the crop. A portion of this requirement is met by the indigenous nutrient supply defined by nutrient contributions from the soil though mineralization, residual effects from a previous crop, crop residues, irrigation water and any other non-fertilizer sources. Since soil chemical analysis, particularly in case of rice soils and nitrogen, is not reliable to quantify indigenous N supply, field specific indigenous nutrient supply is best measured from the yield in N-, P- or K-omission plots.

In case of nitrogen, the current SSNM practice involves four steps to estimate total N rate based on indigenous N supply capacity and target yield:

i. Attainable yield target is taken as 85% of the yield potential

ii. Indigenous N supply is estimated as the yield without fertilizer N

iii. Estimate yield gain due to fertilizer N as the difference between the target yield and the yield without fertilizer N, and

iv. Estimate fertilizer N rate to be applied based on N anticipated N response and agronomic N use efficiency.

The agronomic efficiency defined as the increase in yield per unit of fertilizer N applied, defines the total amount of N needed for each tonne increase in grain yield over no-N control. For rice agronomic efficiency typically ranges from 16 to 25 kg increase in grain yield per kg applied fertilizer N and it amounts to fertilizer N rates of 40 to 60 kg for each tonne increase in grain yield. A realistic estimate of agronomic efficiency at a given site is obtained from yields levels and agronomic efficiencies in previous years. When following SSNM approach if yield without fertilizer N is not available due to non establishment of no-N plots, one can estimate the N response based on information on climatic yield potential and soil fertility. It allows skipping the first two steps in estimating total amount of fertilizer N to be applied.

Guidelines for distribution of fertilizer N during the crop growth season constitute an important aspect of the plant-based SSNM approach. In case of site-specific N management, the main reason of low use efficiency is inefficient splitting of N applications, including the use of N in excess to the requirements. The aim is to synchronize N applications with crop demand. In addition to field-to-field variability, fertilizer N management strategies must be responsive to temporal variations in crop N demands and soil N supply in order to achieve supply-demand synchrony and to minimize N losses. When N application is not synchronized with crop demand, N losses from the soil–plant system are large, leading to low N fertilizer use efficiency. The recovery efficiency of top dressed urea during panicle initiation stage could be as high as 78 per cent. Thus it seems that improving the synchrony between crop N demand and the N supply from soil and/or the applied N fertilizer is likely to be the most promising strategy to increase N use efficiency in rice and wheat.

Site-specific N management is based on periodic assessment of plant N status, and the application of fertilizer N is delayed until N deficiency symptoms start to appear. Thus, a key ingredient for real-time N management is a method for the rapid assessment of leaf N content, which is closely related to photosynthesis rate and biomass production and is sensitive indicator of changes in crop N demand within the growing season. Although estimation of soil N supply or the calculation of a pre-season fertilizer rate is not required, the real-time N management revolves around quick and reliable tools to decide the time when fertilizer N needs to be applied to the crop. Spectral characteristics of radiation reflected, transmitted, or absorbed by leaves can provide an estimate of chlorophyll content. In turn the leaf chlorophyll concentration can help to diagnose N deficiency and, indirectly, to correct N fertilization. During last 2-3 decades some noninvasive, optical methods based on leaf greenness, absorbance, transmittance or reflectance of red, green or near infra red radiation by the intact leaf have been developed. These include chlorophyll meters, leaf color charts, ground-based remote sensors, and digital, aerial, and satellite imageries. Over the last decade, these methods have been extensively tried to improve N use efficiency in cereals grown in different agro-ecosystems and regions of the world.

Total N is distributed in rice in such a way that only a small to moderate amount of fertilizer N is applied within 14 days after transplanting or 21 days after direct sowing. In wheat moderate dose of fertilizer N is applied at sowing. The remaining fertilizer is applied to ensure sufficient N at critical growth stages of the crop – mid-tillering and panicle initiation stages for rice and crown root initiation and tillering stages in wheat. In a study the total N to be applied as: 35, 20, 30 and 15 per cent at transplanting, mid-tillering, panicle initiation and heading stages of transplanted rice, respectively. The actual rate of N top dressing at mid-tillering and panicle initiation stages is adjusted by ± 10 kg N ha^{-1} according to leaf N status measured with a SPAD meter or leaf colour chart. When colour of the leaves is below or above the threshold greenness level, N rate will be increased or decreased by 10 kg N ha^{-1}. Similar fertilizer N application plans are being tested in different regions and for different crops. In rice, fertilizer N can also be applied following a fixed or dynamic threshold values of SPAD meter of LCC irrespective of crop growth stage. After applying a moderate dose of fertilizer N at transplanting, colour of the first fully opened leaf from the top of a rice plant is monitored using LCC or SPAD meter every 7 to 10 days starting from 15 days after transplanting up to initiation of flowering. Whenever leaf colour turns out to be less than the threshold greenness a dose of fertilizer N is applied. The threshold greenness can be fixed or it can be defined as 90 per cent of the SPAD values of a small non-N limiting plot established in a corner of the field. These real-time N management strategies are described by several studies. Using SSNM approach, amounts of fertilizer P and K to be applied to rice and wheat are worked out in way that these are sufficient to overcome deficiencies and ensure profitable farming.

Total amounts of P and K taken up by the crop are determined from the target yield and an established optimal reciprocal internal efficiency (kg nutrient in above-ground dry matter per tonne grain) for each crop. Supply of P or K from crop residues, added organic materials, irrigation water, excess fertilization of the previous crop, and deposited sediment from floods is estimated. A deficit in a nutrient balance (difference between the total amount of P and K taken up for the target yield and the nutrient supply) reflects the amount of added nutrient required to avoid net removal of the nutrient from soil. Fertilizer P and K requirements for different fields are worked out from estimated target yield, nutrient balances, and expected yield gains from added nutrient. These are derived solely from the estimated nutrient balance when yield gain to P or K is negligible. When yield gain to applied P or K is certain, fertilizer P or K requirements are determined from a combination of the nutrient balance and anticipated yield gain to nutrient application.

Sustainable Nutrient Management

Recent research conducted in many Asian countries, including Northwest India, has demonstrated limitations of the current approach of fixed-rate, fixed-time (blanket) fertilizer recommendations being made for large areas. This is mainly because this approach does not take into account the existence of large variability in soil nutrient supply and site-specific crop response to nutrients among farms. This helps to explain why fertiliser N use efficiency is usually poor, the use of P and K fertilisers is often not balanced with crop requirements and other nutrients and, as a result, profitability is not optimized. Based on these conclusions, the original concept of site specific nutrient management (SSNM) to manage among-farm nutrient variability was developed in Asia for rice. A series of on-farm experiment were conducted with rice and wheat crops at 56 farmer fields in Northwest India to test the hypothesis that rice and wheat yields, profit, plant nutrient uptake, and fertiliser efficiencies can be increased significantly through field-specific nutrient management. On farm experiments were conducted from 2002-03 to 2004-05 with irrigated wheat and transplanted rice at 56 sites in six rice-wheat production regions across the three major agro-climatic zones of Punjab. The experimental set-up followed a standard protocol at all sites and included nutrient omission plots (0-N, 0-P, 0-K) to estimate indigenous nutrient supplies, a SSNM treatment plot, and farmer fertilizer practice (FFP) plot in each farmer field. Researchers did not intervene in the FFP plots but managed fertilizer application in the SSNM and nutrient omission plots. Farmers were responsible for all other aspects of general crop and pest management and the choice of variety. Treatments (SSNM and FFP) were compared on 56 farms over a period of 2 cropping years (2003-04 and 2004-05). An estimate of soil indigenous N, P, and K supply was obtained from omission plots situated in each farmer field. Average rice grain yields in nutrient omission plots increased in the order 0-N (3.82) <0-K (5.41) <0-P (5.45 t/ha), while the corresponding values for wheat were 0-N (3.08) <0-K (4.35) <0-P (4.55 t/ha). These data confirm that N deficiency is a general feature of irrigated rice-wheat systems in Punjab, whereas P and K supply are equally limiting factors, especially when considering the average rice and wheat yield goals of 7.9 t/ha in one study and 5.8 t/ha in another study for Punjab. Performance indicators used for the agronomic and economic evaluation of SSNM and FFP were:

» Recovery efficiency of fertilizer N (REN) is the increase in plant N uptake per unit fertilizer N applied (kg plant N/kg fertilizer N).

» Physiological N efficiency (PEN) is the increase in grain per unit increase in plant N uptake from fertilizer (kg grain/kg plant N).

» Agronomic N use efficiency (AEN) is the product of REN and PEN, expressed as the yield increase per unit fertiliser N applied (kg grain yield/kg fertiliser N).

» Gross return over fertiliser costs (Rs/ha/crop) is calculated as revenue (grain yield x farm gate paddy and wheat prices) minus fertiliser cost.

On average, SSNM generated a yield gain of at least 0.9 (17%) and 0.5 t/ha (12%) in rice and wheat crops, respectively, compared with FFP in approximately 48 per cent of the sites studied. At 21 of the total 56 farms studied, rice grain yield increases were >1 t/ha with SSNM compared with FFP, while at 24 of the total 56 farms studied, wheat grain yield increases were >0.8 t/ha, showing the potential of the SSNM approach used. Another interesting fact observed was that the maximum increases in rice and wheat grain yields were obtained at sites with low fertility soils, while the regions with high fertility soils had minimum but significant increases in grain yields of rice and wheat crops. This corroborates our hypothesis that blanket fertilizer recommendations, as is the current norm in Punjab, are of limited use in tackling site-specific soil fertility problems and that the adoption of site-specific strategies can give some impetus to the productivity growth of rice and wheat crops. Average fertilizer N applied to the rice and wheat crops in FFP at all sites in Punjab (148 and 143 kg N/ha, respectively) was relatively higher than the fertilizer N applied in other parts of India.. However, most farmers had no means of adjusting their fertilizer rates according to the actual soil fertility status. Correlation between N rate and indigenous N supply (INS) in wheat was -0.16, clearly outlining why…despite higher N use under FFP…grain yield and N accumulation were low as compared with that under SSNM. Like N, P rates were also not significantly correlated with indigenous P supply (IPS) (r = -0.05 and = 0.01 for wheat and rice, respectively). On the other hand, fertiliser K application in FFP was not much in Punjab probably because of substantial contribution of K (6 to 51 kg K/ha with an average of 29 kg K/ha) from irrigation water. On average, SSNM saved a significant amount (8 and 10% for rice and wheat, respectively) of fertilizer N compared with FFP, clearly bringing out the positive effect of SSNM for N. In contrast, average fertilizer P application significantly increased in rice and remained the same in wheat in both SSNM and FFP treatments, while fertiliser K application was significantly increased with SSNM compared with FFP for both rice and wheat crops. This might be due to the fact that 10 and 30 kg/ha P and K, respectively, were set as the minimum amounts to be applied to replenish net removal of these nutrients from a site and minimise risk of any macronutrient deficiency. Significant increases in N use efficiency were achieved in rice and wheat through the field-specific N management practiced in the SSNM treatment. In general, compared with the FFP, less fertiliser N was applied, and AEN, REN, and PEN were significantly increased with SSNM. On average, AEN was increased by 7.3 kg/kg (83%) and 5.3 kg/kg (63%), REN by 0.10 kg/kg (50%) and 0.10 kg/kg (59%), and PEN by 9.5 kg/kg (27%) and 7.7 kg/kg (26%) in rice and wheat crops, respectively. This increase was attributed to more uniform N applications among sites under SSNM as compared to under FFP. Also, the N applications were spread more evenly through the growing season and avoided heavy single applications at early growth stages of rice and wheat crops when compared with FFP. Site-specific nutrient management led to a small increase in the average fertilizer cost (Rs.370/ha/crop [12%] in wheat and Rs.1,190/ha/crop [52%] in rice), but comparatively a larger increase in gross returns over fertilizer (GRF) (Rs.2,950/ha/crop [13%] in wheat and Rs.3,450/ha/crop [14%] in rice) compared with FFP. Increase in the average fertilizer cost under SSNM was mainly attributed to an increase in K fertilizer use – an important input from the balanced crop nutrition point of view, but one that is generally skipped by farmers in Punjab.

Leaf colour chart

Nitrogen (N) is the most important nutrient for rice but it is the most limiting element in almost all soils. Optimal N supply matching with the actual crop demand is thus vital for improving crop growth and maximizing production. Among the various strategies available for N management, leaf color chart (LCC) for real-time N management in rice is a simple, easy and inexpensive option. It measures leaf color intensity related to leaf N status. LCC is an ideal tool for individual farmers to optimize N use in rice at high yield levels irrespective of the source of N applied, i.e., organic manure, biologically fixed N, or chemical fertilizers. It is also ecologically-friendly. Blanket-fertilizer N recommendations generally take into account crop response to applied N as basis to calculate the amount of N required achieving a targeted yield. These recommendations do not consider variability in soil N supply and changes in crop demand. Farmers generally apply too much N (and little P and K and other nutrients) that results in high incidence of pests and diseases, besides lodging. The consequence of high N application is high pesticide use to control pests, more expenditure on pesticides, and reduced yield and poor grain quality due to lodging. In addition, excess N is leached into water sources that get polluted over time. Several factors influence LCC readings: varietal group, plant or tiller density, variability in solar radiation between seasons, status of nutrients other than N in soil and plant, and biotic and abiotic stresses that induce discoloration of leaves. Users should clearly understand these limitations and know how to tackle them while using the LCC. The Punjab Agricultural University recommended this technology and making efforts to popularize this technology among the farmers.

In the case of nitrogen, findings from IRRI's research on matching site-specific capacities of the soil to supply nutrients and to the demand of crop(s) in the system have been reflected in the development of a leaf color chart (LCC) to help farmers select the right dose and time of application for optimum response in rice. Efforts have also been made to extend the LCC technology to wheat crop by synchronising N application with irrigation practices. The LCC has been widely distributed to tens of thousands farmers in the consortium countries to assess response. LCC technology has the potential to save about 15–20% of N fertiliser application in rice stated in another studies.

Biological nitrogen fixation (BNF)

Augmenting the nitrogen supply to crops through BNF is a viable option for resource-poor farmers of the humid tropics and must be exploited to its fullest potential. The amount of N fixed by legumes can range from 20-200 kg/ha/yr depending on the species, soil type, climate and agro-ecoregion. Some common legumes that can be grown as cover crops to procure mulch and increase BNF. Several perennial shrubs and woody species also can be used to enhance the nitrogen status of the soil.

Concept, Advantages, Components Of Integrated Nutrient Management

Integrated Nutrient Management refers to the maintenance of soil fertility and of plant nutrient supply at an optimum level for sustaining the desired productivity through optimization of the benefits from all possible sources of organic, inorganic and biological components in an integrated manner.

Concepts

- » Regulated nutrient supply for optimum crop growth and higher productivity.
- » Improvement and maintenance of soil fertility.
- » Zero adverse impact on agro – ecosystem quality by balanced fertilization of organic manures, inorganic fertilizers and bio- inoculants.

Determinants

1. Nutrient requirement of cropping system as a whole.
2. Soil fertility status and special management needs to overcome soil problems, if any
3. Local availability of nutrients resources (organic, inorganic and biological sources)
4. Economic conditions of farmers and profitability of proposed INM option.
5. Social acceptability.
6. Ecological considerations.

7. Impact on the environment

Advantages

1. Enhances the availability of applied as well as native soil nutrients
2. Synchronizes the nutrient demand of the crop with nutrient supply from native and applied sources.
3. Provides balanced nutrition to crops and minimizes the antagonistic effects resulting from hidden deficiencies and nutrient imbalance.
4. Improves and sustains the physical, chemical and biological functioning of soil.
5. Minimizes the deterioration of soil, water and ecosystem by promoting carbon sequestration, reducing nutrient losses to ground and surface water bodies and to atmosphere.

Components

Soil source: Mobilizing unavailable nutrients and to use appropriate crop varieties, cultural practices and cropping system.

Mineral fertilizer: Super granules, coated urea, direct use of locally available rock PO_4 in acid soils, Single Super Phosphate (SSP), MOP and micronutrient fertilizers.

Organic sources: By products of farming and allied industries, FYM, droppings, crop waste, residues, sewage, sludge, industrial waste.

Biological sources: Microbial inoculants substitute 15 - 40 Kg N/ha

Integrated Nutrient Management (Fertilizers)

The main objective of Integrated Nutrient Management (INM) Division is to ensure adequate availability of quality fertilizers to farmers through periodical demand assessment and timely supply, promoting integrated nutrient management, which is soil test-based judicious and balanced use of chemical fertilizers in conjunction with organic manures and bio-fertilizers, promotion of organic farming and ensuring quality control of fertilizers through implementation of Fertilizer (Control) Order, 1985.

Fertilizer consumption

India is the third largest producer and consumer of fertilizers in the world after China and the USA. Against 21.65 million tonnes of fertilizer nutrients (NPK) consumed during 2006-07, the nutrient consumption is 22.57 million tonnes during 2007-08. The consumption of major fertilizers namely, Urea, DAP, MOP, SSP and Complexes were 25.96, 7.50, 2.88, 2.29 and 6.57 million tonnes during 2007-08. India is by and large self sufficient in respect of Urea and about 90 per cent in case of DAP. The all India average fertilizer consumption is 116.5 kg/ha of NPK nutrients, though there is wide variation from state to state varying from 212.7 kg/ha in Punjab, 208.2 kg/ha in Andhra Pradesh, 190.9 kg/ha in Haryana to less than 5 kg/ha in States like Arunachal Pradesh and Nagaland etc. Considering the skewed pattern of fertilizer use, Government of India is promoting balanced and integrated use of fertilizer nutrient through various initiatives. As a result, NPK consumption ration has now improved to 5.5:2.1:1 during 2007-08 from 7.0:2.7:1 during 2000-01.

Despite normal south-west monsoon during 2016-17, the pattern of distribution of rainfall was found to be uneven during the period. Uneven distribution of rainfall resulted in lower consumption of fertilisers in some of the major fertiliser consuming states. Total fertiliser nutrient consumption declined by 3 per cent during 2016-17 over the previous year. The supply from indigenous production (N+P) increased nominally. The shortfall between consumption and production was fulfilled through import. Import of urea and DAP declined due to heavy opening inventory of these fertilisers reported in various channels. Import of MOP, however, increased in 2016-17 over the previous year.

Fertiliser consumption during 2016-17

Total nutrient consumption ($N+P_2O_5+ K_2O$) decreased from a total of 26.75 million metric tonne (MMT) during 2015-16 to 25.95MMT during 2016-17. Nitrogen consumption at 16.74 MMT and P_2O_5 at 6.71 MMT registered decline of 3.7 and 3.9%, respectively, during 2016-17 over the previous year. However, K_2O consumption at 2.51 MMT increased by 4.4% during the period. Per hectare use of total nutrients reduced from 134.9 kg in 2015-16 to 130.8 kg in 2016-17. All-India NPK use ratio changed from 7.2:2.9:1 during 2015-16 to 6.7:2.7:1 during 2016-17.

Production of major crops

Total production of food grains touched a record level of 275.7million tonnes in 2016-17, which was 9.6 per cent higher over the previous year's level. During 2016-17, almost all the major crops recorded increase in production except sugarcane. Among principal crops, production of rice at 110.2 MMT, wheat at 98.4 MMT, coarse cereals at 44.2 MMT, pulses at 23 MMT and oilseeds 32.1 MMT recorded increase of 5.5, 6.6, 14.7, 40.4 and 27.1 per cent, respectively during 2016-17 over 2015-16. Similarly, the production of cotton increased by 10.3 per cent, jute and mesta increased by 0.7 per cent during the period. However, the production of sugarcane at 306.7 MMT declined by 12 per cent during the period.

Price of fertilizers

Presently urea is the only fertilizer, which is under Statutory Price Control. To ensure adequate availability of fertilizers to farmers at reasonable rates, subsidy is provided by Government of India. Urea, the most consumed fertilizers, is subsidized under the New Urea Pricing Scheme, whereas P&K fertilizers, which are decontrolled, are covered under the Concession Scheme. The policy for uniform freight subsidy on all fertilizers under the fertilizer subsidy regime is also implemented.

The food grains production of the country during 2016-17 was an all-time high of 275.68 million tonnes. The horticultural production (fruits, vegetables, plantation crops, spices, etc) in the same year surpassed the level of food grains at 300.64 million tonnes. The figures speak volumes of the efforts of all stakeholders in the Indian agriculture sector to have now attained near self-sufficiency from a stage of food scarcity in early 1960s. Supply of inputs like seeds, fertilizers, agro-chemicals, irrigation water and agricultural credit, besides delivery of extension services has definitely played a key role in raising the crop production and productivity over the last five decades.

Fertilizers being important inputs in the farm sector, successive governments have made them available to farmers across the country at affordable prices through concession/subsidy schemes. Subsidy for main nitrogenous fertilizer product, urea is regulated through a policy wherein it is provided to farmers at a statutory maximum retail price (MRP) of Rs 5,360 per tonne (5% extra for the neem coating). The retail price for urea in India is perhaps the lowest anywhere in the world. The Phosphorous and Potassium (P & K) fertilizer grades are covered under the nutrient-based-subsidy (NBS) scheme, in which a fixed amount of subsidy is paid on each grade of subsidized fertilizers (for example, DAP ; SSP ; MOP ; NPK complex, etc) based on their nutrient content and the fertilizer companies have been permitted to fix the MRPs at a 'reasonable' level. In addition to the product subsidy, fertilizer companies are also paid freight subsidy for transporting fertilizers from the plants/ports to various destinations as per as 'agreed supply plan' issued by the Department of Fertilizers of Government of India.

According to updates 29 September, 2018, farmers may have to pay 5-26% more for fertilisers this crop season due to an increase in the global prices of key fertiliser components phosphate and potash, according to analysts and industry executives. They, however, said if the government increases the per unit subsidy for fertiliser companies, the price impact on farmers will be less. Urea prices usually remain steady due to government controls. At Rs 1,400 per bag (50 kg), the price of di-ammonium phosphate (DAP) is 8 per cent higher than what farmers paid in the *kharif* season this year. It can increase by 4 per cent by October; however, the price of muriate of potash (MoP) may increase by 26 per cent to Rs 880 per bag (50 kg) while nitrogen, phosphorous, and potash (NPK) grades may rise by 5-10 per cent to about Rs 960-1,180 per bag (50 kg).

"The rupee depreciation has ensured that the landed cost of raw material has gone up and prices can be increased in the range of 5 to 15 per cent, depending on the fertiliser,".

The industry had ample stock of fertiliser for the upcoming *rabi* season."Potash prices have just increased by $50 a tonne and even phosphate prices have increased by $103 a tonne in the last one quarter. Another rise is expected by October,". The calculation of increase in DAP by 12 per cent and MoP by 26 per cent is based on the premise that the impact of currency and raw material price increase will be passed on by the industry to the farmers to conserve margins.

"In case the government increases the per unit subsidy to fertilisers companies, then the price impact on farmers will be less". According to the agriculture ministry, the fertiliser industry will be able to meet the requirement for *rabi* season of 2018-19. The assessed requirement of urea is 162.74 lakh tonnes, 50.46 lakh tonnes for DAP, 17.28 lakh tonnes for MOP, 52.19 lakh tonne for NPK, and 29.80 lakh tonne for SSP.

India imports about a quarter of the estimated 310 lakh metric tonnes of urea consumed in the country every year. The entire requirement (about 30 lakh tonnes) of MoP is met from imports, since there is no resource of potash in India.

Buffer stocking of P and K fertilizers

A buffer stock of limited quantity of Di-Ammonium Phosphate (DAP) and Muriate of Potash (MOP) is being maintained at strategic locations to met emergent requirements. These stocks are in the nature of rolling stocks and are replenished when depleted. Besides meeting emergent needs, the Buffer Stock also helps to meet requirements of States, which have low demand and hence sometimes find it difficult to induce suppliers to move fertilizers in small quantities.

Fertiliser quality control

Fertilizer is the most critical and costly input for sustaining agricultural production and ensuring food security of the country. The Government ensures the quality of fertilizers through Fertilizer Control Order (FCO), issued under Essential Commodities Act, 1955 to regulate, the trade, price, quality and distribution of fertilizers in the country. The State Governments are the enforcement agencies for implementation of the provisions of FCO, 1985. The Order strictly prohibits the manufacture, import and sale of any fertilizer, which does not meet prescribe standards. The FCO provides for compulsory registration of fertilizer manufacturers, importers and dealers, specifications of all fertilizes manufactured/imported and sold in the country, regulation on manufacture of fertilizer mixtures, packing and marking on the fertilizer bags, appointment of enforcement agencies, setting up of quality control laboratories and prohibition on manufacture/import and sale of non-standard/spurious/adulterated fertilizers. To check the quality of fertilizes sold in the country, there are 71 Fertilizer Quality Control Laboratories at present, which includes 4 Central Government laboratories namely; Central Fertilizer Quality Control & Training Institute, Faridabad and is three Regional Laboratories at Chennai, Navi Mumbai and Kalyani (Near Kolkata). These laboratories have a total annual analyzing capacity of 1.31 lakh samples. The analytical capacity and the number of samples analyzed and found non standard during the last 5 years are as follow:

The percentage of non standard samples on all India basis is hovering around 6% samples during last 5 years. However, there is a large variation from state to state.

Fertilizer control order, 1985

Non-standard samples during the 2004-05 to 2008-09

Year	Number of labs	Annual capacity	Number of samples	Non standard samples (%)
2004-05	67	120315	124730	6.0
2005-06	67	120415	122488	6.0
2006-07	68	125480	129250	6.0
2007-08	68	124778	129331	6.2
2008-09	71	124730	131565	3.9

Analytical annual capacity and the number of samples analysed during 2004-05 to 2008-09

In pursuance of Clause 20 A of FCO, 1985, fertilizer companies namely, Chambal fertilizer Ltd., Sri Ram Fertilizer Ltd., Tata Chemicals, National Fertilizers Ltd and KRIBHCO have been permitted to manufacture neem coated urea as provisional fertilizer for commercial trial upto 3rd August, 2009. In order to encourage use of micronutrients, NRK complex fertilizer 15:15:15:9(S) have been included in FCO, 1985. Two new NPK 100 per cent water soluble fertilizers 28:28:0 and 24:24:0 have also been notified under clause 20A provisionally. The specification of triple super phosphate has been revised under clause 20 of FCO, 1985 to enable easy availability of phosphatic fertilizer. 14 new grades of soil specific and crop specific customised fertilizers have been notified so far.

Promotion of integrated nutrient management

The Government is promoting soil test-based balanced and judicious use of chemical fertilizers, biofertilizers and locally available organic manures like farmyard manure, compost, nadep compost, vermi compost and green manure to maintain soil health and its productivity. A Centrally Sponsored Scheme, "National Project on Management of Soil Health and Fertility" (NPMSF) has been approved during 2008-09 with an outlay of Rs.429.85 crores for the remaining period of XI Plan. The two existing schemes namely: i) centrally sponsored scheme of Balanced and Integrated Use of Fertilizers and ii) Central Sector Scheme "Strengthening of Central Fertilizer and Quality Control & Training Institutes and its Regional Labs. have been subsumed in the new scheme w.e.f. 1.4.2009. The components of the new scheme include setting up of 500 new soil testing laboratories, strengthening of the existing 315 soil testing laboratories, setting up of 250 mobile soil testing laboratories, promotion of organic manures, soil amendments and distribution of micro nutrients, setting up of 20 new fertilizers quality control laboratories and strengthening of 63 existing fertilizer quality control laboratories during 11th Plan.

There are 686 Soil Testing Laboratories (2007-08) in India. These include 560 static and 126 mobile Laboratories maintained by State Governments and fertilizer industry with an annual analyzing capacity of about 7 million soil samples. During 2008-09, an amount of Rs.16.63 crore has been released under NPMSF for 42 new Static Soil Testing Laboratories (STLs), 44 Mobile Soil Testing Laboratories (MSTLs), strengthening of 39 existing soil testing laboratories, 2 new fertilizer quality control laboratories and strengthening of 19 existing fertilizers quality control laboratories (FQCLs) in 16 States.

Goal of integrated nutrient management

Sustainable agricultural production incorporates the idea that natural resources should be used to generate increased output and incomes, especially for low-income groups, without depleting the natural resource base. In this context, INM maintains soils as storehouses of plant nutrients that are essential for vegetative growth. INM's goal is to integrate the use of all natural and man-made sources of plant nutrients, so that crop productivity increases in an efficient and environmentally benign manner, without sacrificing soil productivity of future generations. INM relies on a number of factors, including appropriate nutrient application and conservation and the transfer of knowledge about INM practices to farmers and researchers.

Plant nutrient application

Balanced application of appropriate fertilizers is a major component of INM. Fertilizers need to be applied at the level required for optimal crop growth based on crop requirements and agro climatic considerations. At the same time, negative externalities should be minimized. Over application of fertilizers, while inexpensive for some farmers in developed countries, induces neither substantially greater crop nutrient uptake nor significantly higher yields. Rather, excessive nutrient applications are economically wasteful and can damage the environment. Under application, on the other hand, can retard crop growth and lower yields in the short term, and in the long term jeopardize sustainability through soil mining and erosion. The wrong kind of nutrient application can be wasteful as well. In Ngados, East Java, for example, the application of more than 1,000 kilograms per hectare of chemical fertilizer could not prevent potato crop yields from declining. Yields on these fields decreased more than 50 percent in comparison with yields on fields where improved soil management techniques were used and green manure was applied. The correction of nutrient imbalances can have a dramatic effect on yields. In Kenya the application of nitrogenous fertilizer on nitrogen-poor soils increased maize yields from 4.5 to 6.3 tons per hectare, while application of less appropriate phosphate fertilizers increased yields to only 4.7 tons per hectare. Balanced fertilization should also include secondary nutrients and micronutrients, both of which are often most readily available from organic fertilizers such as animal and green manures. Lastly, balance is necessary for sustainability over time.

Nutrient conservation and uptake

Nutrient conservation in the soil is another critical component of INM. Soil conservation technologies prevent the physical loss of soil and nutrients through leaching and erosion and fall into three general categories. First, practices such as terracing, alley cropping, and low-till farming alter the local physical environment of the field and thereby prevent soil and nutrients from being carried away. Second, mulch application, cover crops, intercropping, and biological nitrogen fixation act as physical barriers to wind and water erosion and help to improve soil characteristics and structure. Lastly, organic manures such as animal and green manures also aid soil conservation by improving soil structure and replenishing secondary nutrients and micronutrients. Improved application and targeting of inorganic and organic fertilizer not only conserves nutrients in the soil, but makes nutrient uptake more efficient. Most crops make inefficient use of nitrogen. Often less than 50 percent of applied nitrogen is found in the harvest crop. In a particular case in Niger, only 20 percent of applied nitrogen remained in the harvest crop. Volatilization of ammonia into the atmosphere can account for a large share of the lost nitrogen. In flooded rice, for example, volatilization can cause 20 to 80 percent of nitrogen to be lost from fertilizer sources. These losses can be reduced, however. Deep placement of fertilizers in soil provides a physical barrier that traps ammonia. The use of inhibitors or urea coatings that slow the conversion of urea to ammonium can reduce the nutrient loss that occurs through leaching, runoff, and volatilization. With innovations of these kinds, better timing, and more concentrated fertilizers, nutrient uptake efficiency can be expected to improve by as much as 30 percent in the developed world and 20 percent in developing countries by the year 2020.

Untapped nutrient sources

If used appropriately, the recycling of organic waste from urban to rural areas is a potential, largely untapped, source of nutrients for farm and crop needs, especially on agricultural lands near urban centers. For example, environmentally undesirable wastewater has been used to irrigate fields and return nutrients and organic matter to the soil. Like organic manure, urban waste sludge is a source of primary nutrients, albeit a relatively poor source in comparison with commercial fertilizers. Stabilized municipal waste sludge typically contains about 3.3 percent nitrogen, 2.3 percent phosphorus, and 0.3 percent potassium, although some concentrations can reach as high as 10 percent nitrogen and 8 percent phosphorus on a dry weight basis. Actual nutrient content, however, varies widely and depends on the source of the waste. Urban waste also has a number of other benefits. Like other organic manures, it helps improve soil structure by adding organic matter to the soil. It is also a source of the secondary nutrients and micronutrients that are necessary in small quantities for proper plant growth. In addition, urban waste transforms material that would otherwise be slated for costly disposal into a useful farm product. Urban waste needs to be treated carefully because it may contain heavy metals, parasites, and other pathogens. The buildup of heavy metal concentrations in the soil can be cause for concern. While trace amounts of some heavy metals play a critical role in plant metabolism, excessive amounts have reduced crop yields and could be dangerous to public and grazing livestock. To minimize these risks the continuous application of urban waste needs to be monitored in order to ensure that heavy metal and overall nutrient concentrations do not reach toxic levels and do not damage the environment through leaching and eutrophication. Urban waste also contains organic compounds such as dyes, inks, pesticides, and solvents that are often found in commercial and industrial sludge. These pathogens have been shown to cause genetic damage, while others, such as bacteria, protozoa, and viruses can cause salmonellosis, amoebic dysentery, and infectious hepatitis. Untreated urban waste can put these pathogens in contact with fruits and vegetables. One option is to compost the sludge. Composting concentrates nutrients and helps to kill disease-causing organisms, slow the release of nitrogen that might otherwise percolate into groundwater, and eliminate aesthetically objectionable odors. Another option is to use ionizing radiation to kill pathogens in and on food without affecting taste. Despite some public concern about the safety of food irradiation, the 17 technique is likely to be adopted more fully in the future in order to protect public health, improve the shelf-life of food, and make it more feasible to apply treated, nutrient-rich urban waste to farmland. Currently, effective use of urban waste is hampered by its high water content, bulkiness, distance from rural areas, contamination with non decomposable household items, and high handling, storage, transport, and application costs. However, given the cost and the lack of availability of inorganic fertilizers in some areas, the relative abundance and benefit of urban waste as a soil amendment, and the rising cost of environmentally safe waste disposal, economies may make urban waste an appropriate fertilizer choice in areas where agricultural lands are near urban centers. Alternative sources of inorganic fertilizers will also be required in the future, particularly in those parts of Africa where the fertility of the soil needs to be rebuilt and high costs and supply constraints limit the application of fertilizer. Soil infertility (particularly phosphorus deficiency) in parts of semiarid West Africa, for example, limits crop production more than the lack of moisture. Phosphorus application of 15-20 kilograms per hectare can substantially improve crop yields. Medium-reactive and partially acidulated, less-reactive phosphate rock found in Mali, Niger, and Senegal are as agronomically effective as commercial superphosphate fertilizers. Low-cost technology needs to be developed so that phosphate fertilizer can be locally produced from these and other untouched or currently uneconomical phosphate rock reserves. Where phosphate deficiency is severe in Sub-Saharan Africa, government assistance in developing low cost technology and in applying phosphate fertilizer should be evaluated. Because phosphorus and phosphorus rock bind to the soil and thereby reduce the opportunity for leaching, and because these fertilizers release nutrients slowly over time, their use to preserve the long-term productive capacity of the land should be considered more of a capital investment than a subsidy or an environmentally undesirable government intervention.

Internal nutrient sources

Although new sources of nutrients can be developed, genetic engineering offers the potential for plants themselves to generate some of the nutrients they require through nitrogen fixation. In this process, rhizobium bacteria infect, invade, and draw energy from leguminous plants, and in return the bacteria convert and store atmospheric nitrogen in a form that the plant can use for growth. Besides helping the plants themselves, cereals grown in rotation with leguminous plants can absorb the nitrates released from the decaying roots and nodules of the leguminous plants. Experiments have shown that rice-legume rotations can result in a 30 percent reduction in chemical fertilizer use. Genetic research has begun to identify the genes responsible for such nitrogen fixation and assimilation. Further research offers the opportunity of altering or developing microorganisms that can fix nitrogen in non leguminous plants, such as cereals. As with leguminous plants, plant nitrogen needs could be partially met by the plant itself, such that farmers would then simply need to top up crops with inorganic nitrogen fertilizers. The task is considerable. Some 17 genes code the enzymes involved in nitrogen fixation. Since these genes, as well as the genes necessary for nodule formation, need to be transferred, the process is complex and its realization will be costly. Furthermore, because the amount of energy required to fix 150 kilograms of nitrogen per hectare could reduce wheat yields by 20-30 percent, an appropriate balance needs to be found between the nutrient-supply-enhancing benefits of nitrogen fixation and the potential reduction in yields.

Nutrient management

Improving and maintaining soil quality for enhancing and sustaining agricultural production is of utmost importance for India's food and nutritional security. Though India is a food surplus nation at present with about 231.5 million tonnes food grain production per annum, it will require about 4-5 million tonnes additional food grains each year if the trend in rising population persists. Due to increasing population pressure, the demand for food, feed, fodder, fibre, fuel, pulses and oilseed products is rapidly increasing. To meet the future demand we would need better planning and resource management as well as intensification of crop production. It is anticipated that in India in the year 2025, total food grain demand will reach 291 million tonnes comprising 109 million tonnes of rice, 91 million tonnes of wheat, 73 million tonnes of coarse grains and 15 million tonnes of pulses against the limitation of expansion of the cultivable land area. One of the alternatives to achieve this goal is to raise crop productivity through improved varieties and the matching production technology to sustain soil fertility and crop productivity in the future. Intensive cultivation and growing exhaustive crops have made the soil deficient in macro as well as in micronutrients. The success of any cropping system depends upon the appropriate management of resources including balanced use of manures and fertilizers.

The rice-wheat cropping system is the most widely adapted system in the world occupying 24 m ha area in India and China alone. In India, the system is covering about 10.5 m ha of land of which about 10 m ha is lying in Indo-Gangatic plains alone. Both the crops are grown in sequence and require high quantity of nutrients to harvest their potential yields. These two crops remove more than 600 kg of NPK nutrients per ha annually when grown under good management conditions. With the onset green revolution i.e introduction of HYV of rice and wheat and increased use of chemical fertilizers, irrigation water, plant protection measures farm implements and appropriate pricing policies these crops replaced the growing area of other cereals and pulses. Consequently, rice-wheat system became more popular even in traditional wheat, maize, pulse or sugar cane based areas of north India. The bumper crop of rice and wheat has been harvested during the period of Green revolution with urea fertilizer alone. Phosphorus and Potassium fertilizers were generally omitted by the farmers due to limited supply of P & K fertilizer or adequacy of nutrients in the soil and poor purchasing power of farmers. But continuous cropping with high yielding varieties and imbalanced use of fertilizer nutrients resulted in declining crop yields and lowering soil fertility over years. The system now has shown many serious weaknesses with plateauning yield, declining factor productivity, lowering ground water and deteriorating soil health. In intensively cropped irrigated area of rice-wheat having high productivity levels, have started showing signs of fatigue and productivity levels of rice and wheat are stagnant for last few years in Punjab, Haryana and UP. The deterioration in productivity is found to be associated with the new and emerging problems of micronutrients such as those of Zn and secondary nutrients like S. Further, imbalanced use of plant nutrients and the resultant mining for particular nutrients has conspicuous in N alone and other imbalanced fertilizer treatments.

The grain yield of rice recorded in 1984 with N, NP, NPK, NPK + Zn was 7.4, 7.1, 7.5 and 7.6 t/ha, respectively, did not differ significantly. But the differences among fertilizer treatments widen gradually and turned out to be statistically significant within few years of rice-wheat cropping and the average grain yield for three years (2004-06) of N, NP, NPK and NPK + Zn had been decline to the level of 3.3, 5.0, 5.3 and 5.4 t/ ha, respectively. The yield gap was still wider when the present day full fertilizer treatment i.e. the recommended NPK + FYM is compared with N or NK alone. The widening of this yield gap (2.8 t/ha as an average of that 2004, 2005 and 2006) was due to continuous use of imbalanced fertilizer as N alone (N alone yield is even less than unfertilized control treatment) resulting from depletion of soil nutrient supply, especially P. The Olsen's P which was 20 kg/ha in 1984 decreased to 12 kg/ha over two decades of rice-wheat cropping with N alone or control treatment. On the contrary, the yield decline was found to be arrested more than 6.0 t/ha with FYM enriched NPK or NPK + Zn treatments. However little declined in grain yield was also noticed in these full fertilizer treatments (NPK + FYM, NPK + Zn + FYM). The grain yield response (kg rough rice/ kg of nutrient applied) in1984 was quit high for N (30 kg rough rice/kg N), low for P (43 kg rough/ kg P) and K (6 kg rough rice/kg K). Now the response to N has come down to only 10 kg of rough rice/kg N applied. While response to P soared up and reached up to 120 kg rough rice/kg P applied. Response to K also showed an increasing trend but comparatively it was small. Response to Zn was variable but need to be applied foliar. Thus, phosphorus turned out to be the second most limiting nutrient element next to N. Deficiency of secondary and micro nutrients are also emerging. Management of Phosphorus under limited supply seems very important.

Balanced use of fertilizers in the country has received a serious setback due to decontrol and escalation of price of P and K fertilizers. Prior to decontrol, the NPK use ratio in the country was 5.4: 2.4: 1, which changed to 9.5: 3.2:1 after decontrol. Thus, farmers were used relatively more N fertilizer which impairing sustainability. Its consequence is not only low yield but also a drop in the efficiency of other inputs. Further, the use of P and K by the farmers is not strictly according to soils needs. It depends on the resources of the farmer. Resourceful farmers use N, P and K, relatively less resourceful N and P, but poor farmers used to apply only N. Thus, precise identification of deficient areas for various nutrients needs to be done and fertilizer used according.

Thus the present levels of fertilizer use efficiency are low and research efforts must be concentrated on increasing nutrient use efficiency. Fertilizer must be used where it is required. In addition all organic sources must be exploited to maximum as a mix of inorganic an organics gives the most sustainable system.

Integrated Nutrient Management Studies

In rice-wheat system

Thus due to increasing cost of chemical fertilizers and sustainability as an important issue, no single source of plant nutrients (chemical, organic or biological) can satisfy the need of all the essential elements that the crop needs for its growth and development. Therefore, different sources of plant nutrients have to be integrated to provide balanced fertilization to a cropping system on a sustainable basis to increase crop productivity and nutrient use efficiency. A judicious mix of chemical fertilizers; organic manures like F.Y.M., compost, animal waste and green manure; bio-fertilizers; crop residues and industrial and city waste has to be worked out for different cropping systems. Research finding indicates that integrated nutrient management with in cropping systems is ideal. However, proven research findings are still limited. Therefore, proper nutrient management for different cropping systems has to be evolved. It needs proper scientific approach based on long term soil fertility experiments coupled with sound soil plant analysis techniques. The various management strategies for the nutrient management on sustainable basis are discussed below

a. Under limited P availability it is suggested that is to be given to wheat crop instead of rice in rice-wheat crop. Application of K and Zn in intensive rice-wheat cropping is also becoming important for obtaining high yields. FYM addition at 5 t/ha on dry weight basis plus recommended NPK + foliar Zinc (0.5% solution of $ZnSO_4$) sustained soil fertility as well as crop productivity which need to be emphasizes farmers to apply.

b. At Pantnagar, another long-term fertility experiment (1988-2008) on the use of organic and inorganic nutrients alone or with their integration, has given valuable information's for food security, ecology as well as soil sustainability. The findings of investigation carried over a decade explains that S*esbania aculiata* or S*esbania rostrata*, green manuring grown with 40-60 kg P_2O_5 and 30-40 kg K_2O/ha alone or S*esbania* + Farmyard Manure (FYM) at 5 t/ha incorporated into soil before transplanting had great meaning for soil fertility and crop productivity. Both the treatments had produced yield significantly higher than recommended NPK (120 kgN; 60kg P_2O_5 and 40 kg K_2O/ha), even more than 150 per cent of recommended NPK ($N_{180}\ P_{25}\ K_{34}$) fertilizers.

c. There are other options which can be applied on sustainable basis for conserving farm resources. Promising among them is: incorporation of crop residues, straw and other wastes instead of burning, Use of Azolla, Neem Cake as such or coated urea (NCU), FYM, Vermi-compost etc. Addition of nutrients should be kept according to removals from the field. Nutrients' budgeting is to be made for individual plot at farm level. RWCS is not old and has been in practiced only during the past four-five decades. But the question of its sustainability has great concerned especially in states which are known for food security of the country. Indo-Gangetic plains still have potential to produce more than 10 t of Rice and Wheat grains per ha per year. However, it needs more attention and strong planning. Therefore, different sources of plant nutrients have to be integrated to provide balanced fertilization to a cropping system on a sustainable basis to increase crop productivity and nutrient use efficiency. A judicious mix of chemical fertilizers; organic manures like F.Y.M., compost, animal waste and green manure; bio-fertilizers; crop residues and industrial and city waste has to be worked out for different cropping systems.

A long term study on the "Impact of organic, integrated and inorganic nutrient management practices on soil health and crop response in different cropping systems" was carried out at MARS, Dharwad during 2007-08. The organic, inorganic and integrated strips of 14 m were laid out with the cropping systems as sub plots of 23 m length. The strips were separated by the hedge rows of subabul and maintained by pruning at a height of 1 m. Across the three main strips of organic, inorganic and integrated nutrient management practices, five cropping systems mainly soybean-wheat, groundnut-soybean, maize-chickpea, potato-chickpea and chilli + cotton were laid out. In organic nutrient management practices -100 per cent recommended dose of nitrogen (RDN) given through 1/3rd Farmyard manure (FYM), 1/3rd vermicompost (VC) and $1/3^{rd}$ Greenleaf manure (GLM). In integrated nutrient management practices 50 per cent RDN given through fertilizers and 50 per cent RDN through $1/3^{rd}$ FYM, $1/3^{rd}$ VC and $1/3^{rd}$ GLM. In inorganic nutrient management practices -100 per cent recommended chemical fertilizers only were added.

The organic carbon of soil was significantly affected by nutrient management practices. Under organic and integrated nutrient management practices, there was significant build up of organic carbon at the end of fourth year. Organic carbon contents of soil changes rapidly with addition of organic manures. On the other hand, under inorganic nutrient management practices, there was decline in organic carbon content from 0.49 per cent (initial value) to 0.44 per cent (end of fourth year). This might be due to use of fertilizers alone for a long period, which depletes the organic matter. The nutrient management practices had significant effect on soil fertility status. The available N, P_2O_5, K_2O and S showed build up in soil over the years under both organic and integrated nutrient management practices. The available nutrients were significantly high under integrated nutrient management practices compared to inorganic nutrient management practices, however was on par with organic nutrient management practices. Under inorganic nutrient management practices there was declining trend in available nutrient and at the end of fourth year. Soil organic matter affects soil fertility and the C and N mineralization capacities of the soil, which determines the availability of plant nutrients. Thus, soil productivity decreases as soil organic matter content decreases. Continuous application of manures increases the level of N, P, K, S, Ca and Mg in the soil over the years. Thus creating a reservoir of soil nutrients for several years after application, use of FYM, vermincompost and GLM attribute to the mineralization of N in soil and due to high enzyme activities in the soil amended with organic manures will increase the transformation of nutrients to available form. Role of FYM, vermicompost and green leaf manures in releasing N and improving N availability in soil was reported by several studies. Similarly, the P availability in soil increased due to use of organics. During decomposition of organic manure, various organic acids will be produced which solubilize phosphatase and other phosphate bearing minerals and thereby lowers the phosphate fixation and increase its availability. It has been reported that available phosphorus content increased due to addition of FYM over initial and control. The builds up of soil available K due to green manuring or FYM application is due to addition of K applied throughout the solubilizing action of certain organic acids produced during FYM decomposition and its greater capacity to hold K in the available form. The builds up of sulphate S content in soil due to use of FYM either alone or in combination with NPK. This improvement in soil fertility was attributed to addition of FYM and other organics which stimulated the growth and activity of microorganisms. They participate in the biological cycling of elements and transformation of the mineral compounds and thus increases the availability of nutrients in soil.

The available nutrients in soil had significant effect of cropping systems. At the end of third year and during fourth year (after harvest of *kharif* and *rabi* crops), these available nutrients differed significantly with cropping systems. Wherever the legume crop was grown both in *kharif* and *rabi* seasons recorded significantly higher available nutrient in soil than with non-legume crops. This is due to nodulation of legume crops which fixes atmospheric N and N content in soil increases. As there is synergistic relation of N with P, K and S, this helps in increasing the nutrient content in soil irrespective of nutrient management practices. The legume cropping helped to increase the available N, P_2O_5 and K_2O content of the soil. On the other hand, chilli+cotton system showed slight build up of nutrients only under integrated and organic nutrient management practices. This ascribed to addition of organic and slow release of nutrients from the organics. Whereas, under inorganic nutrient management practices, nutrient status after fourth year of Long term fertilizer experiment (LTFE) was drastically reduced as compared to the nutrient status at the end of third year of LTFE and at initial LTFE. This is ascribed to exhausting nature of the crop and lack of nutrient source for their replenishment in the soil. Chilli+cotton being a long duration crops, cotton having deep roots and moreover grown as intercrop, there might have be competition for the nutrients which lead to more uptake of release nutrients and their less content in soil.

Among the cropping systems during *kharif*, maize recorded significantly higher uptake of N, P and K. Maize being an exhaustive crop will remove maximum nutrients from soil and also its higher biomass yield will also contribute to higher uptake of nutrients. Maize being an exhaustive crop responded well to the application of FYM and in situ green manuring. Considerable increase in NPK uptake by maize was observed with the conjunctive use of FYM and NPK fertilizers. However, with respect to all *kharif* and *rabi* crops, they responded well to integrated nutrient management practice followed by organic nutrient management practices. The least response was with inorganic nutrient management practices.

The organic nutrient management practices recorded significantly higher equivalent yield of soybean, groundnut and chilli. This was followed by integrated nutrient management practices, whereas equivalent yield of maize and potato were significantly influenced by integrated nutrient management practices and was superior over inorganic nutrient management. However, it was on par with organic nutrient management practices. Due to balanced supply of nutrients to the crops throughout the crop growth period as FYM, vermicompost and GLM undergo decomposition during which series of nutrient transformation takes place and helps in their higher availability to the crops. Higher uptake of nutrients by the crops will result in higher yield. As all the essential elements are released by the organic manures, the released essential elements play a vital functional role in crops and thus ultimately increase the yield with balanced nutrition. As there was build up of soil fertility status in terms of N fractions, available N, P2O5, K2O and S and DTPA extractable micronutrients in organic and integrated nutrient management practices as compared to inorganic nutrient management practices, which might have resulted in better uptake by the crops and responses well in terms of yield. The incorporation of FYM or crop residue in conjunction with inorganic fertilizer significantly enhanced the sustainability and stability with respect to productivity of soybean-wheat system. Application of FYM along with N and P fertilizers resulted in the highest total uptake of N, P and K by soybean which was due to higher availability of these nutrients in soil reservoir besides the additional quantity of nutrients supplied by FYM. Similarly, soybean-wheat was also found as efficient cropping system in terms of productivity and profitability. This was also revealed that the crop removal of N, P and K was highest when organic manures (FYM, vermicompost, crop residues and green manuring) and inorganic fertilizers applied in equal proportion to supply the recommended level of N (150 kg/ha) to maize. This was ascribed to continuous supply of N, P and K throughout the crop growth period as the nutrient from inorganic sources were available to the crop in the early stages and in the later stages of the crop growth, the slow and continuous release of nutrients from the organic source made available. Among different organic manures, application of FYM @ 10 t/ha increased the chilli fruit yield by 19.94 per cent over no FYM treatment. Similarly, application of mulches (gliricidia lopping or organic residues) + organic (FYM) with organic solutions recorded an equivalent yield (860, 830 and 818 kg/ha, respectively) as that of check treatment. The equivalent yield significantly differed with cropping systems. Potato-chickpea system yield was significantly high, followed by maize-chickpea, groundnut-sorghum, soybean-wheat and chilli+cotton . This highest yield in potato-chickpea might be due to benefit of inclusion of legume in the cropping systems and higher market price of the produce of the system. Potato being a vegetable of higher demand its market price might have contributed to higher equivalent yield. It was also highlighted the benefits of inclusion of groundnut (legume crop) in potato-pearl millet system. The higher equivalent yield was maize-chickpea; this was again due to legume-cereal system. The incorporation of FYM or crop residue in conjunction with inorganic fertilizer significantly enhanced the sustainability and stability with respect to productivity of soybean-wheat system. The gross and net returns of the cropping system of 2007-08 were significantly high under organic nutrient management practices when compared to inorganic nutrient management practices. This is ascribed to higher yield of the crops in organic nutrient management practices. Whereas, lower yield of crops in inorganic nutrient management practices might have resulted in significantly lower gross and net returns of the cropping system of 2007-08. Similarly, the B:C ratio of organic nutrient management practices was significantly high, which is be due to higher gross returns and lower cost of cultivation of organic nutrient management practices. The groundnut-sorghum cropping system recorded significantly higher gross and net returns when compared to other cropping systems. This is due to higher yield of the crops. Whereas, the B:C ratio of chilli + cotton was significantly high (3.35) followed by groundnut-sorghum (3.23) and was superior over other systems. This is ascribed to higher market price of chilli and cotton. There was an interaction effect on economics of cropping systems of 2007-08. The groundnut-sorghum cropping system recorded significantly higher gross and net returns under organic nutrient management practices. This might be due to higher market price of groundnut and higher yield of the groundnut and sorghum crops when compared to other combinations. On the other hand, chilli + cotton cropping system recorded significantly higher B:C ratio of 4.05 under organic nutrient management practice, which was superior over rest of the combinations. This is due to lower cost of cultivation, higher gross returns and higher market price of chilli and cotton. It is concluded that integrated practice followed by organic nutrient management practices performed better with respect to soil fertility and crop productivity than inorganic nutrient management practice. Among the cropping systems, legume based system responded better under organic and integrated nutrient management practices.

Thus application of organic manures resulted in significantly higher organic carbon. Whereas integrated application of manure and fertilizers resulted in significantly higher available N, P_2O_5, K_2O and S than chemical fertilizers alone. Significantly higher uptake of N, P and K by *kharif* and *rabi* crops was recorded in integrated practice compared to inorganic nutrient management practice. The available N, P_2O_5, K_2O and S were significantly higher in legume based cropping systems during both the seasons of the study than non-legume system. The B:C ratio of chilli + cotton was found to be significantly higher (3.35) than other cropping systems. The organic nutrient management recorded significantly higher B:C ratio of 2.99 than inorganic (2.38) and integrated (2.59) nutrient management practices.

Integrated nutrient management module for rice-wheat cropping system on an Inceptisol was developed through field experimentation for eight consecutive crop seasons (1999–2003). The treatments consisted of FYM, vermicompost, green manure, *Azotobacter*, phosphate solubilizing bacteria (PSB), blue-green algae (BGA), rice residue incorporation and NPK fertilizers. Significantly higher yields to the tune of 4.3 t/ha for rice and 4.0 t/ha for wheat were recorded when rice-wheat were grown after green manuring of *dhaincha in-situ* or application of FYM (10 t/ha /year) or vermicompost (5 t/ha /year) in *kharif* season along with reduced quantity of fertilizers per hectare per crop (30–90 kg N, 13–20 kg P and 37 kg K) accompanied by microbial cultures (*Azotobacter*, BGA and PSB) as compared to the yield (4.0 – 4.1 t/ha) with recommended dose of NPK (120–26–50)/ha/ crop. Reduction to the tune of 25 per cent in recommended dose of N, P and K fertilizers (30 kg N, 6.5 kg P and 13 kg K/ha/crop) could be made with the application of FYM or vermicompost or green manuring alone without decrease in yield of rice and wheat. Quantity of N and P fertilizers could be further reduced to half of the recommended dose (i.e. 60 kg N and 13 kg P/ha/crop) with the application of *Azotobacter* and PSB along with FYM or vermicompost or green manuring. Application of BGA in paddy and incorporation of rice residue before sowing of wheat further reduced the recommended dose of N fertilizer by 30 kg/ha for both rice and wheat. Cultivation of rice and wheat on the recommended dose of NPK fertilizers alone decreased the organic carbon, available P and K content of the soil. Application of FYM, vermicompost, green manuring, and rice residue incorporation alone or in combination with biofertilizers supplemented by NPK fertilizers improved the soil fertility besides maintaining higher sustainable productivity.

20. Concept And Function Of Integrated Pest Management

PEST PROBLEMS IN AGRICULTURE

Food plants damage by insects and pests are listed below:

i. Agriculture insect and mites= >10,000 Species

ii. Diseases = 1,00,000 species

iii. Nematodes = 1000 Species

iv. Weeds = 30, 000 species

Current global losses (%) by different pests in major crops

Crop	Losses (%)				
	Weeds	Animal pests	Pathogen	Viruses	Total
Wheat	7.7	7.9	10.2	2.4	28.2
Rice	10.2	15.1	10.8	1.4	37.4
Maize	10.5	9.6	8.5	2.7	31.2
Potato	8.3	10.9	14.5	6.6	40.3
Soybean	7.5	8.8	8.9	1.2	26.3
Cotton	8.6	12.3	7.2	0.7	28.8
Average	8.8	10.8	10.0	2.5	32.1

Estimated monetary losses by pests

Estimated losses due to insect pest is about 90.5 billion dollars in damage to 8 principal food crops and cash crops rice, wheat, barley, cotton, maize, potato, soybean and coffee.

Crop losses due to insect-pests in India

Crop	Traditional agriculture (%)	Modern agriculture (%)
Wheat	3	5
Rice	10	25
Maize	5	25
Cotton non-Bt	18	50
Bt cotton	-	10
Sugarcane	10	20

Various problems associated with use of pesticides

1. Pesticide poisoning
2. Pesticide residue
3. Non target organisms
4. Pest resurgence
5. Pest resistance

Modern Agriculture and Pest Problems

IPM refers to an ecological approach in Pesticide management (PM) in which all the available necessary techniques are consolidated in a unified program, so that pest populations can be managed in such a way that economic damage is avoided and adverse side effects are minimized.

Concept of Integrated Pest Management

Integrated pest management (IPM) is a decision support system for the selection and use of pest control tactics, singly or harmoniously coordinated into a management strategy, based on cost/benefit analyses that take into account the interests of and impacts on producers, society and the environment.

Need bio-intensive integrated pest management

» Integrated pest management originated as a reaction to the over use of insecticides and is now the dominant paradigm that guides development of insect pest management technologies all over the world.

» Most IPM programmes still include economic threshold level based application of chemical insecticides as a major input.

» Ample evidence available w.r.t. adverse effects of these synthetic chemicals on the ecosystem, the environment and on human and animal health.

» Therefore, there is an urgent need to curtail and ultimately eliminate the use of chemical pesticides in IPM programmes.

Biointensive IPM: A systems approach to pest management based on an understanding of pest ecology. It begins with steps to accurately diagnose the nature and source of pest problems, and then relies on a range of preventive tactics and biological controls to keep pest populations with in acceptable limits. Reduced-risk pesticides are used if other tactics have not been adequately effective, as a last resort, and with care to minimize risks.

Biointensive IPM Options

1. **Proactive**
    a. Cultural control
    b. Host plant resistance
    c. Transgenic crops

2. **Reactive**
    a. Mechanical control
    b. Biological control
    c. Reduced risk pesticides

Cultural control: The deliberate alteration of the production system, either the cropping system itself or specific crop production practices, to reduce pest populations or avoid pest injury to crops.

The functional mechanisms include:

- » Impediments to pest colonization of the crop.
- » Creation of adverse biotic conditions that reduce survival of individuals or population of the pest
- » Modifications of the crop in such a way that pest infestation results in reduced injury to the crop
- » Enhancement of natural enemies by manipulating the environment.

Cultural control practices

- » Planting time
- » Plant diversity: Trap, barrier and inter crops
- » Tillage
- » Crop sanitation
- » Fertilizer as well as water management
- » Harvesting time and procedures
- » Crop rotation
- » Manipulation of carryover sources

With the awareness of environmental problems, exploitation of different cultural and mechanical practices has been advocated as a vital approach to curb pest populations.

1. **Planting time**
   a. Cotton: Sowing till mid-May lower incidence of PBW, ABW, and Whitefly.
   b. Rapeseed Mustard: Early sown crop suffers lower incidence of Mustard aphid.
   c. Sorghum: Timely (June) sown crop shows lower incidence of shoot fly; Spring and late sown heavily attacked.

2. **Seed rate and spacing**

   a. Fodder Sorghum: Sorghum higher seed rate by 10 per cent compensates for seedling mortality due to shoot fly.

   b. Rice: Closer spacing increases the incidence of planthoppers & leaf folder.

   c. Cotton: Closer spacing increases the incidence of bollworms & jassid.

3. **Tillage**

   a. Rice: Deep poughing after rice helpful in reducing stem borer incidence

   b. Cotton: After cotton, deep ploughing reduces the over wintering population of *Helicoverpaarmigera.*

   c. Sunflower: Deep ploughing before sowing of sunflower reduces the incidence of cutworms.

   d. Groundnut: Deep ploughing in May–June exposes white grub beetles.

4. **Nutrient management**

   a. High N favours population build-up Leaf folder, BPH, WBPH, stemborer (rice) cutworm (wheat) leafhopper (cotton).

   b. High K has depressant effect on insect development Thrips, leaffolder (rice) Jassid (bhindi).

   c. In sugarcane, application of well rotten FYM helps to lower incidence of termites.

5. **Water management**

Required amount at appropriate time is crucial for crop vis-à-vis insects (e.g. Draining of water from fields reduces BPH incidence).

6. **Crop rotation**

Growing rice after groundnut in soil in puddle condition eliminates white grub.

7. **Sanitation**

   a. In sugarcane, removal of stubble and debris of previous crop from field helps to lower incidence of termites.

b. Removal of weed *itsit* in cotton reduces *S.litura* incidence.

8. **Trap crops**

   a. Indian mustard for diamond back moth (DBM) in Cole crops.

   b. Napier grass and Napier millet for stem borer in maize and sorghum.

   c. Marigold for fruit borer in tomato.

9. **Intercrops and crop mixtures**

   a. Ryegrass-berseem mixture lowers the incidence of *T.orichalcea* and *H.armigera*.

   b. A row of sesamum around cowpea field attracts oviposition by hairy caterpillar.

Mechanical Control

   a. Collection and destruction of egg masses of top borer in first and second brood during March and May in sugarcane.

   b. Hand picking of hairy caterpillars, leaf rollers, tobacco caterpillar, cabbage butterfly, white grubs etc.

   c. Clipping of aphid infested twigs in *raya*.

   d. Light traps for mass trapping and management of hairy caterpillars, leaf folder in rice.

   e. Uprooting of infested plants having gregarious phases practiced in various cops to eliminate hairy caterpillars.

   f. Locust invasions-managed by using fire flames and by drum beating during swarming periods.

Host Plant Resistance (HPR)

The production of crop plants with heritable arthropod resistant traits has been recognized for more than 100 years.

Advantages of HPR

   a. Specificity

   b. Cumulative effect

 c. Ecofriendly

 d. Persistent

 e. Ease of adoption

 f. Compatibility

 g. Eliminates or reduces use of pesticides

Economic value of HPR

 a. The multiple insect resistant rice cultivars IR-36 provided about $1billion of additional annual income to Asian growers every year for 20years.

 b. Insect resistant cultivars of alfalfa, corn and beet returned $320 for each $ invested in research.

 c. Wheat cultivars resistant to Hessian fly provided a 120-fold greater return on investment than pesticides.

Disadvantages of plant resistance

 a. Not available for all pests

 b. Preventative in nature

 c. Level of control may not be sufficient

 d. May not be agronomically acceptable

Transgenic crops

 a. Insect pest resistance

 b. Disease resistance

 c. Herbicide resistance

 d. Nutritional improvement

 e. Abiotic stress resistance

 f. Production of edible vaccines

Transgenic crops carrying Bt. genes for insect resistance

Crop	Gene	Target insect pest
Chickpea	*cry1Ac*	*Helicoverpa armigera*
Cotton	*cry!Ab, cry1Ac*	*Helicoverpa armigera, H. zea*
	cry2Ab	*Heliothisvirescens, Pectinophoragossypiella, S. Exigua, T.ni*
Groundnut	*cry1Ac*	*Elasmopalpuslignosellus*
Maize	*cry1Ab*	*Ostrinia nubilalis, Chilo partellus*
	cry1Ac, cry1F	*Busseola fusca, H. Zea, S.frugiperda*
	cry9C, cry3Bb, cry34Ab/cry2Aa	*Diabroticavirgifera*
Rice	*cry1Ab, cry1Ac, cry2Aa*	*Chilo suppressalis, Cnaphalocrocismedinalis, Scipophagaincertulas*

Induced Resistance

Defensive traits that are only expressed in response to herbivory

a. The plant hormone jasmonic acid and related signalling compounds (collectively called jasmonates) activated offense responses.

b. Insect oral secretions combine with plant factors to elicit defence responses in host plants. For example, the fatty-amino acid conjugates found in the oral secretion so flep. Larvae have their fatty acid and A A moieties derived from the insect and the host plant, respectively.

c. Insects from different feeding guilds tend to elicit distinct pattern so f gene expression where as attackers from the same guild evoke very similar responses.

Biologial control

Biological control or biocontrol refers to the use of living organisms to suppress the population density or impact of specific pest organisms, making it less abundant or less damaging than it would otherwise be:

a. Macrobial Control- Birds and Arthropods (Predators & Parasitoids).

b. Microbial control- Entomopathogens (Bacteria, Viruses, Fungi and Nematodes)

Some facts about Biocontrol

a. Classical BC is applied on about 350 million ha and has a phenomenal B:C ratio of 20-500:1.

b. Augmentative BC is applied on 16 million ha and has a B:C ratio of 2-5:1, which is similar to chemical pest control.

c. More than 150 species of NEs including predators, parasitoids and pathogens are currently commercially available.

d. Thousands of NE species have not yet been tested for usefulness in biocontrol programs.

Predators

a. Predaceous insects of potential use in biocontrol are found in Dermaptera, Mantodea, Hemiptera, Thysanoptera, Coleoptera, Hymenoptera and Diptera.

b. More than 30 families of insects are predaceous and of these the Anthocoridae (Piratebugs), Nabidae, Reduviidae, Geocoridae, Carobidae, Coccinellidae (ladybirdbeetles), Nitidulidae, Staphylinidae, Chrysopidae (lacewings), Formicidae, Cecidomyiidae and Syrphidae are commonly important in crops.

Insectivorous birds are Cattle egret, Black drongo, Predatory birds viz., Common myna, Rosy paster, Sparrow, Black drongo- birds perches at 50/ha.

Semiochemicals

» The chemicals that trigger various behavioral responses and mediate the interactions between the organisms are called "Semiochemicals".

"semeon" =a mark or signal.

Pheromones

» A substance that issecreted by an organism to the outside and causes a specific reaction in receiving organism of the same species. (sexpheromones,alarm,aggregationetc.).

Allelochemicals

» A chemical substance that mediate interactions between different species.

Application of Pheromones in IPM

Pheromones of > 800 species of insect pests have been identified, synthesized and formulated.

   a. Detecting and monitoring insect pest populations

   b. Mass trapping

   c. Mating Disruption

Pheromones production and use

   a. Three type of pheromone traps *viz.*, sticky, liquid and dry traps are used in pest management programmes.

   b. The most widely used traps are delta traps for sucking pests in cotton and dry traps for lepidopterous pests in many crops.

   c. In India the pheromones have been used in about 15 species of insects.

   d. The production and supply of pheromone traps is mostly from the private sectors such as Bio-control Research Laboratory and Biopest Management, Bangalore.

Drawbacks of Pheromones

   a. Specificity in action

   b. Difficulties in mass production

   c. Photo degradability

   d. Timing of application

   e. Low persistence

Use of allelochemicals in IPM

   a. There are numerous avenues for use of allelochemicals in IPM programs either

      i. Directly against the insect pests or

      ii. Through enhancement of natural enemies.

   b. Host or prey selection behavior of entomophagous insects is chemically mediated.

- c. Volatiles from herbivorous insects and their host plants may serve as reliable cues for biocontrol agents in search of suitable hosts.

- » Biointensive IPM is considered the desirable path to sustainability in agriculture.

- » There is now wide array of techniques available to replace the use of conventional chemical insecticides in IPM programmes.

- » The challenge before applied entomologists is to develop, validate and disseminate the site-specific biointensive IPM technologies to the farmers.

- » There is also a need to strengthen research in strategic are as like genetically improved bioagents/biopesticides and pest-resistant transgenics.

- » Utilization of molecular approaches will go along way in overcoming the major limitations of biological pest control.

- » With these advancements, biointensive IPM appears poised to play a greater role as an environmentally benign alternative to chemical insecticide based IPM for sustainable insect pest management in future.

Challenges of Established and Emerging Technologies

Green Technology (GT) is environmental healing technology that reduces environmental damages created by the products and technologies for peoples' conveniences. It is believed that GT promises to augment farm profitability while reducing environmental degradation and conserving natural resources. Green technology, also known as sustainable technology, takes into account the long- and short-term impact something has on the environment. Green products are by definition, environmentally friendly. Eco-friendly technologies are recycled, recyclable and/or biodegradable content, plant-based materials, reduction of polluting substances, reduction of greenhouse gas emissions, renewable energy, energy-efficiency, multi-functionality and low-impact manufacturing. Green Technology for Agriculture and Food: The green technology should be efficient, practical, cost effective and free from pollution. The sustainability factor should be looked at the ability of the agricultural land to maintain acceptable levels of production over a long period of time, without degrading the environment. Some define sustainability as the maintenance of productivity under stress conditions. Agricultural sustainability in this context should seek to maximize food production within constraints of profitability. The specific Challenges for green technology in agriculture are Identifying appropriate technology suitable for income generation through sustainable agriculture i.e., ecological agriculture, rural renewable energy, Examining the impact and implications of national policies for making recommendations for the extension of appropriate technology, Diagnosing policy-level impact of such green technology on rural income generation under the sustainable agriculture development framework, Reviewing the challenges and available policy options for the adoption of GT sustainable agriculture integrates three main goals- environmental health, economic profitability, and social and economic equity. Some of the opportunities towards sustainable agriculture are Integrated nutrient management (IPM), Integrated Pest Management (IPM), Farming system, Site specific nutrient management, Organic farming, Carbon farming, Traditional crop rotations, Conservation agriculture, Conservation of natural resources, Crop diversification, Microirrigation, Precision farming, Rotational Grazing, Water quality/wetlands, Cover crops, Crop/ landscape diversity, Nutrient management, Agro-forestry, Agrometeorology, Marketing of green products and reducing environmental degradation in agricultural processes.. The revolution in the challenge is to make applied technology competitive and sustainable nanotechnology innovations in agriculture are expected to solve the problems in the food sector and maximize productivity in agriculture. There is an ever-increasing demand for food and adequate nutrition and nanotechnology will provide solutions through precision farming using nanosensors, nano-pesticides, and inexpensive decentralized water purification. A more advanced nanotechnology solution will be plant gene therapy; creating pest resistant, high yield crops that require less water etc. which also supports a sustainable environment.

Reference

A. Aladjadjiyan 1992. Lessons from Denmark and Austria on the Energy Valorization of Biomass (Contract No: JOU2-CT92-0212, Coordinator for Bulgaria); European Commission: Brussels, Belgium, (1992).

A. Allen, A. Voiland 2017. NASA Earth Observatory, Haze Blankets Northern India (2017). Available online: https://earthobservatory.nasa.gov/images/91240/haze-blankets-northern-india (accessed on 11 July 2018).

A. A. Maynard 2000. Compost: The process and research. The Connecticut agricultural experiment station. Bulletin, (2000): 966, 13.

About Us - Learn More About The Nature Conservancy. Nature.org. 2011-02-23. Retrieved 2011-05 01.

A. Dobermann, P. F. White 1999. "Strategies for nutrient management in irrigated and rainfed lowland rice systems." In Resource management in rice systems: nutrients (1999), pp 1-26.

A. Dobermann, P.F. White 1999. Nutr. Cycling Agroecosyst., 53 (1999):1–18.

Aithal, Sreeramana and Aithal, Shubhrajyotsna 2016. Opportunities & Challenges for Green Technologies in 21st Century. Munich Personal RePEc Archive (MPRA), (2016).

A. J. Ward, P. J. Hobbs, P.J. Holliman, D. L. Jones 1999. Review: Optimization of the anaerobic digestion of agricultural resources. Bioresour Technol., 79(1999): 28–40.

A.K. Shukla, J. K. Ladha, V.K. Singh, B.S. Dwivedi, V. Balasubramanian, R. K. Gupta, S. K. Sharma, Y. Singh, H. Pathak, P. S. Pandey, A. T. Padre, R. L. Yadav 2004. Calibrating the leaf color chart for nitrogen management in different genotypes of rice and wheat in a systems perspective. Agronomy Journal, 96 (2004): 1606–1621.

A. K. Y. N. Aiyer 1942. Mixed Cropping in India, Indian J. Agr. Sci., 19 (1942): 439.

Albuquerque Bernalillo County Water Utility Authority (2009-02-06). "Xeriscape Rebates". Albuquerque, NM. Retrieved 2010-02-02.

Anonymous 2005. India 2005A reference annual. Publ. Division, Ministry of Information and Broadcasting, Gov. of India (2005).

Amy Vickers 2002. Water Use and Conservation. Amherst, MA: water plow Press (2002), pp 434. ISBN 1-931579-07-5.

A. Sarkar, R. L. Yadav, B. Gangwar, P. C. Bhatia 1999. *Crop residues in India*. Modipuram: Project Directorate for Cropping System Research. Tech. Bull (1999).

A. Sinha 2015. Four New Missions to Boost Response to Climate Change; The Indian Express: New Delhi, India (2015).

A. Urja 2016. Generation of Green Energy from Paddy Straw, a Novel Initiative in Sustainable Agriculture Green Energy (2016). Available online: https://mnre.gov.in/file-manager/akshay urja/june-2016/30-33.pdf (accessed on 10 June 2018).

A. Young 1990. Agroforestry environment and sustainability. Outlook on Agriculture 19 (3) (1990): 156-160.

A.Bationo, A. U. Mokwunye 1991. Alleviating soil fertility constraints to increased crop production in West Africa: The experience of the Sahel. In alleviating soil fertility constraints to increased crop production in West Africa, ed. A. Uzo Mokwunye. Dordrecht: Kluwer Academic Publishers (1991).

A. Clark 1998. Managing Cover Crops Profitably, Sustainable Agriculture Network cropping system of the Punjab through zero-tillage.Pakistan J. Agric. Res., 14 (1998): 8-11.

A. Demirbas 2009. Political, economic and environmental impacts of biofuels: A review. Applied Energy 86: S108–S117. doi:10.1016/j.apenergy.2009.04.036. Edit

A.F. Collings, C. Critchley (eds) 2005. Artificial Photosynthesis- from Basic Biology to Industrial Application (Wiley-VCH Weinheim 2005) p ix.

Agricultural Environmental Managemnet, Water Quality Information Center, U.S. Department of Agriculture.

A.Goeppert, M. Czaun, G.S. Prakash, G. A. Olah 2012. Air as the renewable carbon source of the future: an overview of CO_2 capture from the atmosphere. Energy & Environmental Science, 5 (7) (2012): 7833-7853. doi:10.1039/C2EE21586A. Retrieved September 7, 2012. (Review).

Agroforestry Frequently Asked Questions. United States, Department of Agriculture. Retrieved 19 February 2014.

Agroforestry. Science Publishers Inc., Enfield, NH (2002), pp356.

Ag Water Management Summit, NRCS. Nrcs.usda.gov. Retrieved 2013-10-31.

A. H. Khan, M. Ashraf, P. Chandna, G. Singh, M.L. Jat, R. Misra, R. Mann, A. Anjum, M. Gill, B.S. Aithal Sreeramana and Shubhrajyotsna Aithal 2016. Opportunities & Challenges for Green Technologies in 21st Century. Online athttps://mpra.ub.uni-muenchen.de/73661/MPRA Paper No. 73661, posted 13 September (2016) 11:25 UTC.

A.Janaiah, Mahabub Hossain 2003. Farm-level sustainability of intensive rice–wheat system: Socio-economic and policy perspective. Addressing Resource Conservation Issues in Rice-Wheat Systems of South Asia. A Resource Book. Rice-Wheat Consortium for Indo-Gangetic Plains (CIMMYT), March 2003, pp. 39–45

A.L. Bhandari, J.K. Ladha, H. Pathak, A.T. Padre, D. Dawe, R.K. Gupta 2002. Yield and soil nutrient changes in a long-term rice-wheat rotation in India. Soil Science Society of America Journal, 66(1) (2002): 162-170.

Alex Morales 2012. Wind power market rose to 41 Gigawatts in 2011, led by China. Bloomberg News (2012).

American Energy 2006. The Renewable Path to Energy Security. Worldwatch Institute (2006). Retrieved 2007-03-11.

A.M. Johnston, H. S. Khurana, K. Majumdar, T. Satyanarayana 2009. Site-specific nutrient management - concept, current research and future challenges in Indian agriculture. Journal of the Indian Society of Soil Science, 57 (2009): 1-10.

Analysis of Wind Energy in the EU-25 (PDF). European Wind Energy Association. Retrieved 2007-03-11.

A. Prashar, S. Thaman, A. Nayyar, E. Humphreys, S.S. Dhillon, Y. Singh, P.R. Gajri, J. Timsina, D.J. Smith 2004. Performance of wheat on beds and flats in Punjab, India. Proceedings of the Fourth International Crop Science Congress, Brisbane, Australia, 26 September–1 October. Publication No. 65. ASA, CSSA, SSSA, Madison, WI, USA (2004), pp115–172.

A.P. Regmi, J.K. Ladha, E.M. Pasuquin, H. Pathak, P.R. Hobbs, L. Shrestha, D.B. Gharti, E. Duveiller 2002. The role of potassium in sustaining yields in a long-term rice–wheat experiment in the Indo-Gangetic Plains of Nepal. Biol. Fertil. Soils, 36 (2002), pp 240–247

A.P. Regmi, J.K. Ladha, H. Pathak, E. Pasuquin, C. Bueno, D. Dawe, P.R. Hobbs, D. Joshy, S.L. Maskey, S.P. Pandey 2002. Yield and soil fertility trends in a 20-year rice-rice–wheat experiment in Nepal. Soil Sci. Soc. Am. J., 66 (2002), pp 657–867

A. Seth, K. Fischer, J. Anderson, D. Jha 2004. The Rice–wheat Consortium: an institutional innovation in international agricultural research on the rice–wheat cropping systems of the Indo Gangetic plains. The Review Panel Report. Rice–Wheat Consortium office, New Delhi, India (2004).

A. Shukla, J.K. Ladha, V.K. Singh, B.S. Dwivedi, V. Balasubramanian, R.K. Gupta, S.K. Sharma, Y. Singh, H. Pathak, P.S. Pandey, A.T. Padre, R.L. Yadav 2004. Calibrating the leaf colour chart for nitrogen management in different genotypes of rice and wheat in systems perspectives. Agron. J., 96 (2004), pp. 1606-1621.

Aslam Mohammad, N. I. Hashmi, Abdul Majid, P. R. Hobbs 1993. Improving wheat yield in the rice-wheat cropping system of the Punjab through fertilizer management. Pakistan Journal of Agricultural Research 14 (1993): 1-1.

A.U. Mokwunye 1995. Phosphate rock as capital investment. In Use of phosphate rock for sustainable agriculture in West Africa, ed. H. Gerner and A. U. Mokwunye. Miscellaneous Fertilizer Studies No. 11. Lome: International Fertilizer Development Center, Africa (1995).

A.W. Sanford 2011. Ethics, narrative, and agriculture: transforming agricultural practice through ecological imagination. Journal of Agricultural and Environmental Ethics 24(3) (2011): 283-303.

B. M. Jenkins, L. L. Baxter, T. R. Jr. Miles, T. R. Miles 1998. Combustion properties of biomass. Fuel Processing Technology, 54 (1998): 17–46.

B. A. Stewart, C. A. Robinson 1997. Are agroecosystems sustainable in semiarid regions? Advance in Agronomy. 60 (1997): 191-228. Academic Press.

B. Beck-Friis, M. Pell, U. Sonesson, H. Jonsson, H. Kirchmann 2000. Formation and emission of N_2O and CH_4, from compost heaps of organic household waste. Environ. Monit. Assess., 62(2000): 317.

Berlin, Germany, (2015), pp 144, ISBN 978-81-322-2014-5. Available online: http//www.springer.com/978-81-322-2146-3 (accessed on 6March 2019).

B. Gadde, S. Bonnet, C. Menke, S. Garivait 2000. Air pollutant emissions from rice straw open field burning in India, Thailand and the Philippines. Environ. Pollut., 157(2000): 1554–1558.

B. Khanna 2018. Times of India. Haryana Sees Decline in Stubble Burning Cases by 25%.

Available online:https://timesofindia.indiatimes.com/city/chandigarh/haryana-sees-decline-in-stubble burningcases- by-25/articleshowprint/64021934.cms (2018) (accessed on 7 September 2018).

B. N. Reddy, G. Suresh 2008. Crop diversification with oilseeds for higher profitability. Souvenir. National Symposium on 'New Paradigms in Agronomic Research'. Navsari Agricultural University, Navsari, November, 19–21 (2008), pp.33–37.

B. Stinner, J. Blair 1990. Biological and agronomic characteristics of innovative cropping systems In: Edwards CA et al (eds) Sustainable Agricultural Systems. Soil and Water Conservation Society, Ankeng, Iowa (1990), pp 123-40.

B. S. Sidhu, V. Beri 2005. Experience with managing rice residues in intensive rice-wheat cropping system in Punjab. In I. P. Abrol, R. K. Gupta, & R. K. Malik (Eds.), Conservation agriculture: Status and prospects (2005), pp 55–63. New Delhi: Centre for Advancement of Sustainable Agriculture, National Agriculture Science Centre.

B. S. Sidhu, V. Beri, S. K. Gosal 1995. Soil microbial health as affected by crop residue management. In Proceedings of National Symposium on Developments in Soil Science, Ludhiana, India (1995), pp 45–46. New Delhi, India: Indian Society of Soil Science. 2–5 November, 1995.

B. Bumb, C. Baanante 1996. The role of fertilizer in sustaining food security and protecting the environment to 2020. 2020 Vision Discussion Paper 17. Washington, DC: IFPRI (1996).

Ben Sills 2011. Solar May Produce Most of World's Power by 2060, IEA Says. Bloomberg (2011).

Bijay-Singh 2008. Crop demand-driven site specific nitrogen applications in rice (*Oryza sativa*) and wheat (*Triticum aestivum*): some recent advances. Indian Journal of Agronomy, 53 (2008): 157-166.

Bijay-Singh, Yadvinder-Singh, J. K. Ladha, K. F. Bronson, V. Balasubramanian, Jagdeep-Singh, C. S. Khind 2002. Chlorophyll meter- and leaf color chart -based nitrogen management for rice and wheat in northwestern India. Agronomy Journal, 94 (2002): 821–829.

Bijay-Singh, R. K. Sharma, Jaspreet Kaur, M. L. Jat, K. L. Martin, Yadvinder-Singh, Varinderpal-Singh, P. Chandna, O. P. Choudhary, R. K. Gupta, H. S. Thind, Jagmohan-Singh, H. S. Uppal, H. S. Khurana, A. Kumar, R. K. Uppal, M. Vashistha, W. R. Raun, R. Gupta 2010. Assessment of the nitrogen management strategy using an optical sensor for irrigated wheat. Agronomy for Sustainable Development, 31(3) (2011): 589-603.

Biomass Energy Center. Biomassenergycentre.org.uk. Retrieved on 2012-02-28.

Blooming New Energy Finance, UNEP SEFT, Frankfurt School, Global Trends in Renewable Energy Investment (2011). Unep.org. Retrieved 21-11-2011.

B. N.Swetha 2007. Studies on nutrient management through organics in soybean – wheat cropping system. M. Sc. (Agri.) Thesis, Univ. Agric. Sci., Dharwad, India (2007).

Breaking the Biological Barriers to Cellulosic Ethanol: A Joint Research Agenda. June 2006. Retrieved 2010-08-02.

B. Singh, Y. Singh, J.K. Ladha, K.F. Bronson, V. Balasubramanian, J. Singh, C.S. Khind 2002. Chlorophyll meter and leaf color chart-based nitrogen management for rice and wheat in Northwestern India. Agron. J., 94 (2002): 821–829.

C. H. Srinivasarao, B. Venkateswarlu, R. Lal, A. K. Singh, K. Sumanta 2013. Sustainable management of soils of dryland ecosystems for enhancing agronomic productivity and sequestering carbon. Adv. Agron., 121 (2013): 253–329.

C. Kailasam 1988. Intercropping of oilseeds in sugarcane. National Seminar on 'Strategies for making India self-reliant in vegetable oils'. September 5-9, 1988. Indian Society of Oilseeds Research, Hyderabad (1988), pp 325-332.

C. P.Singh, S. Panigrahy 2011. Characterization of residue burning from agricultural system in India using space-based observations. J. Indian Soc. Remote Sens., 39(2011): 423–429.

C. R. Hazra 2001. Crop diversification in India. In: Crop diversification in the Asia-Pacific Region. (Minas K. Papademetriou and Frank J. Dent Eds.). Food and Agriculture Organization of the United Nations. Regional Office for Asia and the Pacific, Bangkok, Thailand (2001), pp 32-50.

C. Schaik, H. Van Murray, J. Lamb, J. Di-Giacomo 2000. Composting reduces fuel and labour costs on family farms. Biocycle, (2000): 41, 72.

C. Venkataraman, G. Habib, D. Kadamba, M. Shrivastava, J. F. Leon, B. Crouzille, O. Boucher, D. G. Streets 2006. Emissions from open biomass burning in India: Integrating the inventory approach with high-resolution Moderate Resolution Imaging Spectroradiometer (MODIS) active-fire and land cover data. Global Biogeochemical Cycles 20(2) (2006): 1–12.

C.V. Reddy, R. K. Malik, A. Yadav 2005. Evaluation of double zero-tillage in rice-wheat cropping system. Project Workshop on. Accelerating the Adoption of Resource

Conservation Technologies in Rice-Wheat Systems of the Indo-Gangetic Plains held on June 1-2, 2005 at Hisar (Haryana), India (2005), pp118-21.

C. W. Deren, G. H. Snyder 1991. Biomass production and biochemical methane potential of seasonally flooded inter-generic and inter-specific saccharum hybrids. J. Bioresour. Technol., 36(1991): 179–184.

C.Adhikari, K. F. Bronson, G. M. Panaullah, A. P. Regmi, P. K. Saha, A. Dobermann, D. C. Olk, P.

R. Hobbs, E. Pasuquin 1999. On-farm N supply and N nutrition in the rice–wheat system of Nepal and Bangladesh. Field Crops Research, 64 (1999): 273–286.

Carolyn Fry 2012. Anguilla moves towards cleaner energy (2012).

C.A. Shapiro 1999. Using a chlorophyll meter to manage nitrogen applications to corn with high nitrate irrigation water. Communications in Soil Science and Plant Analysis, 30 (1999): 1037–1049.

C.B. Christianson, P. L. G. Vlek 1991. Alleviating soil fertility constraints to food production in West Africa: Efficiency of nitrogen fertilizers applied to food crops. In Alleviating soil fertility constraints to increased crop pro- duction in West Africa, ed. A. Uzo Mokwunye. Dordrecht: Kluwer Academic Publishers (1991).

C. Edge 2006. Clean Energy Trends 2009, pp 1-4. 2008-03-26]. http://www. cleanedge.com/reports/pdf/Trends 2008.

C. Graves, S.D. Ebbesen, M. Mogensen, K. S. Lackner 2011. Sustainable hydrocarbon fuels by recycling CO_2 and H_2O with renewable or nuclear energy. Renewable and Sustainable Energy Reviews, 15(1) (2011): 1-23. doi:10.1016/j.rser.2010.07.014. (Review.)

C. Morris, M. Pehnt 2012. German energy transition. Arguments for a renewable energy future, Heinrich Böll Foundation, Berlin (2012).

C. K. Patel, P. P Chaudhari, R. W. Patel, N. H. Patel 2010. Integrated nutrients management in potato based cropping systems in north Gujarath. Potato J., 37(1-2) (2010): 68-70.

Conservation Agriculture, Agriculture and Consumer Protection Department, Food and Agriculture Organization, United Nations.

Conservation Effects Assessment Project, U.S. Department of Agriculture.

Conservation Engineering Division (CED), NRCS. Nrcs.usda.gov. Retrieved 2013-10-31.

Cook Maurice 2011. "Hugh Hammond Bennett: the Father of Soil Conservation". Department of Soil Science, College of Agriculture and Life Sciences. North Carolina State University. Retrieved 30 September 2011.

Council on Foreign Relations 2012. Public Opinion on Global Issues: Chapter 5b: World Opinion Energy Security (2012).

Craig Hooper 2011. Air Force cedes the Green lead–and the lede–to Navy. nextnavy.com. Retrieved December 27, 2011.

C. Witt, R. J. Buresh, S. Peng, V. Balasubramanian, A. Dobermann 2007. Nutrient management. In: Rice: A practical guide to nutrient management (T.H. Fairhurst, C. Witt, R. Buresh and A. Dobermann, Eds.) (2007), pp 1-45. International Rice Research Institute, Los Baños, Philippines and International Plant Nutrition Institute and International Potash Institute: Singapore.

C. Witt, A. Dobermann, S. Abdulrachman, H. C. Gines, G. H. Wang, R. Nagarajan, S. Satawathananont, T. T. Son, P. S. Tan, L. V. Tiem, G. C. Simbahan, D. C. Olk 1999. Internal nutrient efficiencies of irrigated lowland rice in tropical and subtropical Asia. Field Crops Research, 63 (1999):113-138.

DACFW 2017. Minutes of Kharif Campaign. Department of Agriculture Cooperation & Farmers Welfare, the Government of India. (2017). Available online: http://agricoop.nic.in/sites/default/files/Revised_Minutes_of_Kharif_conference_2017.pdf (accessed on 15 November 2018). Int. J. Environ. Res. Public Health 2019, 16, 832 19 of 19

D. Deublein, A. Steinhauser 2008. Biogas from Waste and Renewable Sources: An Introduction;Wiley-VCH Verlag GmbH & Co. KGaA: Weinheim, Germany (2008).

D. G. Streets, K. F. Yarber, J. H. Woo, G. R. Carmichael 2003. An Inventory of Gaseous and Primary Aerosol Emissions in Asia in the Year 2000. J. Geophys. Res., 108(2003): 8809–8823.

D. Hoornweg, P. Bhada-Tata 2012. What a Waste: A Global Review of Solid Waste Management; World Bank: Washington, DC, USA, (2012).

D. H. Wall, U. N. Nielsen, J. Six 2015. Soil biodiversity and human health. Nature, 528(2015): 69–76.

D.J. Andrews, A. H. Kassam 1976. The Importance of Multiple Cropping in Increasing World Food Supplies, In: Papendick, R. 1., et al., Multiple Cropping, ASA Special Publ. No.27 (1976),pp1.

D. M. Hegde 2006. Oilseeds in crop diversification. In: Extended Summaries, National Symposium on 'Conservation Agriculture and Environment', October 26-28, Banaras Hindu University, Varanasi (2006), pp 351–354.

D. M. Hegde, S. Prakash Tiwari, M. Rai 2003. Crop diversification in Indian Agriculture. Agricultural Situation in India. August, (2003), pp 255–272.

D. Pimentel, B. Berger, D. Filiberto, M. Newton, B. Wolfe, E. Karabinakis, S. Clark, E. Poon, E. Abbett, S. Nandagopal 2004. Water resources: agricultural and environmental issues. BioScience, 54(10) (2004): 909-918. doi:10.1641/0006-3568 (2004) 054[0909:WRAAEI]2.0.CO;2.

Daniel Budny, Paulo Sotero, editor 2007. Brazil Institute Special Report: The Global Dynamics of Biofuels. Brazil Institute of the Woodrow Wilson Center (2007).

David Beattie 2011. Wind Power: China Picks Up Pace. Renewable Energy World (2011).

D. Byerlee (1992) Technical change, productivity and sustainability in irrigated cropping systems of South Asia: emerging issues in the post-green revolution era. J. Int. Dev., 4 (1992), 477–496.

D.Carrington 2000. Date set for desert Earth. BBC News (2000). Retrieved 2007-03-31.

D. Dawe, A. Dobermann, J.K. Ladha, R.L. Yadav, L. Bao, R.K. Gupta, P. Lal, G. Panaullah, O. SariRam, Y. Singh, A. Swarup, Q.-X. Zhen. (2003)Do Organic Amendments Improve Yield Trends and Profitability in Intensive Rice Systems?. Field Crop. Res., 83 (2003): 191–213.

Defence-scale supercomputing comes to renewable energy research. Sandia National Laborator Retrieved 2012-04-016.

Denis Lenardic 2010. Large-scale photovoltaic power plants ranking 1 - 50 PVresources.com, 2010.

Department of Energy & Climate Change 2011. UK Renewable Energy Roadmap (PDF), pp35.

D. Duchane, D. Brown 2002. Hot dry rock (HDR) geothermal energy research and development at Fenton Hill, New Mexico. Geo-Heat Centre Quarterly Bulletin, 23 (4) (2002), pp13-19. ISSN 0276-1084. Retrieved 2009-05-05.

D.L. Turcotte, G. Schubert 2002. «4», *Geodynamics* (2 ed.), Cambridge, England, UK: Cambridge University Press (2002), pp 136–137, ISBN 978-0-521-66624-4.

D.Maiti, D. K. Das, T. Karak, M. Banerjee 2004. Management of nitrogen through the use of leaf color chart (LCC) and soil plant analysis development (SPAD) or chlorophyll meter in rice under irrigated ecosystem. The Science World Journal, 4 (2004): 838–846.

DOE Closes on Four Major Solar Projects. Renewable Energy World. 30 September 2011.

DTI, Co-operative Energy: Lessons from Denmark and Sweden, Report of a DTI Global Watch Mission, (2004).

Duxbury, R.K. Gupta and R.J. Buresh, Eds.) (2003), pp 115-147. American Society of Agronomy Special Publication 65. ASA, CSSA, SSSA, Madison, WI, USA.

E. Bruni, A. P. Jensen, I. Angelidaki 2019.Comparative study of mechanical, hydrothermal, chemical and enzymatic treatments of digested biofibers to improve biogas production. J. Bioresour. Technol., 101(2010): 8713–8717. Int. J. Environ. Res. Public Health, 16(2019): 832 16 of 19

E. Franchi, G. Agazzi, E. Rolli, S. Borin, R. Marasco, S. Chiaberge, M. Barbafieri 2016. Exploiting hydrocarbon-degrader indigenous bacteria for bioremediation and phytoremediation of a multi-contaminated soil. Chem. Eng. Technol., 39(2016): 1676–1684.

E.H.Tryon 1948. Effect of charcoal on certain physical, chemical, and biological properties of forest soils. Ecol. Monogr., 18 (1948): 81–115.

E.J. Wals 2000. Integrating sustainability in higher agricultural education: dealing with complexity, uncertainty and diverging world views, Interuniversity Conference for Agricultural and Related Sciences in Europe, 0, Ghent, Belgium.www.ggssc.net/files/pdf/crop_diversification (2000).

Environment Canada 2005. Municipal Water Use, 2001 Statistics (Report). Retrieved 2010-02-02. Cat. No. En11-2/2001E-PDF. ISBN 0-662-39504-2. p. 3.

EPA 2010. How to Conserve Water and Use It Effectively. Washington, DC (2000). Retrieved 2010-02-03.

E.Palmqvist, B. Hahn-Hägerdal 2000. Fermentation of lignocellulosic hydrolysates II: Inhibitors and mechanisms of inhibition. J. Bioresour. Technol., 74(2000): 25–33.

E. Sjöström 1993. Wood Chemistry: Fundamentals and Applications; Academic Press: San Diego, CA, USA, (1993).

Edwin Cartlidge 2011. Saving for a rainy day. Science 334 (2011), pp 922–924.

E. Lantz, M. Hand, R. Wiser 2012. The Past and Future Cost of Wind Energy, National Renewable Energy Laboratory conference paper no. 6A20-54526 (2012), pp 4.

E. M. A. Smaling, A. R. Braun 1996. Soil fertility research in Sub-Saharan Africa: New dimensions, new challenges. Communications in Soil Science and Plant Analysis, 27 (3-4) (1996).

Employment News, 222 (1999): 1–2.

Energy and environment policy case for a global project on artificial photosynthesis. Energy & Environmental Science (RSC Publishing). Retrieved 2013-08-19.

Energy crops. Crops are grown specifically for use as fuel. Biomass Energy Centre. Retrieved 6 April 2013.

Energy for Development: The Potential Role of Renewable Energy in Meeting the Millennium Development Goals, pp 7-9.

Energy Kids. Eia.doe.gov. Retrieved on 2012-02-28.

Energy Sources: Solar. Department of Energy. Retrieved 19 April 2011.

Erica Gies 2010. As Ethanol Booms, Critics Warn of Environmental Effect The New York Times, June 24, 2010.

European Photovoltaic Industry Association 2012. Market Report 2011.

Evergreen Agriculture Project. World Agroforestry Centre. Retrieved 2 April 2014.

Exelon purchases 230 MW Antelope Valley Solar Ranch One from First Solar. Solar Server. 4 October 2011.

E. William 2010. Glassley. Geothermal Energy: Renewable Energy and the Environment CRC Press (2010).

F. Monforti, K. Bódis, N. Scarlat, J.F. Dallemand 2013. The possible contribution of agricultural crop residues to renewable energy targets in Europe: A spatially explicit study. Renew. Sustain. Energy Rev., 19(2013): 666–677.

F. O. Obi, B. O. Ugwuishiwu, J. N. Nwakaire 2016. Agricultural Waste Concept, Generation, Utilization and Management. NIJOTECH, 35 (2016): 957–964.

Federal Crop Insurance Reform and Department of Agriculture Reorganization Act of 1994, 108 Stat. 3223, October 13, 1994.

FAO 2002. Agriculture: towards 2015/2030. Rome (2002), pp 420.

F. Hussain, K. F. Bronson, Yadvinder-Singh, Bijay-Singh, S. Peng 2000. Use of chlorophyll meter sufficiency indices for nitrogen management of irrigated rice in Asia. Agronomy Journal, 92 (2000):875–879.

F. Magdoff 1998. Building soils for better crops: organic matter management. Ohio Agronomy Guide. Bulletin 672 (1998).

Food and Agriculture Organization (FAO). 2006. Agriculture and Consumer Protection Department. Rome, Italy Available from http://www.fao.org/ag/magazine/0110sp.htm(Accessed November 2007).

Food and Agriculture Organization (FAO). 2007. Agriculture and Consumer Protection Department. Rome, Italy Available from http://www.fao.org/ag/ca/ (Accessed November 2007).

FSA Conservation Programs, Farm Service Agency, U.S. Department of Agriculture. M.A. Gill, M.A. Chaudhary, M. Ahmed, A. Mujeeb-ur-Rehman 2002. Water management, cultural practices and mechanization. In: Akhtar, M.S., Nabi, G. (Eds.), National Workshop on Rice–wheat Cropping System Management, Islamabad, Pakistan Agricultural Research Council, Islamabad, Pakistan, 11–12 December (2002), pp10.

Fuel Ethanol Production: GSP Systems Biology Research. U.S. Department of Energy Office of Science. April 19, 2010. Retrieved 2010-08-02.

F. Urban, T. Mitchell 2011. Climate change, disasters and electricity generation (2011).

Future Marine Energy. Results of the Marine Energy Challenge: Cost competitiveness and growth of wave and tidal stream energy, January 2006.

G. Bheemaiah, M. V. R. Subramaniam, Ismail Syed 1994. Compatibility of oilseed crops intercropped with Faidherbia albida under different alley width in dry lands. Journal of Oilseeds Research 11(1) (1994): 94–98

G. A. Thomas, R. C. Dalal, E. J. Weston, A. J. King, C. J. Holmes, D. N. Orange, K. J. Lehane 2011. Crop Rotations for Sustainable Grain Production on a Vertisol in the Semi-Arid Subtropics. J. Sustainable Agri., 35 (2011):2–26

G. Kaschuk, O. Alberton, M. Hungria 2011. Quantifying effects of different agricultural land uses on soil microbial biomass and activity in Brazilian biomes: inferences to improve soil quality. Plant soil 338 (2011):467-81.

G. R. Korwar 1992. Influenc of cutting height of Leucaena hedgerows on alley cropped sorghum and pearl millet. Indian Journal of Dryland Agricultral Research & Development 7(1999): 57–60.

G. R. Korwar, G. Pratibha 1999. Performance of short duration pulses with African winter thorn (Faidherbia albida) in semi-arid India. The Indian Journal of Agricultural Sciences 69 (1999): 560–562.

George Olah CO_2 to Renewable Methanol Plant, Reykjanes, Iceland" (Chemicals-Technology.com) First Commercial Plant» (Carbon Recycling International) Carbon Trust 2006.

Geothermal Energy Association. Geothermal Energy: International Market Update (2010), pp 4-7.

G. Korngold 2000. "The Emergence of Private Land Use Controls in Large-Scale Subdivisions: The Companion Story to Village of Euclid v. Ambler Realty Co." Case W. Res. L. Rev. 51 (2000): 617.

Global Concentrating Solar Power. International Renewable Energy Agency. June 2012. Retrieved 2012-09-08.

Global Market Outlook 2016. Retrieved 2012-11-01.

Global wind energy markets continue to boom - 2006 another record years (PDF).

Great Basin Plant Materials Center | NRCS Plant Materials Program. Plant-materials.nrcs.usda.gov. Retrieved 2013-10-31.

Great Basin Plant Materials Center. USDA NRCS. Retrieved 22 October 2010.

G. R. Conway, E. B. Barbier 1990. After the green revolution: Sustainable agriculture for development. London: Earthscan Publications Ltd (1990).

G.R. Conway, J. N. Pretty 1991. Unwelcome harvest: Agriculture and pollution. London: Earthscan Publications Ltd (1991).

G. Suresh, J. V. Rao 1999. Intercropping sorghum with nitrogen-fixing trees in Semi-arid India. Agroforestry Systems 42 (1999):181–194.

G. Suresh, J. V. Rao 2000. The Influence of nitrogen-fixing trees and fertilizer nitrogen levels on the growth, yield and nitrogen uptake of cowpea on a rainfed alfisol. Experimental Agriculture 36 (2000): 1–10.

Gulf of Mexico Initiative | NRCS. Nrcs.usda.gov. Retrieved 2013-10-31.

GWEC, Global Wind Report Annual Market Update. Gwec.net. Retrieved 2011-11-21.

GWEC Global Wind Statistics 2012. Global Wind Energy Commission. Retrieved 18 February 2013.

H. T. Chandranath 2006. Investigations on nutrient management and planting geometry in companion cropping of sunflower (Helianthus annuus. L.) and Ashwagandha (Withania somnifer Dunal L.). Ph.D thesis, University of Agricultural Sciences, Dharwad (2006).

H. B. Nielsen, I. Angelidaki 2008. Codigestion of manure and industrial organic waste at centralized biogas plants: Process imbalances and limitations. Water Sci. Technol., 58(2008): 1521–1528.

H. Hettiarachchi, R. Ardakanian 2016. Good Practice Examples of Wastewater Reuse; UNU-FLORES: Dresden, Germany (2016); ISBN 978-3-944863-30-6 (web), 978-3-944863-31-3 (print).

H. Hettiarachchi, R. Ardakanian 2016. Environmental Resource Management and Nexus Approach: Managing Water, Soil, and Waste in the Context of Global Change; Springer Nature: Basel, Switzerland (2016); ISBN 978-3-319-28593. © 2019 by the authors. Licensee MDPI, Basel, Switzerland. This article is an open access article distributed under the terms and conditions of the Creative Commons Attribution (CC BY) license (http://creativecommons.org/licenses/by/4.0/).

H. Steppler, B. Lundgreen 1988. Agroforestry: Now and in the future. Outlook on Agriculture, 17 (1988):146-51.

H. B. Moller, S. G. Sommer, B. K. Ahring 2004. Methane productivity of manure, straw and solid fractions of manure. J. Biomass Bioenergy, 26(2004): 485–495.

H. Carrère, C. Dumas, A. Battimelli, D. J. Batstone, J. P Delgenès, J. P. Steyer 2010. Pretreatment methods to improve sludge anaerobic degradability: A review. J. Hazard. Mater., 183 (2010): 1–15.

H. Hettiarachchi, J. N. Meegoda, S. Ryu 2018. Organic Waste Buyback as a Viable Method to Enhance Sustainable Municipal Solid Waste Management in Developing Countries. Int. J. Environ. Res. Public Health, 15(2018): 2483.

H. Jiang, A. L. Frie, A. Lavi, J. Chen, H. Zhang 2019. Brown Carbon Formation from Nighttime Chemistry of Unsaturated Heterocyclic Volatile Organic Compounds. Environ. Sci. Technol. Lett. Artic. ASAP (2019).

H. Khurana, S.B. Phillips, Bijay-Singh, A. Dobermann, A. S. Sidhu, Yadvinder-Singh, S. Peng. 2007. Agron. J., 99 (2007):1436-1447.

H. Khurana, S.B. Phillips, Bijay-Singh, M.M. Alley, A. Dobermann, A.S. Sidhu, Yadvinder-Singh, S. Peng. 2008. Nutr. Cycling Agroecosyst,.82 (2008):15-31.

H. Jorgensen, J. B. Kristensen, C. Felby 2007. Enzymatic conversion of lignocellulose into fermentable sugars: Challenges and opportunities. J. Biofuels Bioprod. Bioref., 1(2007): 119–134.

H.L.S. Tandon 1997. Experiences with balanced fertilization in India. Better Crops, 11(1) (1997): 20-21.

H. Zhang, D. Hu, J. Chen, X. Ye, S. X. Wang, J. Hao, L. Wang, R. Zhang, A. Zhi 2011. Particle Size Distribution and Polycyclic Aromatic Hydrocarbons emissions from Agricultural Crop Residue Burning. Environ. Sci. Technol., 45(2011): 5477–5482.

H. B. Babalad 1999. Integrated nutrient management for sustainable production in soybean based cropping system. Ph. D. Thesis, Univ. Agric. Sci., Dharwad, India (1999).

H. Gerner, C. Baanante 1995. Economic aspects of phosphate rock application for sustainable agriculture in West Africa. In Use of phosphate rock for sustainable agriculture in West Africa, ed. H. Gerner and A. U. Mokwunye. Miscellaneous Fertilizer Studies No. 11. Lome: International Fertilizer Development Center Africa (1995).

H. Gloystein 2011. Renewable energy becoming cost competitive, IEA says. Reuters. International Renewable Energy Agency 2012. Renewable Power Generation Costs in 2012: A Overview.

History of PV Solar. Solarstartechnologies.com. Retrieved 2012-11-01.

H. Kirchmann, G. Thorvaldsson 2000. Challenging Targets for Future Agriculture. European Journal of Agronomy, 12 (3-4) (2000): 145-161.

H.K. Rai, A. Sharma, U.A. Soni, S.A. Khan, K. Kumari, N. Kalra 2004. Simulating the impact of climate change on growth and yield of wheat J. Agrometeorol., 6 (1) (2004): 1–8.

H. L. S. Tandon 1992. Fertilizers, organic manures, recyclable wastes and biofertilizers: Components of integrated plant nutrition. New Delhi: Fertilizer Development and Consultation Organization (1992).

How Does A Wind Turbine's Energy Production Differ from Its Power Production? Ccgrouppr.com. 1999-10-06. Retrieved 2013-08-19.

H. Pathak, A. Bhatia, Shiv Prasad, S. Singh, S. Kumar, M.C. Jain, U. Kumar 2002. Emission of nitrous oxide from soil in rice–wheat systems of Indo-Gangetic Plains of India J. Environ. Monitor. Assessm., 77 (2) (2002): 163–178.

H. Pathak, J.K. Ladha, P.K. Aggarwal, S. Peng, S. Das, Y. Singh, B. Singh, S.K. Kamra, B. Mishra, S.R.A.S. Asastri, H.P. Aggarwal 2003. Trends of climatic potential and on-farm yield of rice and wheat in the Indo-Gangetic Plains. Field Crops Research, 80(3) (2003): 223-234.

H. S. Khurana, S. B. Phillips, Bijay-Singh, A. Dobermann, A. S. Sidhu, Yadvinder-Singh, S. Peng 2007. Performance of site-specific nutrient management for irrigated, transplanted rice in northwest India. Agronomy Journal, 99(6) (2007): 1436-1447.

I. Marjanovic 2016. The Best Practices for Using Plant Residues, Agrivi. 2016. Available online: http://blog.agrivi.com/post/the-best-practices-for-using-plant-residues (2016). (accessed on 15 November 2018).

I. S. Arvanitoyannis, P. Tserkezou 2008. Wheat, barley and oat waste: A comparative and critical presentation of methods and potential uses of treated waste. Int. J. Food Sci. Technol., 43(2008): 694–725.

ICAR 1999. ICAR – Vision 2020, Indian Council of Agricultural Research, New Delhi, India (1999).

I. de Carvalho Macedo, M.R. Leal, J.E. da Silva (2002). Greenhouse gas emissions in the production and use of ethanol in brazil: present situation (2002).

IEA says biofuels can displace 27% of transportation fuels by 2050 Washington. Platts. 20 April 2011.

I. M. Chhiba, D.K. Bembi, G.S. Hira 2006. Sustenance of soil and water resources for food security and healthy environment.(Edited Book). Centre of Advanced studies, Department of Soils, PAU, Ludhiana (2010).

References

Industry Statistics: Annual World Ethanol Production by Country. Renewable Fuels Association. Archived from the original on 2008-04-08. Retrieved 2008-05-02.Intergovernmental Panel on Climate Change.

International Energy Agency 2007. Contribution of Renewables to Energy Security IEA Information Paper, pp 5.

International Energy Agency 2007. Renewables in global energy supply: An IEA facts sheet (PDF), OECD (2007), pp 3.

International Energy Agency 2012. Energy Technology Perspectives (2012).

Intergovernmental Panel on Climate Change.

International Programs | NRCS. Nrcs.usda.gov. Retrieved 2013-10-31.

International Technical Assistance | NRCS | NRCS. Nrcs.usda.gov. Retrieved 2013-10-31.

Invasive Species: Animals - European Gypsy Moth (Lymantria dispar). Invasivespeciesinfo.gov. Retrieved 2013-10-31.

Invasive Species: Animals - Sirex Woodwasp (Sirex noctilio). Invasivespeciesinfo.gov. 2013-09-24. Retrieved 2013-10-31.

Invasive Species: Animals - Wild Boar (Sus scrofa). Invasivespeciesinfo.gov. 2013-10-17. Retrieved 2013-10-31.

I. P. Abrol, K. Bronson, J.M. Duxbury, R.K. Gupta 2000. Long-Term Fertility Experiments in IndiaSoil Sci. Soc. Am. J., 66 (2002): 162–170.

I. P. Abrol, S. Sangar 2006. Sustaining Indian agriculture- conservation agriculture the way forward. *Current Sci* 91 (2006): 1020-25.

IPCC Special Report on Land Use, Land-Use Change and Forestry, 2.2.1.1 Land Use FAO Land and Water Division retrieved 14 September 2010.

I. R. E. W. Party 2002. Renewable Energy into the mainstream (2002), pp 9.

I. R. E. W. Party 2002. Renewable Energy into the mainstream. Sittard: Novem, Netherlands (2002), pp54.

IRRI (International Rice Reseach Institute) 1996. Use of Leaf Color Chart (LCC) for N management in rice. Crop Resource Management Network Technology Brief No. 1. International Rice Research Institute, Manila, Philippines (1996).

IRRI (International Rice Research Institute) 2010. Site-specific nutrient management (2010). www.irri.org/ssnm

J. F. Organ, V. Geist, S.P. Mahoney, S. Williams, P.R. Krausman, G.R. Batcheller, T.A. Decker, R. Carmichael, P. Nanjappa, R. Regan, R.A. Medellin, R. Cantu, R.E. McCabe, S. Craven, G.M. Vecellio, D.J. Decker 2012. The North American Model of Wildlife Conservation.. The Wildlife Society Technical Review 12-04. (Bethesda, Maryland: The Wildlife Society) (2012). ISBN 978-0-9830402-3-1.

J. Amonette, S. Joseph 2009. Characteristics of biochar: Micro-chemical properties. In Biochar for Environmental Management: Science and Technology; Lehmann, J., Joseph, S., Eds.; Earth Scan: London, UK (2009), pp 33–52.

J. B. Holm-Nielsen, T. Al Seadi, P. Oleskowicz-Popiel 2009. The future of anaerobic digestion and biogas utilization. Bioresour. Technol., 1000(2009): 5478–5484.

J. C. Gilbert, D. J. G. Gowing, P. R. G. Higginbottom, R. J. Godwin 2000. The habitat creation model: A decision support system to assess the viability of converting arable land into semi natural habitat. Computers and Electronics in Agriculture 28 (2000):67-85.

J. Decker 2009. Going Against the Grain: Ethanol from Lignocellulosics, Renewable Energy World, January 22, 2009.

J. Gaunt, A. Cowie 2009. Biochar greenhouse gas accounting and emission trading. In Biochar for Environmental Management: Science and Technology; Lehmann, J., Joseph, S., Eds.; Earthscan: London, UK (2009), pp 317–340.

J. K. Lynam 1994. Opportunities, Use and Transfer of Systems Research Methods in agriculture in developing countries, (eds: Goldsworthy, P. and Penning de Vries, F.W.T) (1994).

J. K. Lynam, R. W. Herdt 1989. Sense and sustainability: sustainability as an objective in international agricultural research. Agricultural Economics, 3 (4) (1989): 381-98.

J. Lehmann, S. Joseph 2009. Biochar systems. In Biochar for Environmental Management: Science and Technology; Lehmann, J., Joseph, S., Eds.; Earthscan: London, UK (2009), pp 147–168.

J. Makower, R. Pernick, C. Wilder 2009. Clean energy trends. Clean Edge (2009), pp1-4.

J. N. Meegoda, B. Li, K. Patel, L. B. Wang 2018. A Review of the Processes, Parameters, and Optimization of Anaerobic Digestion. Int. J. Environ. Res. Public Health, 15(2018): 2224.

J. Perez, J. M. Dorado, T. D. Rubia, J. Martinez 2002. Biodegradation and biological treatment of cellulose, hemicellulose and lignin: An overview. J. Int. Microbiol., 5(2002): 53–56.

J. Schmaltz, A. Voiland 2017. NASA Earth Observatory, Stubble Burning in Punjab, India (2017). Available online: https://earthobservatory.nasa.gov/images/86982/stubble-burning-in-punjab-india (accessed on 11 July 2018).

J. Singh, R. S. Sidhu 2004. Factors in declining Crop diversification. Economic and Political Weekly, 39 (52) (2004).

J. Sood 2015. Not aWaste untilWasted, Down to Earth (2015). Available online: https://www.downtoearth.org.in/coverage/not-a-waste-until-wasted-40051 (accessed on 7 September 2018).

Jitendra and Others 2017. India's Burning Issues of Crop Burning Takes a New Turn, Down to Earth. (2017). Available online: https://www.downtoearth.org.in/coverage/river-of-fire-57924 (accessed on 7 September 2018).

Jacobson Nemz, Z. Mark, M.A. Delucchi 2009. A Path to Sustainable Energy by 2030. Scientific American 301 (5) (2009): 58-65.

Jacobson Nemz, Z. Mark, M.A. Delucchi 2011. Providing all global energy with wind, water, and solar power, Part I: Technologies, energy resources, quantities and areas of infrastructure, and materials. Energy Policy 39 (3) (2011): 1154.

James Russell 2010. Record Growth in Photovoltaic Capacity and Momentum Builds for Concentrating Solar Power Vital Signs, June 3, 2010.

J.D.T. Kumwenda, S. R. Waddington, S. S. Snapp, R. B. Jones, M. J. Blackie 1996. Soil fertility management research for the maize cropping systems of smallholders in southern Africa: A review. Natural Resources Group Paper 96-02. Mexico City: International Maize and Wheat Improvement Center (CIMMYT) (1996).

J.F. Teboh 1995. Phosphate rock as a soil amendment: Who should bear the cost? In Use of phosphate rock for sustainable agriculture in West Africa, ed. H. Gerner and A.U. Mokwunye. Miscellaneous Fertilizer Studies No. 11. Lome: International Fertilizer Development Center, Africa (1995).

J.K. Ladha, D. Dawe, H. Pathak, A.T. Padre, R.L. Yadav, B. Singh, Y. Singh, Y. Singh, P. Singh, A.L. Kundu, R. Sakal, N. Rame, A.P. Regmi, S.K. Gami, A.L. Bhandari, R. Amin, C.R. Yadav, E.M. Bhattarai, S. Das, H.P. Aggarwal, R.K. Gupta, P.R. Hobbs (2003)

How extensive are yield declines in long-term rice wheat experiments in Asia? Field Crop. Res., 81 (2-3) (2003): 159–180.

J.K. Ladha, J. Hill, R.K. Gupta, J. Duxbury, R.J. Buresh (Eds.), Improving the Productivity and Sustainability of Rice–Wheat Systems: Issues and Impact. ASA Special Publication 65, ASA, Madison, WI, USA (2003): 77–96.

J.K. Ladha, K.S. Fischer, M Hossain, P.R. Hobbs, B. Hardy (Eds.) 2000. Improving the productivity and sustainability of rice–wheat systems of the Indo-Gangetic plains: a synthesis of NARS-IRRI partnership research. IRRI Discussion Paper Series No. 40, IRRI, Los Banos, Philippines (2000).

J.M. Pogodzinski, T.R. Sass 1991. Measuring the effects of municipal zoning regulations: a survey. Urban Studies, 28(4) (1991):597-621.

J. Nemzer 1998. Geothermal heating and cooling. Archived from the original on (1998), pp01-11.

John Timmer 2013. Cost of renewable energy's variability is dwarfed by the savings: Wear and tear on equipment costs millions, but fuel savings are worth billions. Ars Technica. Condé Nast. Retrieved 26 September 2013.

J. Palmer 2008. Hope dims that Earth will survive Sun's death. New Scientist (2008). Retrieved 2008-03-24.

J.R. Freney 1996. Efficient use of fertilizer nitrogen by crops. In Appropriate use of fertilizers in Asia and the Pacific, ed. S. Ahmed. Taipei: Food and Fertilizer Technology Center (1996).

J.R. Goldberger 2011. Conventionalization, civic engagement, and the sustainability of organic agriculture. Journal of Rural Studies, 27(3) (2011): 288-296.

J.R. Nolon 1992. Local Land Use Control in New York: An Aging Citadel Under Siege (July/Aug. 1992). New York State Bar Journal, p. 38, July–August 1992.

J.S. Nelson 1995. Residential Zoning Regulations and the Perpetuation of Apartheid. UCLA L. Rev., 43 (1995): 1689.

J.S. Singh, V.C. Pandey, D.P. Singh 2011. Efficient soil microorganisms: a new dimension for sustainable agriculture and environmental development. Agriculture, Ecosystems & Environment, 140(3-4) (2011): 339-353.

J. Timsina, D.J. Connor (2001) Productivity and management of rice–wheat cropping systems: issues and challenges. Field Crop Res., 69 (2001): 93–132.

J.V.D.K. Kumar Rao, C. Johansen, T.J. Rego (Eds.) 1998. Residual Effects of Legumes in Rice and Wheat Cropping Systems of the Indo-Gangetic Plain. Patancheru 502 324, International Crops Research Institute for the Semi-Arid Tropics, Andhra Pradesh, India (1998), pp. 207–225.

K. Bullis 2012. In the Developing World, Solar Is Cheaper than Fossil Fuels-Technology Review. Technology Review (2012).

K. G. Cassman, A. Dobermann, P. C.Sta.Cruz, H. C. Gines, M. I. Samson, J. P. Descalsota, J. M. Alcantara, M. A. Dizon, D. C. Olk 1996a. Soil organic matter and the indigenous nitrogen supply of intensive irrigated rice systems in the tropics. Plant and Soil 182 (1996a): 267-278.

K. G. Cassman, H. C. Gines, M. Dizon, M. I. Samson, J. M. Alcantara 1996b. Nitrogen-use efficiency in tropical lowland rice systems: contributions from indigenous and applied nitrogen. Field Crops Research 47 (1996b): 1-12.

K. Gupta Raj, M. S. Zia 2003. Reclamation and management of alkali soils. In: Addressing Resource Conservation Issues in Rice-Wheat Systems of South Asia: A Resource Book. Rice-Wheat Consortium for the Indo-Gangetic Plains. International Maize and Wheat Improvement Center, New Delhi, India (2003), pp 30.

K. Hayashi, K. Ono, M. Kajiura, S. Sudo, S. Yonemura, A. Fushimi, K. Saitoh, Y. Fujitani, K. Tanab 2014. Trace gas and particle emissions from open burning of three cereal crop residues: Increase in residue moistness enhances emissions of carbon monoxide, methane, and particulate organic carbon. Atmos. Environ., 95(2014): 36–44.

K. N. Tiwari 2001. Phosphorus needs of Indian soils and crops. Better Crops, 15(2) (2001): 6-10.

K. N. Tiwari 2008. Future of plant nutrition research in India. 26th J. N. Mukherjee-ISSS Foundation lecture, Indian Society of Soil Science, UAS, Bangalore, November 27, 2008.

Korea Joong ang Daily: Turning tides. Tidal power (PDF), retrieved 2010-03-20. Jinangxia Tidal Power Station, pp 194, retrieved 2010-03-21.

K.V.S.Badarinath, T. R. Chand Kiran 2006. Agriculture crop residue burning in the Indo-Gangetic Plains-A study using IRSP6 WiFS satellite data. Current Science, 91(8) (2006): 1085–1089.

K. P. R. Vittal, G. Ravindra Chary, C. A. Rama Rao, G. R. Maruti Sankar 2007. Oilseeds in crop diversification in rainfed regions. In : Challenging global vegetable oils scenario: Issues and Challenges before India. Hegde, D.M (Ed) (2007). Indian Society of Oilseeds Research, Hyderabad (2007),. pp 175–200.

K Sayre 2000. Effects of tillage, crop residue retention and nitrogen management on the performanc of bed-planted, furrow irrigated spring wheat in northwest Mexico.. Paper presented at the Conference of the International Soil Tillage Research Organization, 15; Fort Worth, Texas, USA; 2-7 (Jul, 2000).

K. Sidhu, V. Kumar, T. Singh 2009. Diversification through Vegetable Cultivation. Journal of Life Sciences, 1(2) (2009): 107-113.

K .S. Lackner, S. Brennan, J. M. Matter, A. H. A. Park, A. Wright, B. Van Der Zwaan 2012. The urgency of the development of CO_2 capture from ambient air. Proceedings of the National Academy of Sciences, 109(33) (2012), pp 13156-13162. Bibcode: 2012. PNAS. 10913156L. doi:10.1073/pnas.1108765109. Retrieved September 7, 2012.

K. Govindan, V. Thirumurugan 2002. Organic manure for sustaining productivity in soybean. Finan. Agri., 34 (2002): 23-26.

K. K. Manna, B. S. Brar, N. S. Dhillon 2006. Influence of long term use of FYM and inorganic fertilizers on nutrient availability in a Typic Ustochrept. Indian J. Agric. Sci., 76 (8) (2006): 77-480.

K. K. M. Nambiar, I. P. Abrol 1989. Long-term fertilizer experiments in India: An overview, Ferti. News 34 (4): 1-20, 26. National Academy of Agricultural Sciences, New Delhi, India (1989).

K. Kris Hirst 2013. The Discovery of Fire. About.com. Retrieved 15 January 2013.

K K. Singh, M.Khan, M.S. Shekhawat 2000. Green Revolution-How Green it is? Yojana-Delhi 44.6 (2000): 26-28.

K. Kurihara 1984. Urban and Industrial wastes as fertilizer materials. In *Organic matter and rice*. Los Banos, Laguna, Philippines: International Rice Research Institute (1984).

K.P. Schröder, R. Connon Smith 2008. Distant future of the Sun and Earth revisited. Monthly Notices of the Royal Astronomical Society, 386(1) (2008), pp155-163.

L. E. Hatch, W. Luo, J. F. Pankow, R. J. Yokelson, C. E. Stockwell, K. C. Barsanti 2015. Identification and Quantification of Gaseous Organic Compounds Emitted from Biomass

Burning using Two-Dimensional Gas Chromatography-time-of-flight Mass Spectrometry. Atmos. Chem. Phys., 15(2015): 1865–1899.

L, H., Jr. Allen, T. R. Sinclair, E. R. Lemon, 1976. Radiation and Microclimate Relationships in Multiple Cropping Systems. *In:* Papendick, R. I., et al., Multiple *Cropping*, ASA Special Publ. No. 27 (1976), pp171.

Lidong Bi, Bin Zhang, Guangrong Liu, Zuzhang Li, Yiren Liu, Chuan Ye, Xichu Yu, Tao Lai, Jiguang Zhang, Jianmin Yin and Yin Liang 2009. Long-term effects of organic amendments on the rice yields for double rice cropping systems in subtropical China. Agriculture, Ecosystem and Enviornment 129 (2009):534-41.

L. R. Brown 2001. Eco-Economy: Building an Economy for the Earth, WW Norton &Co., New York (2001).

L. Wu, L. Q. Ma, G. A. Martinez 2000. Comparison of methods for evaluating stability and maturity of biosolids compost. J. Environ. Q., 29 (2000): 424.

Lars Kroldrup 2010. Gains in global wind capacity reported. Green Inc (2010).

L.G. Horlings, T.K. Marsden 2011. Towards the real green revolution? Exploring the conceptual dimensions of a new ecological modernisation of agriculture that coul feed the world'. Global Environmental Change, 21(2) (2011): 441-452.

L. Pingali, Prabhu (Eds.) 1999. Sustaining Rice–Wheat Production Systems: Socio-economic and Policy Issues. Rice–Wheat Consortium Paper Series 5, Rice–Wheat Consortium for the Indo-Gangetic Plains, New Delhi, India (1999), pp43-60.

L. R. Varalakshmi, C. A. Srinivasamurthy, S. Bhasakar 2005. Effect of integrated use of organic manures and inorganic fertilizers on organic carbon, available N, P and K in sustaining productivity of groundnut-Finger millet cropping system. J.Indian Soc. Soil Sci., 53 (8) (2005): 315-318.

Lucas Mearian 2013. U.S. flips switch on massive solar power array that also stores electricity: The array is first large U.S. solar plant with a thermal energy storage system, October 10, 2013. Retrieved October 18, 2013.

L.W. Harrington, S. Fujisaka, M. L. Morris, P.R. Hobbs, H.C. Sharma, R.P. Singh, M.K. Chaudhary, S. D. Dhiman 1992. Wheat and rice in Karnal and Kurukshetra districts, Haryana, India: farmers' practices, problems and an agenda for action. Exploratory Surveys. CIMMYT, Haryana Agricultural University, Indian Council of Agricultural Research and IRRI (1992).

L.W. Harrington, P.R. Hobbs, D.B. Tamang, C. Adhikari, B.K. Gyawali, G. Pradhan, B. K.Batsa, J.D. Ranjit, M. Ruckstuhl, Y.G. Khadka, M.L. Baidya 1993. Wheat and rice in the hills: farming systems, production techniques and research issues fro rice–wheat cropping pattern in the mid-hills of Nepal. Report on an exploratory survey conducted in Kabhre district. Nepal Agricultural Research Council and CIMMYT (1993).

M.A.Altiere, W. H. Whitcomb 1978-79. "Manipulation of Insect Populations Through Seasonal Disturbance of weed Communities. Protection Ecology, 1 (1978-79):185.

M.A. Alteiri, M. Z. Liebman 1986. Incset, weed and plant disease management in multiple cropping systems. In Francis CA (ed.) Multiple cropping systems, Mac Milan New York (1986), pp183-218.

M. Ali Khan 2007. The Geysers Geothermal Field, an Injection Success Story. Annual Forum of the Groundwater Protection Council (2007). Retrieved 2010-01-25.

Mahoney, Shane 2004. The North American Wildlife Conservation Model. *Bugle* (Rocky Mountain Elk Foundation) 21 (3) (2004).

M. D. Eisaman, K. Parajuly, A. Tuganov, C. Eldershaw, N. Chang, K. A. Littau 2012. CO_2 extraction from seawater using bipolar membrane electrodialysis. Energy & Environmental Science, 5(6) (2012): 7346-7352. doi:10.1039/C2EE03393C. Retrieved July 6, 2013.

M. F. Amador 1980. Comportamiento de tres Especies (Maiz, Frijol, Calabaza) en Policultivos en la Chontalpa, Tabasco, Mexico,"Tesis Professional, Colegio Superior de Agricultural Tropical, H Cardenas, Tabasco, Mexico (1980).

M. J. Taherzadeh 1999. Ethanol from Lignocellulose: Physiological Effects of Inhibitors and Fermentation Strategies. Ph.D. Thesis, Biotechnology, Chemical Reaction Engineering, Chalmers University of Technology, Gothenburg, Sweden, (1999).

Ministry of New and Renewable Energy (MNRE). 2011. Strategic Plan for New and Renewable Energy Sector for the Period 2011–2017. In Energy; Ministry of New and Renewable Energy: New Delhi, India (2011).

Ministry of New and Renewable Energy (MNRE) 2015. Annual Report, 2015–2016. In Energy; Ministry of New and Renewable Energy: New Delhi, India (2015).

Ministry of Agriculture 2000. National Agricultural Policy 9 National Research Council (1999) Our Common Journey: a transition toward Sustainability, National Academy of Sciences, USA (1999), pp363.

M. K. Devarajaiah, M. S. Nataraju 2009. Procedures and Practices of Organic Certification in India- A Step towards Sustainable Agriculture. *Financing Agriculture* -A National Journal of Agriculture & Rural Development (2009).

M.L.McCallum, J. L. McCallum, S. E. Trauth. 2009. Predicted climate change may spark box turtle declines. Amphibia-Reptilia 30 (2009): 259-264.

M.L.McCallum, G.W. Bury 2013. Google search patterns suggest declining interest in the environment. Biodiversity and Conservation (2013). DOI: 10.1007/s10531-013-0476-6.

M. M. Alam, R. j. Buresh, J. K. Ladha, S. Foyjunnessa 2006. Optimization of phosphorus and potassium management in lowland rice in Bangladesh through site specific nutrient management approach. 18th World Congress of Soil Science, 9 – 15 July 2006, Philadelphia, Pennsylvania, USA. Published on CD (2006).

M. M. V. Srinivasa Rao, G. Bheemaih 2001. Response of groundnut alley cropped with nitrogen fixing tree species to application of organic and inorganic sources of nitrogen. Journal of Oilseeds Research 18 (2001): 147-149.

M. P. McHenry 2009. Agricultural biochar production, renewable energy generation and farm carbon sequestration in Western Australia, Certainty, uncertainty and risk. Agric. Ecosyst. Environ., 129(2009): 1–7.

M. S. Gill, I. P. S. Ahlawat 2006. Crop diversification- its role towards sustainability and profitability. Indian Journal of Fertilizers 2 (9) (2006):125–138, 150.

M. Sirhindi 2018.Punjab Witnesses 38% Fall Stubble Burning Instances, Times of India. Available online:http//timesofindia.indiatimes.com/city/chandigarh/punjab-witnesses-38-fall-stubble-burning instances/articleshowprint/64018735.cms (accessed on 7 September 2018).

M. Tuomela, M. Vikman, A. Hatakka, M. Itavaara 2000. Biodegradation of lignin in a compost environment: A review. Bioresour. Technol., (2000): 72, 169.

M. Tvaronavičienė 2012. Contemporary perceptions of energy security: policy implications. Journal of Security & Sustainability 1(4) (2012).

M.A. Razzaque, M. Badaruddin, C.A. Meisner (Eds.) 1995. Sustainability of rice–wheat systems in Bangladesh. In: Proceedings of the National Workshop, Dhaka, Bangladesh, 14–15 (November 1994). Bangladesh and Australia Wheat Research Centre, Dhaka, Bangladesh.

Mark Tran 2011. UN calls for universal access to renewable energy. The Guardian (London) (2011).

M. Greenberg, F. Popper, B. West, D. Krueckeberg 1994. Linking city planning and public health in the United States. Journal of Planning Literature, *8*(3) (1994): 235-239.

M. Greenley, J. Farrington 1989. Potential implications of agriculture for the Third World. In Agriculture biotechnology: Prospects for the 3rd World, ed. J. Farrington. London: Overseas Development Institute (1989).

Mitigation, C.C., 2011. IPCC special report on renewable energy sources and climate change mitigation (2011).

M. Jacobson, S. Kar 2013. Extent of agroforestry extension programs in the United States. Journal of Extension, 51(4) (2013).

M. Lipton, R. Longhurst 1989. New seeds and poor people. London: Unwin Hyman (1989).

M. Singh, A. Ankit. A Bulletin on needs and future of precision farming in Punjab.

M. Z. Jacobson, M. A. Delucchi 2011. Providing all global energy with wind, water, and solar power, Part I: Technologies, energy resources, quantities and areas of infrastructure, and materials. Energy Policy, 39(3) (2011): 1154-1169.

Nagendran, R. Agricultural Waste and Pollution. Waste 2011, 341–355.

N. C. Brady, R. R. Weil 1996. The Nature and Properties of Soils, 14th ed.; Prentice Hall: Upper Saddle River, NJ, USA, (1996).

N.H.Rao 2002. Sustainable Agriculture: Critical Challenges Facing the Structure and Function of Agricultural Research and Education in India. National Workshop on Agricultural Policy, April 2002. nhrao@naarm.ernet.in

N. Jain, A. Bhatia, H. Pathak 2014. Emission of Air Pollutants from Crop Residue Burning in India. Aerosol Air Qual. Res., 14(2014): 422–430.

NPMCR.(2019) Available online: http://agricoop.nic.in/sites/default/files/NPMCR 1.pdf (accessed on 6 March 2019).

NRG Energy Completes Acquisition of 250- Megawatt California Valley Solar Ranch from SunPower. MarketWatch, 30-10-2011.

Nuclear Energy and the Fossil Fuels (PDF). Retrieved 2012-11-01.

N. P. Singh, R. S. Sachan, P. C. Pandey, P. S. Bisht 1999. Effect of decade long –term fertilizer and manure application on soil fertility and productivity of rice wheat system. In mollisoi. J. Indian Soc. Soil Sci. 47 (1) (1999): 72-80.

NARA. Records of the Natural Resources Conservation Service. Retrieved 2008-01-10.

National Agricultural Technology Project 2002. Annual Report, 2001–2002. Irrigated Agro-Ecosystem. Directorate of Maize Research, IARI, New Delhi, India (2002).

National Agroforestry Center. USDA National Agroforestry Center (NAC). Retrieved 2 April 2014.

Natural Resource Group Paper, CIMMYT, Mexico (1996), pp. 96–101

New Standard Encyclopedia 1992. Standard Educational Operation. Chicago, Illinois (1992), pp A-141, C-546.

NRCS Biography of Hugh Hammond Bennett. Retrieved 2008-01-10.

NRCS Conservation Programs, Natural Resources Conservation Service, U.S. Department of Agriculture.

NRCS National Conservation Practice Standards. National Handbook of Conservation Practices. Accessed 2009-06-05.

NRCS Natural Resources Conservation Service. U.S. Washington, DC. «Soil Survey Programs"Accessed 2009-06-05.

N. S. S. Rao 1993. Biofertilizers in agriculture and forestry, 3rd ed. New York: InternationalScience Publishers (1993).

OECD (Organization for Economic Co-operation and Development). 2001. Available online: https://stats.oecd.org/glossary/detail.asp? ID=77 (accessed on 10 November 2018).

O. Masek 2009. Biochar Production Technologies. Available online: http://www.geos.ed.ac.uk/sccs/biochar/documents/BiocharLaunch-OMasek (2009).pdf (accessed on 6 March 2019).

Opinion of the EEA Scientific Committee on Greenhouse Gas Accounting in Relation to Bioenergy. Retrieved 01-11-2012

O. Oenema, L. van Liere, L. O. Schoumans 2005. Effects of lowering nitrogen and phosphorus surpluses in agriculture on the quality of groundwater and surface water in the Netherlands. Journal of Hydrology, 304(1-4) (2005): 289-301.

P. Agamuthu 2009. Challenges and Opportunities in Agro-waste Management: An Asian Perspective. In Proceedings of the Meeting of First Regional 3R Forum in Asia, Tokyo, Japan, 11–12 November (2009).

P. Gkorezis, M. Daghio, A. Franzetti, J. D. Van Hamme, W. Sillen, J. Vangronsveld 2016. The Interaction between Plants and Bacteria in the Remediation of Petroleum Hydrocarbons: An Environmental Perspective. Front. Microbiol., 7(2016): 1836.

P. K. Gupta, S. Sahai, N. Singh, C. K. Dixit, D. P. Singh, C. Sharma 2004. Residue burning in rice-wheat cropping system: Causes and implications. Curr. Sci. India, 87(2004): 1713–1715. Int. J. Environ. Res. Public Health, 16(2019): 832 17 of 19

P. Kumar, D. M. Barrett, M. J. Delwiche, P. Stroeve 2009. Methods for pre-treatment of lignocellulosic biomass for efficient hydrolysis and biofuel production. J. Ind. Eng. Chem., 48(2009): 3713–3729.

P. R. Shukla 2007. Biomass Energy Strategies for Aligning Development and Climate Goals in India; Environmental Assessment Agency: The Hague, The Netherlands (2007).

P. Weiland 2003. Production and energetic use of biogas from energy crops and wastes in Germany. Appl. Biochem. Biotechnol., 109(2003): 263–274.

P. Kumar, S. Kumar, L. Joshi 2015. The extend and management of crop residue stubbles. In Socioeconomic and Environmental Implications of Agricultural Residue Burning: A Case Study of Punjab, India; Kumar, P., Kumar, S., Joshi, L., Eds.; Springer Briefs in Environmental Science: CSO (Central Statistics Office) 2014. Energy Statistics. In Ministry of Statistics and Program Implementation Office; CSO: New Delhi, India, (2014).

P. Sequi 1996. The role of composting in sustainable agriculture. In The Science of Composting; Bertoldi, M., Sequi, P., Lemmens, B., Papi, T., Eds.; Blackie Academic & Professional: London, UK, (1996), pp 23–29.

Pratap Singh D, R. Prabha 2017. Bioconversion of Agricultural Wastes into High Value Biocompost: A Route to Livelihood Generation for Farmers. Adv. Recycl. Waste Manag, (2017): 137.

P. Abrol, S. P. Palaniappan 1987. Green manure crops in irrigated and rainfed lowland rice based cropping system in South Asia. In: Green Manure In rice farming. Proc. Of a Symposium on "Sustainable Agriculture", Los Banos, Philippines (1987), pp71.

Paul Gipe 2013. 100 Percent Renewable Vision Building. Renewable Energy World (2013).

P. Biradar, Y. R. Aladakatti, T. N. Rao, K. N. Tiwari 2006. Site-specific nutrient management for maximization of crop yields in northern Karanataka. Better Crops 90(3)(2006): 33-35.

P. K. Ghosh, A. Das, I. R. Saha, E Kharkrang, A. K. Tripathi, G. C. Munda, S. V. Ngachan 2010. Conservation agriculture towards achieving food security in North East India. Current Sci., 99 (7) (2010): 15-21.

Power for the People, pp 3.

Power from Sunshine: A Business History of Solar Energy May 25, 2012. S. G. Sommer, P. Dahl 1999. Nutrient and carbon balance during the composting of deep litter. J. Agric. Eng. Res., 74(1999): 145.

Paul A. Wojtkowski 1998. The Theory and Practice of Agroforestry Design. Science Publishers Inc., Enfield, NH (1998), pp282.

P. A, Wojtkowski 2002. Agroecological perspectives in agronomy, forestry, and agroforestry (2002).

P. Schroeder 1994. Carbon storage benefits of agroforestry systems. Agroforestry Systems, *27*(1) (1994): 89-97.

P. Hobbs, M. L. Morris 1996. "Meeting South Asia's future food requirements from rice-wheat cropping systems: priority issues facing researchers in the post-Green Revolution era." (1996): viii-46.

P. Kristjanson, H. Neufeldt, A. Gassner, J. Mango, F.B. Kyazze, S. Desta, G. Sayula, B. Thiede, W. Forch, P.K. Thornton, R. Coe 2012. "Are food insecure smallholder households making changes in their farming practices? Evidence form East Africa". Food Security 4 (3) (2012): 381–397. doi:10.1007/s12571-012-0194-z.

P. Kumar, W.M. Rosegrant (1994) Productivity and sources of growth for rice in India. Econ. Polit. Weekly, 29 (53) (1994): 183–188.

P.K. Gupta, S. Shivraj, S. Nahar, C.K. Dixit, D.P. Singh, C. Sharma, M.K. Tiwari, R.K. Gupta, S.C. Garg (2004) Residue burning in rice–wheat cropping system: causes and implications. Curr. Sci., 87 (12) (2004): 1713–1717.

P.K. Kataki (Ed.) 2001. The Rice–Wheat Cropping Systems of South Asia: Efficient Production Management, Food Products Press, New York, USA (2001), pp. 87–131.

P. Kumar, Mrithyunjaya 1992. Measurement and analysis of total factor productivity growth for wheat in India. Indian J. Agric. Econ. 47(3) (1992): 451–458.

P.L. Pingali, R.V. Gerpacio 1997. Towards reduced pesticide use for cereal crop in Asia. Economics Working Paper No. 97-04. CIMMYT, Mexico, D.F., Mexico (1997).

P.L. Pingali, P. Heisey 1996. Cereal crop productivity in developing countries: Past trends and future prospects. In: Conference Proceedings of Global Agricultural Science on Policy for the 21st Centuary, Melbourne, Australa, Victoria Department of Natural Resources and Environment, Melbourne, Australia, 26–28 August (1996), pp61–94.

P.L. Pingali, M. W. Rosegrant 1994. Confronting the environmental consequences of the Green Revolution in Asia. EPTD Discussion Paper No.2. Washington, DC: IFPRI.

P.L. Pingali, M.W. Rosegrant 2001. Intensive food systems in Asia: can the degradation problems be reversed? In: Lee, D.R., Barrett, C.B. (Eds.), Tradeoffsa or Synergies? Agricultural Intensification, Economic Development, And The Environment 2001, Based on Papers Presented at an International Conference held in Salt Lake City, Utah, July–August 1997.

P.L. Pingali, M. Shah, M 1999. Rice–wheat cropping system in the Indo-Gangetic Plains: policy re-directions for sustainable resource use. In: Pingali, P.P. (Ed.), Sustaining Rice–Wheat Production Systems: Socio-Economic and Policy Issues, Rice–Wheat Consortium Paper Series. 5 (1999), Rice–Wheat Consortium for Indo-Gangetic Plains, CIMMYT-Outreach in India, New Delhi.

Population Reference Bureau. 2007. Washington, D.C. Available from http://www.prb.org/Journalists/FAQ/WorldPopulation.aspx. (Accessed December 2007).

P.R. Hobbs, Ken Sayre, R. Gupta 2008. The role of conservation agriculture in sustainable agriculture. Philosophical transactions of the Royal society of London Series B, Biological Science 363 (1491) (2008): 543-555.

P. S. Bisht, P.C. Pandy, D.K. Singh 2006. Monitoring of long-term fertility experiment after two decades of rice–wheat cropping. Abstracts, 2nd Int. Rice Congress, 2006. Science technology and trade far peace and prosperity. Oct 9-12, 2006, New Delhi (2006), pp 400.

Purpose of the CTA Program | NRCS. Nrcs.usda.gov. Retrieved 2013-10-31.

Q.M. Ali, S. Tunio 2002. Effect of various planting patterns on weed population and yield of wheat. Asian Journal of Plant Sciences (2002).

R. Chandra, H. Takeuchi, T. Hasegawa 2012. Methane production from lignocellulosic agricultural crop wastes: A review in context to second generation of biofuel production. Renew. Sustain. Energy Rev., 16(2012): 1462–1476.

R. C.Izaurralde, N. J. Rosenberg, R. Lal 2001. Mitigation of climate change by soil carbon sequestration: Issues of science, monitoring, and degraded lands. Adv. Agron., 70 (2001): 1–75.

REN21 2008. Renewables 2007. Global Status Report (2008), pp 18.

REN21 2010. Renewables 2010 Global Status Report (2010), pp 12.

R. V.Misra, R. N. Roy, H. Hiraoka 2003. On Farm Composting Methods; Food and Agricultural Organization of the United Nations: Rome, Italy, 2003. Int. J. Environ. Res. Public Health, 16(2019): 832 18 of 19

R. A.Washenfelder, A. R. Attwood 2015. Biomass burning dominates brown carbon absorption in the rural southeastern United States. Geophys. Res. Lett., 42(2015): 653–664.

Ramesh Chand, Sonia Chauhan 2002. Socio-economic factors in agricultural diversification in India. Agricultural Situation in India. Feb. (2002), pp 523–529.

R. P. Gupta, S. K. Tewari 1985. Factors affecting crop diversification: A critical Analysis. Indian Journal of Agricultural Economics 40(3) (1985): 304–309.

R. L. Yadav, N. D. Shukla 2002. Diversification in cropping systems for sustainable production of oilseeds (2002). In : Oilseeds and oils: Research and Development Needs. Rai, Mangala., Singh Harvir and Hegde, D.M. Eds. Indian Society of Oilseeds Research, Hyderabad (2002), pp101-111.

R.Hasan 2002. Potassium status of soils in India. Better Crops 16(2) (2002): 3-5.

Rice–Wheat Consortium (RWC), 2004. Progress reports. In: The 12thRegional Technical Coordination Committee Meeting, 7–9 February, Islamabad, Pakistan. RWC, New Delhi (2004), pp 135.

R. J. Buresh 2010. Nutrient best management practices for rice, maize, and wheat in Asia. 19th World Congress of Soil Science, Soil Solutions for a Changing World 1 – 6 August 2010, Brisbane, Australia. Published on DVD (2010).

R. J.Buresh, M. F. Pampolino, C. Witt 2010. Field-specific potassium and phosphorus balances and fertilizer requirements for irrigated rice-based cropping systems. Plant and Soil 335 (2010): 35-64.

R. J. Pearson, M. D. Eisaman, J. W. Turner, P. P. Edwards, Z. Jiang, V. L. Kuznetsov, S. G. Taylor 2012. Energy Storage via Carbon-Neutral Fuels Made From CO_2, Water, and Renewable Energy. Proceedings of the IEEE, 100(2) (2012), pp 440-460. doi:10.1109/JPROC.2011.2168369. Retrieved September 7, 2012. (Review)

R. Nagarajan, S. Ramanathan, P. Muthukrishnan, P. Stalin, V. Ravi, M. Babu, S. Selvam, M. Sivanatham, A. Dobermann, C. Witt 2004. Site-specific nutrient management in irrigated rice systems of Tamil Nadu, India. In: Increasing productivity of intensive rice systems through site-specific nutrient management (A. Dobermann, C. Witt, and D. Dawe, Eds.) (2004), pp 101–123. Science Publishers, Inc., Enfield, N.H., U.S.A. and International Rice Research Institute, Los Baños, Philippines.

RWC-CIMMYT 2003. Addressing resource conservation issues in rice-wheat systems of South Asia resource book. New Delhi, India, Rice-Wheat Consortium for the Indo-Gangetic Plains – International Maize and Wheat Improvement Centre (2003), pp305.

Renewables. eirgrid.com. Retrieved 22 November 2010.

REN 21 2008. Renewables 2007 Global Status Report (2007), pp 12.

REN 21 2010. Renewables 2010 Global Status Report (2010), pp 15. http://www.iea.org/publications/freepublications/publication/cooking.pdf

REN 21 2010. Renewables 2010 Global Status Report (2010), pp 53.

REN21 2011. Renewables 2011: Global Status Report (2011), Pp 11.

REN 21 2011. Renewables 2011: Global Status Report (2011), pp 13–14.

REN 21 2011. Renewables 2011: Global Status Report (2011), pp 14.

REN21 2011. Renewables 2011: Global Status Report (2011), pp15.

REN 21 2011. Renewables 2011: Global Status Report (2011), pp 17, 18.

REN 21 2012. Renewables Global Status Report (2012), pp 17.

REN 21 Renewables Global Status Report (2006 - 2012). Ren21.net. Retrieved 2012-10-21.

REN 21 2013. Renewables Global Status Report (PDF). Retrieved 30-01-2014.

R.G. Muschler 2001. Shade improves coffee quality in a sub-optimal coffee-zone of Costa Rica. Agroforestry Systems 85 (2001):131-139.

R.Gupta, P.R. Hobbs, Ken Sayre 2007. The role of conservation agriculture in sustainable agriculture. The Royal Society (2007), pp1-13.

R.H. Nelson 1977. Zoning and property rights: An analysis of the American system of land-use regulation. Cambridge, MA: MIT Press (1977), pp 15.

Rice–Wheat Consortium (RWC) 2000. Developing an Action Program for Farm-level Impact in Rice–wheat Systems of the Indo-Gangetic Plains. RWC New Delhi, India (2000).

Rice–Wheat Consortium (RWC) 2004. Progress reports. In: The 12th Regional Technical Coordination Committee Meeting, 7–9 February (2004), Islamabad, Pakistan. RWC, New Delhi, pp135.

Rice–Wheat Consortium (RWC) 2005. Progress reports. In: The 13th Regional Technical Coordination Committee Meeting, 6–8 February (2005), Dhaka, Bangladesh. RWC, New Delhi, pp 115.

R.J.I. Ortiz Monasterio, S.S. Dhillon, R.A. Fischer 1994. Date of sowing effects on grain yield components of irrigated spring wheat cultivars and relationships with radiation and temperature in Ludhiana, India. Field Crop Res., 37 (1994): 169–184.

R. Khosla 2008. Precision Agriculture: Challenges and Opportunities in a Flat World. Department of Soil & Crop Sciences, Colorado State University, USA (2008).

R.K. Malik (1996) Herbicide resistant weed problems in developing world and methods to overcome them. Proceedings of the Second International Malikional Weed Science Congress, Copenhagen, Denmark, Food and Agriculture Organization, Rome, Italy (1996), pp 665–673.

R.K. Malik, G. Gill, P.R. Hobbs 1998. Herbicide resistance—a major issue for sustaining wheat productivity in rice–wheat cropping system in the Indo-Gangetic plains. Rice–Wheat Consortium Paper Series 3 (1998): 32.

R.K. Malik, S. Singh (1993) Evolving strategies for herbicide use in wheat: resistance and integrated weed management. In Proc. Indian Soc. Weed Sci. Int. Symp. Hisar, India 1993 Nov 18 (Vol. 1, pp 225-238).

R.K. Malik, S. Singh 1995. Littleseed canarygrass (Phalaris minor) resistance to isoproturon in India. Weed Technology, 9(3) (1995): 419-425.

R. Lal, P. Hobbs, N. Uphoff, D.O. Hansen (Eds.) 2004. Sustainable Agriculture and the Rice–wheat System, Ohio State University, Marcel Dekker, Columbus, OH, New York, USA (2004), pp101–119.

R.L. Yadav 2001. On farm experiments on integrated nutrient management in rice–wheat cropping systems. Expl. Agric., 37 (2001): 99–113.

R.L. Yadav, B.S. Dwivedi, P.S. Pandey 2000. Rice–wheat cropping system: assessment of sustainability under green manuring and chemical fertilizer inputs. Field Crop Res., 65 (2000): 15–30.

R.Muschler 1999. Árboles en Cafetales. Materiales de Enseñanza No. 45, CATIE, Turrialba, Costa Rica (1999), pp139.

Robert Hart 1996. Forest Gardening. Forest gardening, in the sense of finding uses for and attempting to control the growth of wild plants, is undoubtedly the oldest form of land use in the world (1996), pp 124.

R.S. Mehla, J.K. Verma, R.K. Gupta, P.R. Hobbs 2000. Stagnation in the productivity of wheat in the Indo-Gangetic plains: zero-till-seed-cum-fertilizer drill as an integrated solution. Rice–Wheat Systems of the Indo-Gangetic Plains. RWC Paper Series No.8, New Delhi, India (2000), pp9.

R.S. Paroda, T. Woodhead, R.B. Singh 1994. Sustainability of Rice–Wheat Production Systems in Asia. RAPA Publication 1994/11. FAO, Bangkok, Thailand (1994)

R. Wassman, R.S. Lantin, H.U. Neue (Eds.) 2001. Methane Emissions from Major Rice Ecosystems in Asia. Developments in Plant and Soil Sciences, vol. 91 Kiuwer Academic Publishers, Dordrecht (2001), pp 416.

RWC-CIMMYT 2003. Addressing Resource Conservation Issues in Rice Wheat Systems of South Asia: A Resource Book. Rice Wheat Consortium for Indo-Gangetic Plains, International Maize and Wheat Improvement Center, New Delhi (2003), pp305.

R. Zehra 2017.. How Clean Is the Air around You. Available online: https://fit.thequint.com/health-news/clean-your-air-as-per-who-standards-2 (2017). (accessed on 10 June 2018).

S. Dumontet, H. Dinel, S. B. Baloda 1999. Pathogen reduction in sewage sludge by composting and other biological treatments: A review. Biol. Agric. Hortic., 16(1999): 409.

S. Shilev, M. Naydenov, V. Vancheva, A. Aladjadjiyan 2006. Composting of Food and Agricultural Wastes. Utilization of By-Products and Treatment of Waste in the Food Industry; Oreopoulou, V., Russ, W., Eds.; Springer: New York, NY, USA, (2006), pp 283–301.

S. S. Verma 2014. Technologies for stubble use. J. Agric. Life Sci., (2014): 1, 2.

S. Kim, B. E. Dale 2004. Cumulative energy and global warming impacts from the production of biomass for biobased products. Journal of Industrial Ecology, 7(3–4) (2004): 147–162.

S. P. Palaniappan, A. Jeyabal 2002. Crop diversification with oilseeds in non-traditional areas and seasons. Oilseeds and oils: Research and Development Needs. Rai, Mangala., Singh, Harvir and Hegde, D.M. (Eds) 2002. Indian Society of Oilseeds Research, Hyderabad (2002), pp 94–100.

S. P. Palaniappan, A. Jeyabal, S. Chelliah 1999. Evaluation of integrated nutrient management techniques in summer sesame. Sesame and Safflower Newsletter 14 (1999): 32–35.

S. Singh, Renu Batra, M. M. Mishra, K. K. Kapoor, Sneh Goyal 1992. Decomposition of paddy straw in soil and the effect of straw incorporation in the field on the yield of wheat. Journal of Plant Nutrition and Soil Sciences, 155(4) (1992): 307–311.

S. Garg 2017. Bioremediation of Agricultural, Municipal, and Industrial Wastes. Handb. Res. Inventive Bioremediat. Tech. (2017).

S. Geerts, D. Raes 2009. Deficit irrigation as an on-farm strategy to maximize crop water productivity in dry areas. Agric. Water Manage 96 (9) (2009): 1275–1284. doi:10.1016/j.agwat.2009.04.009.

S. K. Lohan, H. S. Jat, A. K. Yadav, H. S. Sidhu, M. L. Jat, M. Choudhary, P. Jyotsna Kiran, P.C. Sharma 2018. Burning issues of paddy residue management in north-west states of India. Renew. Sustain. Energy Rev., 81(2018): 693–706.

S. K. Sharma, I. M. Mishra, M. P. Sharma, J. S. Saini 1988. Effect of particle size on biogas generation from biomass residues. J. Biomass, 17(1988): 251–263.

S. Jeff, M. Prasad, P. Agamuthu 2017. Asia Waste Management Outlook. UNEP Asian Waste Management Outlook; United Nations Environment Programme: Nairobi, Kenya, (2017).

S. John, C. Mini 2005. Development of an okra based cropping system. Indian J Horti.,612 (2005).

S. K. Mittal, K. Susheel, N. Singh, R. Agarwal, A. Awasthi, P. K. Gupta 2009. Ambient air quality during wheat and rice crop stubble burning episodes in Patiala. Atmos. Environ., 43(2009): 238–244.

S. Mittal, E. Ahlgren, P. Shukla 2017. Barriers to biogas Dissemination in India: A review. Energy Policy, 112 (2017): 361–370.

S. Ross 2018. Countries That Produce the Most Food, Investopedia. 2018. Available online: https://www.investopedia.com/articles/investing/100615/4-countries-produce-most-food (2018). asp#ixzz5WRqV85mY (accessed on 10 November 2018).

S. Sahai, C. Sharma, S. K. Singh, P. K. Gupta 2011. Assessment of Trace Gases, Carbon and Nitrogen Emissions from Field Burning of Agricultural Residues in India. Nutr. Cycl. Agroecosyst., 89(2011): 143–157.

S.C.E. Jupe, A. Michiorri, P.C. Taylor 2007. Increasing the energy yield of generation from new and renewable energy sources. Renewable Energy 14 (2) (2007): 37–62.

S. C. Panda, S. C. Panda 2004. Cropping and Farming systems. Agrobios (India) (2004).

S. C. Panda. A Textbook of Agronomy.

S. D. Basavaraju 2007. Integrated nitrogen management in maize in a Vertisol of Malaprabha command area. M. Sc. (Agri.) Thesis, Univ. Agric. Sci., Dharwad, India (2007).

S.D. Billore, A. K. Vyas, A. Ramesh, O.P. Joshi, I.R. Khan 2008, Sustainability of soybean (*Glycine max*) – wheat (*Triticum aestivum*) cropping system under integrated nutrient management. Indian J. Agric. Sci., 78(4) (2008): 358-361.

Simon Gourlay 2008. Wind farms are not only beautiful, they're absolutely necessary. The Guardian (UK) (2008). Retrieved 17 January 2012.

Solar Integrated in New Jersey. Jcwinnie.biz. Retrieved 2013-08-20.

Solar Energy Perspectives: Executive Summary (2011). International Energy Agency. 2011. Archived from the original on 2011-12-03.

Solar Fuels and Artificial Photosynthesis. Royal Society of Chemistry 2012. http://www.rsc.org/ScienceAndTechnology/Policy/Documents/solar-fuels.asp (accessed 11 March 2013).

Solar Expected to Maintain its Status as the World's Fastest-Growing Energy Technology. Socialfunds.com. 2009-03-03. Retrieved 2011-11-21.

Solar Thermal Projects Under Review or Announced. Energy.ca.gov. Retrieved 2011-11-21.

S.S. Dhillon, P.R. Hobbs, J.S. Samara 2000b. Investigations on bed planting system as an alternative tillage and crop establishment practice for improving wheat yields sustainably. In: Proceedings of the 15th Conference of International Tillage Research Organization, Fort Worth, TX, USA, 2–7 July (2000).

S.S. Dhillon, P.R. Hobbs, J.S. Samara 2000a. Investigations on bed planting system as an alternative tillage and crop establishment practice for improving wheat yields sustainably. In: Proceedings of the 15th Conference of the International Soil Tillage Research Organisation, Fort Worth, TX, 2–7 July 2000, (CDROM).

S. Fujisaka, L. Harrington, P. Hobbs 1994. Rice–wheat in South Asia: systems and long-term priorities established through diagnostic research. Agric. Syst., 46 (1994): 169–187.

S.K. Gami, J.K. Ladha, H. Pathak, M.P. Shah, E. Pasuquin, S.P. Pandey, P.R. Hobbs, D. Joshy, R. Mishra 2001. Long-term changes in yield and soil fertility in a twenty-year rice–wheat experiment in Nepal. Biol. Fertil. Soils, 34 (2001): 73–78.

S.K. Ghosh, K.M.D. Murthy, G. Ramesh, S.P. Palaniappan 1999. Green revolution.

Steve Leone 2011. U.N. Secretary-General: Renewable Can End Energy Poverty. Renewable Energy World (2011).

S.K. Sinha, G.B. Singh, M. Rai 1998. Decline in Crop Productivity in Haryana and Punjab: Myth or reality? Indian Council of Agricultural Research, New Delhi, India (1998), pp89.

S. Peng, R. Buresh, J.L. Huang, J.C. Yang, Y.B. Zou, X.H. Zhong, G.H. Wang, F.S. Zhang 2006.

Strategies for overing low agronomic nitrogen use efficiency in irrigated rice systems in China. Field Crops Research, 96 (2006): 37-47.

S. Peng, K. G. Cassman 1998. Upper thresholds of nitrogen uptake rates and associated nitrogen fertilizer efficiencies in irrigated rice. Agronomy Journal, 90 (1998): 178–185.

S. Peng, F.V. Garcia, R.C. Laza, A. L. Sanico, R. M. Visperas, K. G. Cassman 1996. Increased nitrogen use efficiency using a chlorophyll meter in high-yielding irrigated rice. Field Crops Research, 47 (1996): 243–252.

S. Peng, R.J. Buresh, J. Huang, X. Zhong, Y. Zou, J. Yang, G. Wang, Y. Liu, R. Hu, Q. Tang, K. Cui, F. Zhang, A. Dobermann 2010. Improving nitrogen fertilization in rice by site-specific N management – A review. Agronomy for Sustainable Development, 30 (2010): 649-656.

S. Portch, A. Hunter 2002. A systematic approach to soil fertility evaluation and improvement. Modern Agriculture and Fertilizers, Special publication no. 5. IPNI China program, Beijing, China (2002).

S. Portch, M. D. Stauffer 2005. Soil testing: A proven diagnostic tool. Better Crops with Plant Food, 89(1) (2005): 28-32.

S. Singh, A. Yadav, R.K. Malik, H. Singh 2002b. Long-term effect of zero-tillage sowing technique on weed flora and productivity of wheat in rice–wheat cropping zones of Indo-Gangetic Plains. In: Proceedings of the International Workshop on Herbicide Resistance Management and Zero Tillage in Rice–Wheat Cropping System, CCS Haryana Agricultural University, Hisar, 4–6 March (2002b), pp. 155–157.

S. Singh Sidhu, R.K. Gupta, Y. Hanmin 2004. Resource Conserving Technologies for Rice Production. Travelling Seminar Report Series 5. Rice–Wheat Consortium for the Indo-Gangetic Plains, New Delhi, India (2004), pp36.

Steve Leone 2011. Billionaire Buffett Bets on Solar Energy. Renewable Energy World (2011).

Soil Conservation and Domestic Allotment Act, P.L. 74-46, 49 Stat. 163, 16 U.S.C. § 590(e), April 27, 1935.

South Asia. Rice Wheat Consortium, Paper Series 6. RWC, New Delhi (2000), pp171.

Sustainable Agriculture and Natural Resource Management (SANREM CRSP)

Sweet sorghum for food, feed and fuel New Agriculturalist, January 2008.

T.A. Volk, L.P. Abrahamson, E.H. White, E. Neuhauser, E. Gray, C. Demeter, C. Lindsey, J. Jarnefeld, D.J. Aneshansley, R. Pellerin and S. Edick 2000. "Developing a Willow Biomass Crop Enterprise for Bioenergy and Bioproducts in the United States. Proceedings of Bioenergy (2000). Adam›s Mark Hotel, Buffalo, New York, USA: North East Regional Biomass Program. OCLC 45275154. Retrieved 2006-12-16.

T. C. Mendoza, B. C. Mendoza 2016. A review of sustainability challenges of biomass for energy, focus in the Philippines. Agric. Technol., 12 (2016): 281–310.

The Central Pollution Control Board (CPCB) 2013. Consolidated Annual Review Report on Implementation of Municipal SolidWastes (Management and Handling) Rules. In Ministry of Environment Forests and Climate Change; Board, C.P.C., Ed.; The Central Pollution Control Board: New Delhi, India, (2013).

The Foreign Assistance Act of 1961, as amended (PDF). Retrieved 2011-05-01.

The Hindu Crop Residue-CoalMix to Nix Stubble Burning 2018. Available online: http//www.thehindu.com/news/national/other-states/ntpc-to-mix-crop-residue-with-coal-to-curb-crop-burning/article20492123.ece (accessed on 25 June 2018).

Time for universal water metering? Innovations Report. May 2006.

T. Pettinger 2016. Polluter Pays Principle (PPP), (2016). Available online: http://www.economicshelp.org/blog/ 6955/economics/polluter-pays-principle-ppp (accessed on 23 November 2018).

T. Phonbumrung, C. Khemsawas 1998. Agricultural Crop Residue. In Proceedings of the Sixth Meeting of Regional Working Group on Grazing and Feed Resources for Southeast Asia, Legaspi, Philippines, 5–9 October (1998), pp. 183–187.

TWS Final Position Statement. Retrieved 2011-04-04.

The Encyclopedia of Alternative Energy and Sustainable Living. Retrieved 15 January 2013.

The myth of renewable energy | Bulletin of the Atomic Scientists". Thebulletin.org. 2011-11-22. Retrieved 2013-10-03.

The surprising history of sustainable energy. Sustainablehistory.wordpress.com. Retrieved 2012-11-01.

T.N. Chaudhary 1997. Water management in Rice for Efficient Production. Directorate of Water Management Research. Indian Council of Agricultural Research, Patna, India (1997).

T. Rice (1988). Re-Evaluating the Balance Between Zoning Regulations and Religious and Educational Uses. Pace L. Rev. 8 (1988): 1.

Turning the tide on farm productivity in Africa: an agroforestry solution. July 8, 2009. Retrieved 2 April 2014.

United Nations Environmental Programme Global Trends in Sustainable Energy Investment 2007. Analysis of Trends and Issues in the Financing of Renewable Energy and Energy Efficiency in OECD and Developing Countries (2007), pp 3.

United Nations. Glossary of Environment Statistics, Studies in Methods; Series F, 67; Department for Economic and Social Information and Policy Analysis, Statistics Division: New York, NY, USA, 1997; Volume 96. Int. J. Environ. Res. Public Health 2019, 16, 832 15 of 19

UN Land Degradation and Land Use/Cover Data Sources ret. 26 June 2007.

UN Report on Climate Change retrieved 25 June 2007 [from Web archive]

UNEP, Bloomberg, Frankfurt School, Global Trends in Renewable Energy Investment 2011.

USDA. NRCS - Natural Resources Conservation Service (in (Greek)). Nrcs.usda.gov. Retrieved 2013-10-31.

USDA. NRCS - Natural Resources Conservation Service (in (Spanish)). Nrcs.usda.gov. Retrieved 2013-10-31.

USDA. NRCS - Natural Resources Conservation Service. Nrcs.usda.gov. Retrieved 2013-10-31.

USDA. NRCS - Natural Resources Conservation Service. Nrcs.usda.gov. 2013-09-30. Retrieved 2013-10-31.

U.S. Environmental Protection Agency (EPA) 2002. Cases in Water Conservation (Report) (2002). Retrieved 2010-02-02. Document No. EPA-832-B-02-003.

U.Sonesson, A. Bjorklund, M.Carlsson, M. Dalemo 2000. Environmental and economic analysis of management systems for biodegradable wastes. Resour. Conserv. Recycl., (2000): 28, 29.

Varinderpal-Singh, Yadvinder-Singh, Bijay-Singh, Baldev-Singh, R. K. Gupta, Jagmohan-Singh, J. K. Ladha, V. Balasubramanian 2007. Performance of site specific nitrogen management for irrigated transplanted rice in northwestern India. Archives of Agronomy and Soil Science, 53(2007): 567-579.

V. Balasubramanian, A. C. Morales, R. T. Cruz, S. Abdulrachman 1999. On-farm adaptation of knowledge-intensive nitrogen management technologies for rice systems. Nutrient Cycling in Agroecosystems, 53 (1999): 93–101.

V. Balasubramanian, A. C. Morales, R. T. Cruz, T. M. Thiyagarajan, R. Nagarajan, M. Babu, S. Abdulrachman, L. H. Hai 2000. Adaptation of the chlorophyll meter (SPAD) technology for real-time N management in rice: a review. International Rice Research Notes, 25(1) (2000): 4–8.

V.Balasubramanian, J. K. Ladha, R. K. Gupta, R. K. Naresh, R. S. Mehla, Y. Singh, B.Singh 2003. Technology options for rice in rice–wheat systems in Asia. ASA, Special Publication No. 65. ASA (2003).

V. Balasubramanian, J. K. Ladha, R. K. Gupta, R. K. Naresh, R. S. Mehla, Bijay-Singh, Yadvinder-Singh 2003. Technology options for rice in the rice-wheat system in South Asia. In: Improving the productivity and sustainability of rice-wheat systems: Issues and impacts (J.K. Ladha, J.E. Hill, J.M.

V. Balasubramanian, J.K. Ladha, R.K. Gupta, R.K. Naresh, R.S. Mehla, Y. Singh, B. Sing, Bulletin E-2646.

V. Barnett R. Payner, R. Steiner 1995. Agricultural Sustainability: Economic, Environmental and Statistical Considerations, John Wiley and Sons, UK (1995), pp 266.

V. Barnett, R. Payner, R. Steiner 1995. Agricultural Sustainability: Economic, environmental and statistical Considerations, John Wiley and Sons, UK (1995), pp 19

V. Beri, B.S. Sidhu, A. K. Bhat, B. P. Singh 1992. Nutrient balance and soil properties as affected by management of crop residues. In: M.S. Bajwa et. al. (Eds.), Nutrient management for sustained productivity (1992), pp 133–135. Proceedings of International Symposium (vol. II). Ludhiana, India: Department of Soil, Punjab Agricultural University.

V. Kumar, A. Yadav, R. K. Malik 2005. Effect of planting methods and herbicides in transplanted rice. Project Workshop Proceedings on Accelerating the Adoption of Resource Conservation Technologies in Rice-Wheat Systems of the Indo-Gangetic Plains held on June 1-2, 2005 at Hisar (Haryana), India (2005), pp 122-126.

V. Kumar, Y. S. Shivay 2010. Integrated nutrient management: An ideal approach for enhancing agricultural production and productivity. Indian J. Fert., 6(5) (2010):41-57.

Water conservation « Defra. *defra.gov.uk*. 2013. Retrieved January 24, 2013.

World Conservation Strategy (PDF). Retrieved 2011-05-01.

Water - Use It Wisely. U.S. multi-city public outreach program. Park & Co., Phoenix, AZ. Accessed 2010-02-02.

Whatever Happened to Wind Energy? LiveScience. 14 January 2008. Retrieved 17 January 2012.

W. C. Leighty, J. H. Holbrook 2012. Running the World on Renewables: Alternatives for Transmission and Low-cost Firming Storage of Stranded Renewables as Hydrogen and Ammonia Fuels via Underground Pipelines. In ASME 2012 International Mechanical Engineering Congress and Exposition (American Society of Mechanical Engineers) November 9-15 (2012), pp 513-522.

Wildlife Conservation. Conservation and Wildlife. Retrieved 1 June 2012. Fromhttp://animals.about.com/od/animalswildlife101/a/threats.htmMcCallum, M.L. 2010. Future climate change spells catastrophe for Blanchard's Cricket Frog (*Acris blanchardi*). Acta Herpetologica 5 (2012):119 -130.

W. Jackson, L. L. Jackson 1999. Developing high seed yielding perennial polycultures as a mimic of mid-grass prairie. In Agriculture as a Mimic of Natural Ecosystems, (ed.) Lefroy E C, Hobbs R J, O'Connor M H, Pate J S, Dordrecht N L: Kluwer Academic Publishers (1999), pp 1-38.

WWF in Brief. Wwf.panda.org. Retrieved 2011-05-01.

Water Management | NRCS. Nrcs.usda.gov. Retrieved 2013-10-31.

W.C. Wheaton 1993. Land capitalization, Tiebout mobility, and the role of zoning regulations. Journal of Urban Economics, 34(2) (1993): 102-117.

W. G. Jaeker, A. J. Plantinga 2007. How have Land-use regulations Affected Property Values in Oregon? OSU Extension (2007).

W. H. Pawley 1963. Possibilities of Increasing World Food Production. Food and Agriculture Organization of the United Nations. Rome, Italy. pp 98.

W. G. Jaeker, A. J. Plantinga 2007. How have Land-use regulations Affected Property Values in Oregon? OSU Extension (2007).

Wildlife Habitat Incentive Program (WHIP) | NRCS. Nrcs.usda.gov. Retrieved 2013-10-31.

Wind Power: Capacity Factor, Intermittency, and what happens when the wind doesn't blow? Retrieved 24 January 2008.

Working Lands for Wildlife | NRCS. Nrcs.usda.gov. 2012-09-17. Retrieved 2013-10-31.

World Energy Assessment 2001. Renewable energy technologies (2001), pp 221.

World watch Institute 2012. Use and Capacity of Global Hydropower Increases (2012).

World Wind Energy Report 2010. Report. World Wind Energy Association. February 2011. Retrieved 30 April 2011.

X. Liu, L. Lynch 2011. Do zoning regulations rob rural landowners' equity?. *American* Journal of Agricultural Economics, 93(1): 1-25.

Y. Venkanna 2008. Studies on the effect of mulches, organics and organic solutions on growth, yield and quality of chilli Cv.Byadagi Dabbi in Northern Transition Zone of Karnataka. M. Sc. (Agri.) Thesis, Univ. Agric. Sci., Dharwad, India (2008).

Yadvinder-Singh, Bijay-Singh, J. K. Ladha, J. S. Bains, R. K. Gupta, Jagmohan Singh V. Balasubramanian 2007. On-farm evaluation of leaf color chart for need-based nitrogen management in irrigated transplanted rice in northwestern India. Nutrient Cycling in Agroecosystems 78 (2007): 167-176.

Y.Singh, H. S. Sidhu 2014. Management of cereal crop residues for sustainable rice-wheat production system in the Indo-gangetic plains of India. Proc. Indian Natl. Sci. Acad., 80(2014): 95–114.

Y. Singh, R. K. Gupta, J. Singh, G. Singh, G. Singh, J. K. Ladha 2010. Placement effects on paddy residue decomposition and nutrient dynamics on two soil types during wheat cropping in paddy-wheat system in north western India. Nutr. Cycl. Agroecosyst., 88(2010): 471–480.

Z. Lei, J. Chen, Z. Zhang, N. Sugiura 2010.Methane production from rice straw with acclimated anaerobic sludge: Effect of phosphate supplementation. J. Bioresour. Technol., 101(2010): 4343–4348.